Lecture Notes in Computer Science　14642

Founding Editors

Gerhard Goos
Juris Hartmanis

Editorial Board Members

The series Lecture Notes in Computer Science (LNCS), including its subseries Lecture Notes in Artificial Intelligence (LNAI) and Lecture Notes in Bioinformatics (LNBI), has established itself as a medium for the publication of new developments in computer science and information technology research, teaching, and education.

LNCS enjoys close cooperation with the computer science R & D community, the series counts many renowned academics among its volume editors and paper authors, and collaborates with prestigious societies. Its mission is to serve this international community by providing an invaluable service, mainly focused on the publication of conference and workshop proceedings and postproceedings. LNCS commenced publication in 1973.

Ioanna Miliou · Nico Piatkowski ·
Panagiotis Papapetrou

Editors

Advances in Intelligent Data Analysis XXII

22nd International Symposium on Intelligent Data Analysis, IDA 2024
Stockholm, Sweden, April 24–26, 2024
Proceedings, Part II

 Springer

Editors
Ioanna Miliou ⓘ
Stockholm University
Kista, Sweden

Nico Piatkowski ⓘ
Fraunhofer IAIS
Sankt Augustin, Germany

Panagiotis Papapetrou ⓘ
Stockholm University
Kista, Sweden

ISSN 0302-9743   ISSN 1611-3349 (electronic)
Lecture Notes in Computer Science
ISBN 978-3-031-58555-5   ISBN 978-3-031-58553-1 (eBook)
https://doi.org/10.1007/978-3-031-58553-1

This Springer imprint is published by the registered company Springer Nature Switzerland AG
The registered company address is: Gewerbestrasse 11, 6330 Cham, Switzerland

Paper in this product is recyclable.

Preface

We are delighted to introduce the proceedings of the 22nd International Symposium on Intelligent Data Analysis, held from April 24–26, 2024, in Stockholm, Sweden. Originating in 1995, the symposium was organized biennially until 2009. Starting from 2010, the symposium shifted its focus towards encouraging submissions that present groundbreaking and innovative ideas, even if they might not be as fully developed as those presented at other conferences. The 2024 edition of IDA maintained this tradition, inviting submissions that, while possibly considered preliminary in other contexts, promise significant research advancements. This edition also marked the return of the Industrial Challenge track, encouraging both academic and industrial researchers to tackle a machine learning challenge focused on predicting the imminent failure of specific vehicle engine components. Participants worked with data from Scania trucks operating under demanding conditions. The IDA Symposium welcomes a broad range of modeling and analysis methodologies from all disciplines, aiming to be an interdisciplinary forum that fosters discussions on intelligent data analysis that span various fields. We invite contributions that address intelligent support for modeling and analyzing data from complex, dynamic systems.

Within this context, IDA 2024 extended its support for data analysis beyond the conventional algorithmic solutions typically discussed in academic literature. Only those submissions that integrated established technologies within intelligent data analysis frameworks or applied such technologies in innovative ways to the analysis and/or modeling of complex systems were considered. The traditional review process, which tends to favor small, incremental improvements over existing research, might have deterred the submission of the innovative research that IDA 2024 aimed to attract. To counter this, reviewers and senior PC members were encouraged to favor innovative ideas to complex solutions and extensive experimental evaluations. The outcome was a highly compelling program. We received 94 submissions, from which 40 were accepted as regular papers and 3 as short papers, including those from the industrial challenge. Each submission underwent a rigorous single-blind review by three members of the program committee and one senior member.

We were pleased to have the following distinguished invited speakers at IDA 2024:

- Dimitrios Gunopulos, National and Kapodistrian University of Athens, Greece on the topic "Computing counterfactuals with feasibility and compactness guarantees"
- Dino Pedreschi, University of Pisa, Italy on the topic "Social Artificial Intelligence: Challenges of the Human-AI Ecosystem"
- Danica Kragic, KTH Royal Institute of Technology, Sweden on the topic "Representation learning and foundation models in robotics".

The conference was held at the Department of Computer and Systems Sciences of Stockholm University in Stockholm.

We wish to express our gratitude to all authors of submitted papers for their intellectual contributions; to the Program Committee members and advisors and additional reviewers for their effort in reviewing, discussing, and commenting on the submitted papers and to the members of the IDA Steering Committee for their ongoing guidance and support. We thank Zed Lee for running the conference website. Special thanks go to the industrial challenge chair Tony Lindgren for handling the submission and reviewing process of the industrial challenge papers, as well as to Luis Galárraga for handling the PhD Poster Track. We gratefully acknowledge those who were involved in the local organization of the symposium: Sindri Magnusson, Ali Beikmohammadi, Franco Rugolon, and Maria Bampa. We thank our Frontier Prize chair, Jaakko Hollmen, our advisory chairs Matthijs van Leeuwen and Siegfried Nijssen for their precious guidance during the preparation of IDA, and our Social Media chair, Zhendong Wang who together with Ioanna Miliou took care of all strategic social media communications related to IDA. Finally, we are grateful to our sponsors: Stockholm University, Scania AB, Digital Futures, Springer, and The Artificial Intelligence Journal. We are especially indebted to KNIME, who funded the IDA Frontier Prize for the most visionary contribution presenting a novel and surprising approach to data analysis in the understanding of complex systems. Last but not least we thank the City of Stockholm for hosting IDA's reception evening at the Stockholm City Hall.

April 2024   Ioanna Miliou
                                                                      Nico Piatkowski
                                                              Panagiotis Papapetrou

Organization

General Chair

Panagiotis Papapetrou   Stockholm University, Sweden

Program Committee Chairs

Ioanna Miliou   Stockholm University, Sweden
Nico Piatkowski   Fraunhofer IAIS, Germany

Senior Program Committee

Hendrik Blockeel   KU Leuven, Belgium
Henrik Bostrom   KTH Royal Institute of Technology, Sweden
Peggy Cellier   INSA Rennes and IRISA, France
Bruno Cremilleux   Université de Caen Normandie, France
Wouter Duivesteijn   TU Eindhoven, Netherlands
Benoît Frénay   University of Namur, Belgium
Elisa Fromont   Université Rennes 1 and IRISA/Inria, France
Esther Galbrun   University of Eastern Finland, Finland
Joao Gama   INESC TEC - LIAAD, Portugal
Sibylle Hess   TU Eindhoven, Netherlands
Frank Höppner   Ostfalia University of Applied Science, Germany
Arno Knobbe   Leiden University, Netherlands
Georg Krempl   Utrecht University, Netherlands
Charlotte Laclau   Polytechnique Institute and Télécom Paris, France
Siegfried Nijssen   Université Catholique de Louvain, Belgium
Celine Robardet   INSA Lyon, France
Arno Siebes   Universiteit Utrecht, Netherlands
Stephen Swift   Brunel University London, UK
Maryam Tavakol   TU Eindhoven, Netherlands
Allan Tucker   Brunel University London, UK
Matthijs van Leeuwen   Leiden University, Netherlands
Veronica Vinciotti   University of Trento, Italy
David Weston   Birkbeck University of London, UK

Program Committee

Pedro Henriques Abreu	CISUC, Portugal
Thiago Andrade	University of Porto and INESC TEC, Portugal
Ali Ayadi	University of Strasbourg, France
Paulo Azevedo	Universidade do Minho, Portugal
Maria Bampa	Stockholm University, Sweden
Mitra Baratchi	LIACS - University of Leiden, Netherlands
José Borges	INESC TEC - FEUP, Portugal
Tassadit Bouadi	Universite de Rennes 1, France
Paula Brito	University of Porto and INESC TEC - LIAAD, Portugal
Dariusz Brzezinski	Poznan University of Technology, Poland
Mirko Bunse	TU Dortmund, Germany
Sebastian Buschjäger	TU Dortmund, Germany
Rui Camacho	University of Porto, Portugal
Paulo Cortez	University of Minho, Portugal
Thi-Bich-Hanh Dao	University of Orléans, France
Maria Demidik	DESY, Germany
Hadi Fanaee-T	Halmstad University, Sweden
Brígida Mónica Faria	Polytechnic Institute of Porto and LIACC, Portugal
Ad Feelders	Universiteit Utrecht, Netherlands
Sébastien Ferré	Univ Rennes and CNRS Inria IRISA, France
Carlos Ferreira	INESC TEC, Portugal
Françoise Fessant	Orange, France
Mikel Galar	Universidad Pública de Navarra, Spain
Luis Galárraga	Inria, France
Benoit Gauzere	INSA Rouen, France
Rui Gomes	Universidade de Coimbra, Portugal
Zhijin Guo	University of Bristol, UK
Thomas Guyet	Inria Centre de Lyon, France
Barbara Hammer	CITEC and Bielefeld University, Germany
Alberto Fernandez Hilario	University of Granada, Spain
Tomáš Horváth	ELTE, Hungary
Dino Ienco	Irstea, France
Szymon Jaroszewicz	Polish Academy of Sciences, Poland
Baptiste Jeudy	Laboratoire Hubert Curien, France
Bo Kang	Ghent University, Belgium
Frank Klawonn	Ostfalia University, Germany
Jiri Klema	Czech Technical University, Czech Republic
Maksim Koptelov	University of Caen Normandy, France

Alejandro Kuratomi	Stockholm University, Sweden
Christine Largeron	Université Jean Monnet Saint-Etienne, France
Nada Lavrač	Jožef Stefan Institute, Slovenia
Zed Lee	Stockholm University, Sweden
Vincent Lemaire	Orange Innovation, France
Sindri Magnusson	Stockholm University, Sweden
Joao Mendes-Moreira	INESC TEC, Portugal
Vera Miguéis	University of Porto, Portugal
Rita Nogueira	INESC TEC, Portugal
Slawomir Nowaczyk	Halmstad University, Sweden
Andreas Nürnberger	Magdeburg University, Germany
Kaustubh Patil	Forschungszentrum Jülich, Germany
Ruggero Pensa	University of Turin, Italy
Pedro Pereira Rodrigues	University of Porto, Portugal
João Pimentel	Dalhousie University, Canada
Marc Plantevit	EPITA, France
Luboš Popelínský	Masaryk University, Czech Republic
Filipe Portela	UMINHO, Portugal
Luis Reis	University of Porto, Portugal
Justine Reynaud	University of Caen Normandy, France
Rita Ribeiro	University of Porto, Portugal
Duncan Ruiz	Pontifícia Universidade Católica do Rio Grande do Sul, Brazil
Amal Saadallah	TU Dortmund, Germany
Akrati Saxena	Leiden University, Netherlands
Jørg Schløtterer	University of Marburg and University of Mannheim, Germany
Roberta Siciliano	University of Naples Federico II, Italy
Paula Silva	University of Porto and INESC TEC, Portugal
Amina Souag	Canterbury Christ Church University, UK
Arnaud Soulet	University of Tours, France
Myra Spiliopoulou	Otto-von-Guericke-University Magdeburg, Germany
Jerzy Stefanowski	Poznan University of Technology, Poland
Shazia Tabassum	INESC TEC, Portugal
Sónia Teixeira	INESC TEC, Portugal
Alicia Troncoso	Pablo de Olavide University, Spain
Peter van der Putten	Leiden University, Netherlands
Bruno Veloso	University of Porto and INESC TEC - FEP, Portugal
Tom Viering	Delft University of Technology, Netherlands

João Vinagre	Joint Research Centre - European Commission, Spain
Sheng Wang	University of Bristol, UK
Hilde Weerts	Eindhoven University of Technology, Netherlands
Pascal Welke	TU Wien, Austria
Zhaozhen Xu	University of Bristol, UK
Paul Youssef	University of Marburg, Germany
Leishi Zhang	Canterbury Christ Church University, UK
Albrecht Zimmermann	Université de Caen Normandie, France

Sponsors

We thank our sponsors for their support:

- Stockholm University
- The City of Stockholm
- Digital Futures
- Scania
- KNIME
- Springer

Contents – Part II

Optimization

XAI

Industrial Challenge

Contents – Part I

Applications

Natural Language Processing

Temporal and Sequence Data

Kernel Corrector LSTM

Rodrigo Tuna[1]([✉])[iD], Yassine Baghoussi[1,4][iD], Carlos Soares[1,2,3][iD],
and João Mendes-Moreira[1,4][iD]

[1] Faculdade de Engenharia, Universidade do Porto, Porto, Portugal
up201904967@edu.fe.up.pt, {baghoussi,csoares,jmoreira}@fe.up.pt
[2] Artificial Intelligence and Computer Science Lab. (LIACC – member of LASI LA),
Universidade do Porto, Porto, Portugal
[3] Fraunhofer AICOS Portugal, Porto, Portugal
[4] INESC TEC, Porto, Portugal

Abstract. Forecasting methods are affected by data quality issues in
two ways: 1. they are hard to predict, and 2. they may affect the model
negatively when it is updated with new data. The latter issue is usually
addressed by pre-processing the data to remove those issues. An alternative approach has recently been proposed, Corrector LSTM (cLSTM),
which is a Read & Write Machine Learning (RW-ML) algorithm that
changes the data while learning to improve its predictions. Despite
promising results being reported, cLSTM is computationally expensive,
as it uses a meta-learner to monitor the hidden states of the LSTM. We
propose a new RW-ML algorithm, Kernel Corrector LSTM (KcLSTM),
that replaces the meta-learner of cLSTM with a simpler method: Kernel Smoothing. We empirically evaluate the forecasting accuracy and
the training time of the new algorithm and compare it with cLSTM and
LSTM. Results indicate that it is able to decrease the training time while
maintaining a competitive forecasting accuracy.

Keywords: Time series forecasting · Recurrent Neural Networks ·
Data-Centric AI

1 Introduction

In many fields, including energy, healthcare, management, and climate research,
time series forecasting is a crucial task that can be accomplished using machine
learning or statistical methods [8]. As data becomes widely available, more precise forecasting models are expected. However, data quality issues like outliers,

This work was partially funded by projects AISym4Med (101095387) supported
by Horizon Europe Cluster 1: Health, ConnectedHealth (n.º 46858), supported
by Competitiveness and Internationalisation Operational Programme (POCI) and
Lisbon Regional Operational Programme (LISBOA 2020), under the PORTU-
GAL 2020 Partnership Agreement, through the European Regional Development
Fund (ERDF); NextGenAI - Center for Responsible AI (2022-C05i0102-02), sup-
ported by IAPMEI, and also by FCT plurianual funding for 2020–2023 of LIACC
(UIDB/00027/2020_UIDP/00027/2020) and SONAE IM Labs@FEUP.

I. Miliou et al. (Eds.): IDA 2024, LNCS 14642, pp. 3–14, 2024.
https://doi.org/10.1007/978-3-031-58553-1_1

missing values, and changes in the underlying data generation process might impact predictive techniques.

Traditional machine learning (ML) models are often considered read-only models, capable of learning from data but neglecting the feedback loop for correcting the data during the learning process. This approach, while efficient in many cases, lacks proper adaptation of preprocessing techniques and the ML model itself, as the model's feedback is often overlooked.

To address this limitation, the concept of Read-Write Machine Learning (RW-ML) has emerged. RW-ML models, such as Corrector LSTM (cLSTM) [1], not only learn from data but also have the capability to change the data during the learning process. cLSTM is a time series forecasting method designed to improve forecasting accuracy by dynamically adjusting the data. It utilizes a meta-model of the Hidden State Dynamics obtained with SARIMA to detect data quality issues and employs a greedy heuristic to correct them. cLSTM has demonstrated superior predictive performance compared to traditional LSTM models. However, the computational cost associated with the meta-learning component of cLSTM is significant.

In this paper, we propose a computationally less expensive variant of cLSTM, named Kernel Corrector LSTM (KcLSTM), which replaces the meta-learner with a simpler method: Kernel Smoothing. We empirically compare KcLSTM with both cLSTM and LSTM models. Results reveal that KcLSTM achieves better predictive performance than LSTM and cLSTM, while also being faster than cLSTM, although the computational efficiency improvement is not as substantial as expected.

The main contributions of this paper are:

- Introducing a variant of cLSTM, KcLSTM, which is computationally less expensive while maintaining high predictive accuracy.
- An empirical study comparing KcLSTM with LSTM and cLSTM in terms of predictive performance and training time.

This paper is structured as follows: we first provide an overview of the state-of-the-art forecasting method, LSTM. Then, we delve into the concept of RW-ML and its significance in time series forecasting. Next, we introduce the proposed algorithm, KcLSTM. Finally, we describe the experimental setup, present the results, and discuss their implications.

2 Related Work

In this section, we first present the Long Short-Term Memory. The algorithm that our proposed algorithm is built on and the one it will be compared to. We then define Data-Centric AI and provide examples of Data-Centric models built for time series forecasting.

2.1 LSTM

The Long Short-Term Memory (LSTM) [16] is a Recurrent Neural Network (RNN), that can capture long-term dependencies in the input and it is used for processing sequential data. RNNs differ from feed-forward networks through recurrent connections, allowing them to learn from sequential data. Back Propagation is applied to RNNs by taking advantage of the fact that for every recurrent network, there exists an equivalent feed-forward network with identical behavior for a finite number of steps [21], training it using Back Propagation Through Time (BPTT) [23] (Fig. 1).

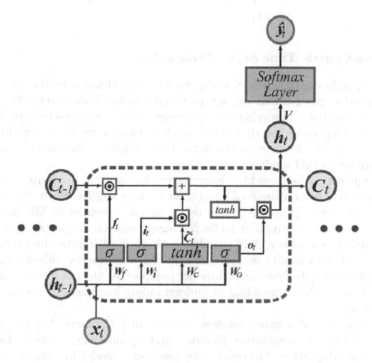

Fig. 1. Cell unit of the LSTM recurent neural network [6].

RNNs have some well-known limitations. First, they have problems capturing long-term dependencies, being limited to only bridge between 5–10 steps [23]. This occurs because RNNs are sensible to the exploding/vanishing gradient problem [15]. The LSTM solves this problem through the use of a gating mechanism.

An LSTM network consists of blocks, with each block containing an input gate, forget gate, output gate, and memory cell (Eqs. (1b) to (1d)). The input gate controls which inputs are relevant; the forget gate learns which information should be kept in memory; and the output gate controls which information should be passed to the next block. The information is retained through the use of two states called the cell state, (Eq. (1e)), and the hidden state, (Eq. (1f)). The forward pass concatenates the input with the hidden state from the last block, while the backward pass derives the error and updates the gates using the chain rule of derivatives. The gate derivatives are multiplied by the hidden output to obtain the gradient deltas that update the gates.

$$i_t = \sigma(W_i \cdot h_{t-1} + V_i \cdot x_t + b_i) \tag{1a}$$

$$o_t = \sigma(W_o \cdot h_{t-1} + V_o \cdot x_t + b_o) \tag{1b}$$

$$f_t = \sigma(W_f \cdot h_{t-1} + V_f \cdot x_t + b_f) \tag{1c}$$

$$\hat{C}_t = tanh(W_c \cdot h_{t-1} + V_c \cdot x_t + b_c) \tag{1d}$$

$$C_t = i_t \cdot \hat{C}_t + f_t \cdot C_{t-1} \tag{1e}$$

$$h_t = o_t \cdot tanh(C_t) \tag{1f}$$

$$z_t = h_t \tag{1g}$$

2.2   Data-Centric Time Series Forecasting

Anomalies, including outliers, missing values, and changes in the underlying data generation process can impact predictive tasks. This affects the predictions of such methods, hindering their performance [17], conversely to traditional machine learning methods, that build models using a fixed dataset. In Data-Centric AI [25], the focus is on the data. Data quality is increased to improve the performance of AI models.

The exploration into machine learning models capable of learning and correcting data has been a topic of interest in various studies. Both [22] and [20] delve into this concept, with [22] focusing on the potential of ML models to memorize sensitive information while [20] emphasize the importance of model interpretability and safety. Additionally, authors in [4] further underscore the significance of data quality in enhancing model performance, advocating for a data-centric approach. Providing a broader perspective, the work in [14] discusses the role of probabilistic modeling in understanding learning and uncertainty in machine learning.

Moving to neural network models, authors in [11] discuss highly interconnected networks for associative memory and optimization, with a focus on learning and adaptation. Moreover, [10] propose a model for neural networks that learn temporal sequences through selection, employing synaptic triads and a local Hebbian learning rule. Furthermore, [9] introduce predictive-corrective networks for action detection in videos, which utilize top-down predictions and bottom-up observations for adaptive computation and simplified learning. These models collectively demonstrate the potential of neural networks to learn and correct data across various applications.

Similarly, recurrent neural network (RNN) models have been developed to address the challenge of learning and correcting data. In [13], an attempt is made to introduce a learning algorithm for the recurrent random network model, employing gradient descent of a quadratic error function. Later, authors in [2] propose a recurrent network architecture for modeling dynamical systems, which can learn from multiple temporal patterns and cope with sparse data. More recently, the research in [5] demonstrates that tree-structured recursive neural networks can learn logical semantics, including entailment and contradiction.

In the context of time series forecasting, some data-centric approaches have been employed. In dLSTM [19], the authors train the model on non-anomalous data and use the predictive errors to detect anomalies. The deviation from the normal state is measured through delayed prediction errors. The normal state can then be restored from several candidate values. Following the idea of using the prediction errors to improve the quality of the data Pastprop was introduced [3]. The responsibility for the training error is shared between the model parameters and the training data. The backpropagation of the derivatives is applied to the input, indicating the part of the input that caused the training error.

2.3 Corrector LSTM

cLSTM [1] is an architecture that improves its predictive performance by reconstructing the data of the model. The architecture of the algorithm is based on the LSTM and a data correction component. This data correction component uses a meta-learner, SARIMA, to identify problems in the hidden states of an LSTM model. This is achieved by predicting the hidden states using SARIMA and if the difference between the predicted and the real hidden states is over a certain threshold they are considered anomalous. The anomalies detected in the hidden states are assumed to be caused by the data which is then reconstructed. The reconstruction of the data points is such that the difference between the predicted and the real hidden states falls under a certain threshold. The authors showed that analyzing the Hidden State Dynamics [24] of an LSTM can be used to detect anomalies in the training data and consequently improve the forecasting performance of the model. However, the data correction relies on a meta-learner which makes the algorithm computationally expensive.

3 Kernel Corrector LSTM

The architecture of the Kernel Corrector LSTM (KcLSTM) is the same as the cLSTM architecture, and the meta-learner used to detect problems in the learning is substituted by a simpler approach, kernel smoothing.

3.1 Training

The KcLSTM utilizes the hidden states learned during the training process to find and correct data points of the series. The training of the KcLSTM is divided into three distinct phases. The first phase consists of training the data on a standard LSTM. This allows the hidden states to capture the information of the time series and to be indicative of problems in the data. We then perform the correction, which is comprised of a detection and a correction component. These two components find and correct errors in the data respectively, this phase is thoroughly explained in Sect. 3.2. Finally, the LSTM is trained on the new data, learning a corrected time series, that can improve the predictions of the model.

3.2   Data Correction

The Data Correction phase of the algorithm is divided into two different components: the correction and the detection. These two components aim to find data
points that worsen the learning of the model and change the data so that the
learning process is improved and a better model is obtained. Each phase has a
threshold δ_d and δ_c. The hidden states of the last iteration of the first training
phase, $H = h_0, ..., h_n$ are used to find errors in the training data. cLSTM uses a
meta-model that is computationally expensive to compute; our goal is to assess
if a simpler method can obtain competitive results with less cost; the method
selected for this purpose is Kernel Smoothing because states are estimated rather
than predicted which makes it computationally. A new set of estimated hidden
states $H' = h'_0, ..., h'_n$ is calculated using Gaussian Kernel Smoothing of H as
described in Eq. (2) (Fig. 2).

$$h'_i = \frac{\sum_{j \in [i-W/2, i+W/2], i \neq j} h_j * K(h_i, h_j)}{\sum_{j \in [i-W/2, i+W/2], i \neq j} K(h_i, h_j)} \tag{2}$$

where $K(h_i, h_j)$ is:

$$K(h_i, h_j) = e^{\frac{\|h_i - h_j\|^2}{2\sigma^2}} \tag{3}$$

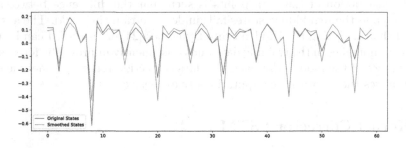

Fig. 2. Gaussian Smoothing of the Hidden States, represented as a series.

The goal of error detection is to discriminate between data points that
need reconstruction and those that do not. A point needs reconstruction if the
Dynamic Time Warp similarity between the hidden state from which the point
originated h_i and the corresponding estimated hidden state h'_i is greater than a
given threshold δ_d. This relation is depicted in Eq. (4).

$$DTW(h'_i, h_i) > \delta_d \tag{4}$$

In the error correction, we reconstruct the detected points such that the
Dynamic Time Warp similarity of the hidden state and estimated hidden state
is less or equal to a given threshold δ_c Eq. (5). Early stopping is employed, and
if a maximum number of iterations is reached, the original value for the point is
restored.

$$DTW(h'_i, h_i) \leq \delta_c \tag{5}$$

4   Experimental Setup

The goals of the empirical validation are to investigate if the proposed algorithm is faster than the original one, without a significant decrease in forecasting accuracy.

A straightforward holdout method was used to estimate forecasting performance, used when there is a temporal dependency in the dataset [7]. The model is trained on the first s samples and assessed on the succeeding $n - s$ samples. The data used for evaluation is always the original one. Using corrected data for the evaluation would likely lead to inadequate optimistic estimates of the forecasting performance of the corresponding method.

The hyperparameters of LSTM and KcLSTM are chosen using hyperparameter tuning using grid search. The learning rate was varied between: 0.0001, 0.001, 0.01, 0.1; and the batch size was varied between: 1, 2, 4, 8. For cLSTM we do not perform hyper-parameter tuning due to the high computational cost. Instead, we use the results described in the original paper [1]. The thresholds for KcLSTM are fixed with values of 0.6 for the detection and 0.5 for the correction. The thresholds for cLSTM are described in the original paper [1], 0.6 for the detection and 0.2 for the correction. We chose to maintain the same detection threshold and increased the correction threshold. The kernel estimates of the hidden states are smoother; thus, a small correction threshold would significantly alter the hidden states, and the information learned in the previous phase would be lost.

4.1   Datasets

Table 1. Statistical description of the dataset.

	Monthly
Timeseries	200
Average Length	366
Mean	4222
Standard Deviation	1160

We have used the M4 Competition Dataset [18] comprising six subsets. From one subset, Monthly, we evaluate the performance of the algorithms on the first 199 time series. To evaluate the time taken to train the models, we use the first 20 time series of that subset (Table 1).

4.2   Evaluation Metrics

This study focuses on both the predictive performance of the algorithm as well as its training time. To quantify the error of forecasts, we focus on the Mean Absolute Scaled Error (MASE) in Eq. 6 because it allows for the averaging of results across different time series as opposed to the Rooted Mean Squared

Error (RMSE). The MASE measures the appropriateness of a forecast against the naive forecast of predicting the previous value. To assess if the differences are statistically different, we use the Mariano-Diebold Test [12].

$$MASE = \frac{\frac{1}{n-s}\sum_{i=s+1}^{n} \hat{y}_i - y_i}{\frac{1}{n-1}\sum_{i=2}^{n} |y_i - y_{i-1}|} \tag{6}$$

The execution time is measured in seconds and the experiments were run on an Intel(R) Core(TM) i7-1065G7 CPU @ 1.30 GHz processor.

5 Results

Investigate if the proposed algorithm is faster than the original one, without significantly decreasing forecasting accuracy. To illustrate the usefulness of the proposed algorithm, we first evaluate its forecasting capabilities, comparing it with LSTM Sect. 5.2 and cLSTM Sect. 5.1.

5.1 Comparison with cLSTM

Results of MASE for the algorithms presented in Table 2 indicate KcLSTM outperforms cLSTM, but this may be explained by the hyper-parameter tuning that was performed for the KcLSTM and not the cLSTM. As such comparison between these two algorithms, can not be performed directly.

The Mariano-Diebold Test for cLSTM and KcLSTM resulted in 40 wins for cLSTM, 111 wins for KcLSTM, and 48 draws. This shows an improvement in forecasting accuracy by substituting the meta-learner with Kernel Smoothing. Again the uneven conditions do not allow us to reach clear conclusions about these two methods.

Results for the training time presented in Table 2 indicate that KcLSTM is indeed faster than cLSTM, significantly. Nonetheless, the gain is not as large as would be expected. Estimating the states with Kernel Smoothing is less computationally expensive than predicting the states with SARIMA. However, this is a cruder method that results in estimated states that are farther away than from the original states when compared with. Consequently, more points are considered anomalies that will be corrected, which will cause the training time to increase.

Table 2. Comparison of each algorithm.

	Mean	Median	Standard Deviation	Average time (s)
LSTM	3.48	0.74	6.07	20.47
cLSTM	8.77	1.04	36.96	56.15
KcLSTM	4.64	0.83	11.96	48.77

However, when performing the Mariano-Diebold test to compare LSTM and KcLSTM at the significance level of $\alpha = 0.05$ we get 102 wins for KcLSTM, 50 wins for LSTM, and 47 draws. We can conclude that KcLSTM is superior to LSTM as it wins more often, although when it loses it is by a greater margin. This is confirmed by the values of the standard deviation of MASE for LSTM and KcLSTM and explains the (apparent) superiority of LSTM when analyzing only the MASE. We see examples of a series with clear outliers that KcLSTM is able to correct and as such increase their predictions in Fig. 3. Conversely, an example of a series without outliers made KcLSTM wrongfully alter the data which results in disastrous predictions in Fig. 4. These two examples reflect the different behaviors mentioned before.

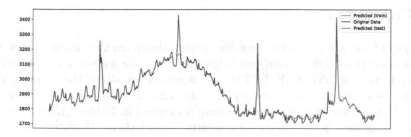

Fig. 3. Example where data reconstruction was successful.

5.2   Comparison with LSTM

Results of MASE for the algorithms presented in Table 2 indicate that LSTM has an overall better performance than KcLSTM with lower values for both the median and the mean for the MASE. However, when performing the Mariano-Diebold test to compare LSTM and KcLSTM at the significance level of $\alpha = 0.05$ we get 102 wins for KcLSTM, 50 wins for LSTM, and 47 draws. We can conclude that KcLSTM is superior to LSTM as it wins more often, although when it loses

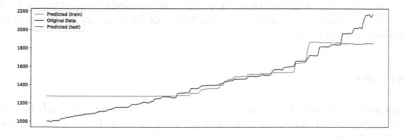

Fig. 4. Example where the data reconstruction destroyed the data which caused the model not to capture the information of the series.

it is by a greater margin. This is confirmed by the values of the standard deviation of MASE for LSTM and KcLSTM and explains the (apparent) superiority of LSTM when analyzing only the MASE. We see examples of a series with clear outliers that KcLSTM is able to correct and as such increase their predictions in Fig. 3. Conversely, an example of a series without outliers made KcLSTM wrongfully alter the data which results in worse predictions in Fig. 4. These two examples reflect the different behaviors mentioned before.

Results for the training time presented in Table 2 indicate that KcLSTM is significantly slower than LSTM. This to be expected as KcLSTM has a data correction component that is responsible for most of the execution time of the algorithm.

6 Conclusion

The goal of our work is to create a forecasting algorithm that reconstructs data that is faster than current solutions in the literature. We present a new algorithm: Kernel Corrector LSTM (KcLSTM). This algorithm alters the training data to improve its forecasting accuracy. Like in cLSTM, this is done by analyzing the Hidden States Dynamics and finding anomalies in hidden states to detect anomalies in data points and consequently correct them. However, the meta-learner of cLSTM was replaced by the Gaussian Kernel Smoothing of the hidden states to decrease the training time of cLSTM.

We empirically compare our algorithm with LSTM and cLSTM both in terms of predictive performance and training time. Results show that KcLSTM obtains a competitive forecasting accuracy surpassing both the LSTM and cLSTM in the number of statistically significant wins. However, KcLSTM is more sensitive to its training data and more prone to making worse forecasts than the baseline, which caused the average MASE of LSTM to be inferior to the average MASE of KcLSTM. The measured training times also show that KcLSTM indeed improves on cLSTM in terms of computational cost, but the margin is smaller than expected because KcLSTM detects more points as anomalies than cLSTM. The estimated hidden states by KcLSTM are more distant from the real hidden states than the predicted states of cLSTM. The empirical study showed that KcLSTM is a faster algorithm that corrects its training data than cLSTM and that those corrections improve the forecasts by being superior to LSTM. Future work comprises the possibility of implementing the algorithm with different estimators.

References

1. Baghoussi, Y., Soares, C., Mendes-Moreira, J.: Corrector LSTM: built-in training data correction for improved time series forecasting. In: Proceedings of the 8th SIGKDD International Workshop on Mining and Learning from Time Series–Deep Forecasting: Models, Interpretability, and Applications, Washington DC, USA, pp. 1–8. ACM (2022)

2. Bailer-Jones, C., MacKay, D.J.C., Withers, P.J.A.: A recurrent neural network for modelling dynamical systems. Network **9**(4), 531–47 (1998). https://api.semanticscholar.org/CorpusID:653765

3. Baptista, A., Baghoussi, Y., Soares, C., Mendes-Moreira, J., Arantes, M.: Pastprop-RNN: improved predictions of the future by correcting the past (2021)

4. Bhowmik, P., Partha, A.S.: A data-centric approach to improve machine learning model's performance in production. Int. J. Eng. Adv. Technol. (2021). https://api.semanticscholar.org/CorpusID:240328155

5. Bowman, S.R., Potts, C., Manning, C.D.: Recursive neural networks can learn logical semantics. In: Workshop on Continuous Vector Space Models and their Compositionality (2014). https://api.semanticscholar.org/CorpusID:15618372

6. Castro, J., Achanccaray Diaz, P., Sanches, I., Cue La Rosa, L., Nigri Happ, P., Feitosa, R.: Evaluation of recurrent neural networks for crop recognition from multitemporal remote sensing images (2017)

7. Cerqueira, V., Torgo, L., Mozetič, I.: Evaluating time series forecasting models: an empirical study on performance estimation methods. Mach. Learn. **109**(11), 1997–2028 (2020). https://doi.org/10.1007/s10994-020-05910-7

8. Cerqueira, V., Torgo, L., Soares, C.: Machine learning vs statistical methods for time series forecasting: size matters (2019)

9. Dave, A., Russakovsky, O., Ramanan, D.: Predictive-corrective networks for action detection. In: 2017 IEEE Conference on Computer Vision and Pattern Recognition (CVPR), pp. 2067–2076 (2017). https://api.semanticscholar.org/CorpusID:2466592

10. Dehaene, S., Changeux, J.P., Nadal, J.P.: Neural networks that learn temporal sequences by selection. Proc. Natl. Acad. Sci. USA **84**(9), 2727–31 (1987). https://api.semanticscholar.org/CorpusID:7423734

11. Denker, J.S.: Neural network models of learning and adaptation. Phys. D Nonlinear Phenom. **2**, 216–232 (1986). https://api.semanticscholar.org/CorpusID:119988262

12. Diebold, F., Mariano, R.: Comparing predictive accuracy. J. Bus. Econ. Stat. **13**(3), 253–63 (1995). https://EconPapers.repec.org/RePEc:bes:jnlbes:v:13:y:1995:i:3:p:253-63

13. Gelenbe, E.: Learning in the recurrent random neural network. Neural Comput. **5**, 154–164 (1992). https://api.semanticscholar.org/CorpusID:38667978

14. Ghahramani, Z.: Probabilistic machine learning and artificial intelligence. Nature **521**, 452–459 (2015). https://api.semanticscholar.org/CorpusID:216356

15. Hochreiter, S.: The vanishing gradient problem during learning recurrent neural nets and problem solutions. Int. J. Uncertain. Fuzziness Knowl.-Based Syst. **6**, 107–116 (1998). https://doi.org/10.1142/S0218488598000094

16. Hochreiter, S., Schmidhuber, J.: Long short-term memory. Neural Comput. **9**, 1735–80 (1997). https://doi.org/10.1162/neco.1997.9.8.1735

17. Kanarachos, S., Christopoulos, S.R.G., Chroneos, A., Fitzpatrick, M.E.: Detecting anomalies in time series data via a deep learning algorithm combining wavelets, neural networks and hilbert transform. Expert Syst. Appl. **85**, 292–304 (2017). https://doi.org/10.1016/j.eswa.2017.04.028. https://www.sciencedirect.com/science/article/pii/S0957417417302737

18. Makridakis, S., Spiliotis, E., Assimakopoulos, V.: The m4 competition: results, findings, conclusion and way forward. Int. J. Forecast. **34**(4), 802–808 (2018). https://doi.org/10.1016/j.ijforecast.2018.06.001. https://www.sciencedirect.com/science/article/pii/S0169207018300785

19. Maya, S., Ueno, K., Nishikawa, T.: dLSTM: a new approach for anomaly detection using deep learning with delayed prediction. Int. J. Data Sci. Anal. **8**, 137–164 (2019). https://doi.org/10.1007/s41060-019-00186-0

20. Otte, C.: Safe and interpretable machine learning: a methodological review (2013). https://api.semanticscholar.org/CorpusID:56899177

21. Sherstinsky, A.: Fundamentals of recurrent neural network (RNN) and long short-term memory (LSTM) network. Phys. D Nonlinear Phenom. **404**, 132306 (2020). https://doi.org/10.1016/j.physd.2019.132306. https://www.sciencedirect.com/science/article/pii/S0167278919305974

22. Song, C., Ristenpart, T., Shmatikov, V.: Machine learning models that remember too much. In: Proceedings of the 2017 ACM SIGSAC Conference on Computer and Communications Security (2017). https://api.semanticscholar.org/CorpusID:2904063

23. Staudemeyer, R.C., Morris, E.R.: Understanding LSTM – a tutorial into long short-term memory recurrent neural networks (2019)

24. Strobelt, H., Gehrmann, S., Pfister, H., Rush, A.M.: LSTMVis: a tool for visual analysis of hidden state dynamics in recurrent neural networks (2017)

25. Zha, D., et al.: Data-centric artificial intelligence: a survey (2023)

Unsupervised Representation Learning for Smart Transportation

Thabang Lebese[2,3](✉) [iD], Cécile Mattrand[1], David Clair[1],
Jean-Marc Bourinet[2], and François Deheeger[3]

[1] Université Clermont Auvergne, CNRS, Institut Pascal, Clermont-Ferrand, France
[2] Université Clermont Auvergne, CNRS, LIMOS, Clermont-Ferrand, France
[3] Manufacture Française des Pneumatiques Michelin, Clermont-Ferrand, France
thabang.lebese@sigma-clermont.fr

Abstract. In the automotive industry, sensors collect data that contain valuable driving information. The collected datasets are in multivariate time series (MTS) format, which are noisy, non-stationary, lengthy, and unlabeled, making them difficult to analyze and model. To understand the driving behavior at specific times of operation, we employ an unsupervised representation learning method. We present Temporal Neighborhood Coding for Maneuvering (TNC4maneuvering), which aims to understand maneuverability in smart transportation data via a use-case of bivariate accelerations from three operation days out of 2.5 years of driving. Our method proves capable of extracting meaningful maneuver states as representations. We evaluate them in various downstream tasks, including time-series classification, clustering, and multi-linear regression. Moreover, we propose methods for pruning the sizes of representations along with a window-size optimizing algorithm. Our results show that TNC4maneuvering has the capacity to generalize over longer temporal dependencies, although scalability and speedup present challenges.

Keywords: Multivariate Time-series · Representation learning · Classification · Clustering · Regression

1 Introduction

Modern transportation is now equipped with more sensors than ever before, making the term "smart transportation" more appropriate. This improves efficiency, security, and helps keep up with ever-changing environmental and government regulations. The sensors collect large amounts of data during operational hours from various parts of the vehicle, including but not limited to engine performance, external conditions, and tire states. These sensors measure different driving behaviors and states as a function of operational time or mileage. For example, the Global Positioning System (GPS) collects geographical data, while

I. Miliou et al. (Eds.): IDA 2024, LNCS 14642, pp. 15–27, 2024.
https://doi.org/10.1007/978-3-031-58553-1_2

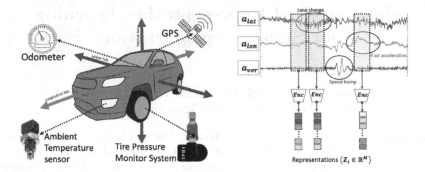

Fig. 1. Left: Smart vehicle with multiple sensors. Right: Encoding multivariate acceleration signals.

sensors inside the odometer read mileage coverage. Ambient temperature sensors measure external driving temperature conditions, and Tire Pressure Monitoring Systems (TPMS) measure the temperature and pressure inside the tires over time as depicted in Fig. 1 (left). The collected data is often high sampled, lengthy, noisy, and impractical to label. As a result, it is challenging to relate underlying behaviors/states to other datasets. This highlights the need for representation learning methods, which can output vectorial summaries from multi-sensory inputs of variables over a given time window. The resulting vectors are descriptors of latent behaviors of the physical system as illustrated with the three input accelerations in Fig. 1 (right). These accelerations are expected to describe different physical maneuvers of a vehicle, rendering them indirectly related. Vehicle maneuvers are rather directly related to driving behavior because driving generally involves three main actions: controlling the steering wheel, stepping on the accelerator, and pressing the brake pedal.

The three accelerations in Fig. 1 (right) are lateral acceleration (a_x), longitudinal acceleration (a_y), and vertical acceleration (a_z), which pertain to steering actions, accelerator and brake pedal usage, and up-and-down movements experienced by a vehicle, respectively. Following our work in [10] on simulated datasets, we here present Temporal Neighborhood Coding for Maneuvering (TNC4maneuvering), an unsupervised representation learning method to extract states for understanding maneuverability. TNC4maneuvering is robust enough to identify and locate temporal transitions between states without any prior knowledge about labels of the states. It employs contrastive learning for its ability to handle long, noisy, and non-linear MTS datasets without the need for reconstruction, significantly reducing computation costs. Furthermore, as an improvement we propose two offline pruning methods for optimizing the sizes of learned representations as well as a window-size selection algorithm. These are useful in the absence of expert knowledge. We evaluate the obtained latent representations by assessing key performance indicators (KPIs) of downstream tasks, namely clustering, classification, and multi-linear regression based on three different driving days to observe the generalization and scalability of our method. To sum up, our contribution is three-fold: 1) TNC4maneuvering, an unsupervised representation method for understanding maneuverability in smart transportation, 2) an offline window-size selection and optimization method that avoids treating it as an additional hyper-parameter, and 3) two offline representation pruning strategies for optimizing dimensions of representations.

2   Related Work

Unsupervised representation learning has excelled in various MTS tasks, but its application to smart transportation MTS datasets is generally limited if any. Existing attempts, such as the application of Bag of Words (BoW) model in [1], led to a representation like output with focus on classifying aggressive driving maneuvers only. Such approaches do not generalize well making them incapable of other alternative subsequent tasks. Recent works explore contrastive learning for representation learning by contrast of similar and dissimilar instances. Examples include [3,4,6,7,9,13,15,18,19]. Notable exceptions are [17], which disentangles seasonal-trend features using time and frequency domains, and [2], which jointly learns contextual, temporal, and transformation consistencies, later applying them to classification, forecasting, and anomaly detection tasks. To the best of our knowledge this is the first work reporting the use of pure unsupervised representation learning on acceleration MTS, specifically for understanding vehicle maneuvering with capabilities of multitask downstream.

3   Method

In [10], we explored three state-of-the-art approaches [6,13,15] that use contrastive learning on simulated MTS datasets to extract underlying states. Building on these findings, we further enhanced TNC by incorporating offline window-size selection, latent space tuning by pruning, and an exponentially dilated convolutional neural network (CNN) encoder. In our extension, here dubbed TNC4maneuvering, introduces an unsupervised representation learning framework to extract underlying driver maneuver states from acceleration signals of a vehicle. Our encoder is specifically designed to efficiently extract maneuver states.

TNC4maneuvering: The backbone of our method is a non-linear composition function encoder (Enc), typically a deep neural network, taking a static window $W_t \in \mathbb{R}^{F \times \delta}$ centered at time t with sub-length δ and F number of features. A tuple of samples, an anchor (W_t), a positive (W_l) and negative (W_k) windows are sampled from input MTS where each window $W_{t,l,k} \in \mathbb{R}^{F \times \delta}$ generates a representation vector $Z_{t,l,k} \in \mathbb{R}^M$, where $M << F \times \delta$ is the size of the representation vector. W_l and $W_t \in N_t$ share the same neighborhood centered at t, while $W_k \in \overline{N}_t$ is at a distant non-neighbourhood. The semantic similarities and dis-similarities between windows is controlled by the temporal neighborhood around W_t. This region is defined as a region where acceleration signals

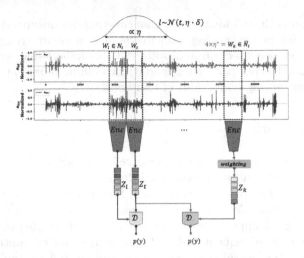

Fig. 2. Overall TNC4maneuvering framework: Encoder: $Enc(W_t) \in \mathbb{R}^{F \times \delta}$, outputs representations $Z_t \in \mathbb{R}^M$, with Discriminator: $\mathcal{D}(Z_t, Z_{l \vee k}) \in [0, 1]$.

are relatively stationary compared to their pre and post-windows, they are therefore assumed to be generated from the same underlying maneuvering state. The objective function (1) is a partial contrastive loss that learns signals via encoding and evaluates them using a Discriminator (D) that identifies representations with similar underlying maneuverings.

$$\mathcal{L} = -\mathbb{E}_{W_t \sim X} \left[\mathbb{E}_{W_l \sim N_t} \left[\log \left(\mathcal{D}(Z_t, Z_l) \right) \right] \right.$$

$$\left. + \mathbb{E}_{W_k \sim \overline{N}_t} \left[w_t \log \left(\mathcal{D}(Z_t, Z_k) \right) + (1 - w_t) \log \left(1 - \mathcal{D}(Z_t, Z_k) \right) \right] \right] \quad (1)$$

The unit root test, Augmented Dickey-Fuller (ADF)[1] is used for determining relative stationarity regions. Furthermore, the objective function is weighted with (w_t) and $(1 - w_t)$, an ideas from Positive-Unlabeled (PU) learning to counter potential sampling bias in the contrastive objective. This compensates for negative samples drawn from outside of the neighborhood which may in fact be similar to those of an anchor window. The overall framework is depicted in Fig. 2, details on this framework can be found in [10,15].

4   Experiments

TNC4maneuvering extends [10], it is implemented in PyTorch framework (v1.12.1) and the source code is available on GitHub[2]. All experiments are conducted using a single Nvidia Tesla P40 GPU with CUDA 11.2.152. All datasets

---

[1] arch.unitroot.ADF.
[2] https://github.com/ThabangDLebese/tnc4maneuvering.

Table 1. Selected different driving days showing mileage coverage, time taken and corresponding length of observations.

Operational day	Mileage (Km)	Time (Mins)	No. observations
2018/10/23 (One D_s)	20	584	1957
2019/11/28 (One D_l)	665	932	16568
2018/10/24-31 (Eight D)	499	5260	19273

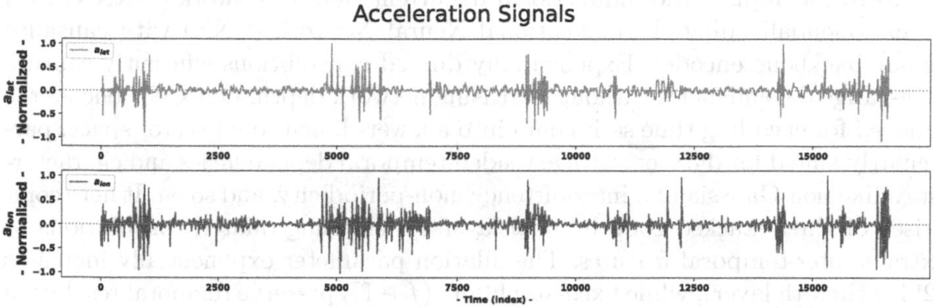

Fig. 3. Normalised bivariate (a_{lat}, a_{lon}) acceleration for One D_l operation day.

are normalized, and the evaluations include three downstream tasks: clustering, classification, and multi-linear regression.

4.1 Vehicle Acceleration Datasets

Vehicle maneuvering is an automotive problem that is central to understanding driving behavior from sensory signals. Our use-case vehicle is a Peugeot 208 model used as a fleet car, where operation time is accumulated as an amount of time where driving activities are collected by different sensors. In this particular work we focus only on the two accelerations, namely the lateral (a_{lat}) which is the effective measure of cornering (negative is for right turning, 0 is straight line or breaking and positive is left turning) and the longitudinal acceleration (a_{lon}) where the straight line acceleration (negative braking, 0 is constant speed and positive is accelerating).

Both accelerations are reported as a fraction of the gravitational acceleration (ms^{-2}). Analysing different vehicle acceleration behaviours on different driving days can help to understand different maneuvering behaviors of a vehicle over time. We consider three bivariate sample signals with different dates, signal lengths, total covered time and mileage covered as depicted in Table 1.

As a pre-processing stage, we perform a data normalization to avoid statistical biases that can lead to misinterpretation of the encoded results. Both input features are normalized such that each $X_i = x_i/x_{\max} \in [-1, 1]$, for $x_{\max} = \max|x_i|, i = \{1, 2\}$, preserving the zero values on each feature. Our bivariate acceleration datasets are extracted from different driving days from the overall 2.5 years of driving. From these datasets, we extract one short day (One D_s), one long day (One D_l) and eight days (Eight D) that is inclusive of (One D_s). These days correspond to separate driving dates with different mileage coverage, time overall required time and total corresponding observation lengths. Figure 3 depicts the normalized accelerations of the One D_l operational day.

4.2 Encoder Details

From [10], we replace the Bidirectional Recurrent Neural Network (BiRNN) with an exponentially dilated Convolutional Neural Network (CNN) with causality as our backbone encoder. Exponentially dilated convolutions efficiently capture long-range dependencies without increasing network depth. Our CNN encoder is tailored for encoding time series data into a lower-dimensional vector space, particularly suited for datasets with extended temporal dependencies and characteristics like non-Gaussianity, intermittency, non-periodicity, and so on. It here comprises of three stacked convolutional layers, each using dilated convolutions to extract inter-temporal features. The dilation parameter exponentially increases (2^i for the i-th layer), while fixed-size filters ($f \in \mathbb{N}$) preserve temporal resolution and alignment. The output undergoes global max pooling, compressing temporal information into a fixed-size vector. This result is flattened and processed by a linear layer, further reducing the dimensionality to produce an encoding of size M, serving as a compressed representation based on a window size W_t.

Our encoder design offers flexibility by allowing customizable encoder sizes (M), this is to say it incorporates a classification component for compatibility with subsequent classification tasks. This design choice provides several advantages, including enhanced generalization for downstream tasks and easy pruning options. Each exponentially dilated convolution layer encodes data through a convolution operation with dilation defined as:

$$F(s) = (W_t \star_d f)(s) = \sum_{i=0}^{k-1} f(i) W_t^{s-d \cdot i}, \qquad (2)$$

where $F(s)$ represents the computed output on each layer for samples $s \in W_t$ ($\in \mathbb{R}^{F \times \delta}$), with a dilation rate of d, filter size k, and ($s - d \cdot i$) accounting for historical direction. Other hyperparameters include a batch size of 5, a learning rate of 1×10^{-5}, and a weight decay of 1×10^{-4}, using the Adam optimizer. We perform a train/test data split without validation, training epochs are limited to 30, 20, and 10 epochs for datasets One D_s, One D_l, and Eight D, respectively.

4.3 Hyperparameter Tuning

In [10], we recognized the need for further tuning two hyperparameters: the window size (W_t) and the latent space dimension (M), while keeping the PU learning parameter fixed at $w_t = 0.05$.

Window Selection: An appropriate window size should capture important information about maneuvering states without being too wide or too narrow.

Table 2. Cross data performances on multi-task downstream before pruning.

TNC4maneuvering (Before pruning)							
Operational day	W_t	Classification		Clustering		Regression	
		AUPCR	Accuracy	Silhouette	DBI	R^2	Loss
One D_s	250	0.988	98.82	0.715	0.492	−0.290	2.075
One D_l	250	0.936	84.86	0.372	1.014	0.326	1.023
Eight D	250	0.976	84.47	0.320	1.202	0.288	1.255

Determining a suitable window size can be achieved by relying on expert knowledge or by treating it as a hyperparameter. We here combine two offline methods for selecting a suitable window-size, 1) we examine numerical first order derivatives of acceleration signals of the window-size. If the derivative is constant, this indicates no state change; otherwise, a different state. This is simultaneously applied on both a_{lat} and a_{lon}, with windows non-overlapping. The numerical gradients within each window are approximated using numpy.gradient[3] approximation, where interior[4] and end[5] points are approximated differently as the window size increases until a predefined necessary condition is satisfied. For sufficiency, 2) is employed using the Augmented Dicky-Fuller (ADF)[6] test with a p-value threshold similar to that in TNC4maneuvering encoder. As a result, it was found that window sizes shorter than 250 do not contain enough non-stationarity, especially in One D_l and Eight D. Therefore, we determined that window size of $W_t = 250$ ($\equiv$ 4.2 min of driving) is suitable. We use this window size in all experiments, for instances where the window size is larger than the sampled size, padding with zeros is applied. On the other hand, a downside of this window selection method is that gradients are prone to total samples evaluated compared to statistical variance.

Optimizing Representation Size: Determining the optimal size M for $Z_t \in \mathbb{R}^M$ is a challenging open question in representation learning. A larger encoding size captures more information but risks adding irrelevant details, affecting interpretability. Conversely, a smaller size may lead to insufficient encoding and reduced generalizability. Achieving the right balance is crucial. We propose two methods for selectively removing unnecessary details from representations, a technique referred to as latent space pruning. Initially setting $M \in \mathbb{N}$, we obtain the optimal pruned $m \leq M, m \in \mathbb{N}$ using two proposed methods: 1) Pearson Cor-

---

[3] numpy.gradient package.
[4] Interior points:$(f(x+h) - f(x-h))/2h$, for evenly spaced ($h = 1$).
[5] End points:$(f(x+h) - f(x))/h$ and $(f(x) - f(x-h))/h$, for evenly spaced ($h = 1$).
[6] ADF(W_t), if p-value > 0.01 signals is non-linear, else linear..

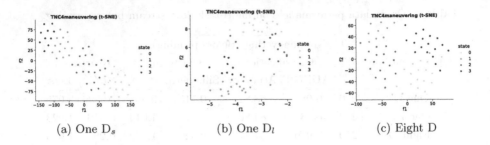

(a) One D_s (b) One D_l (c) Eight D

Fig. 4. t-SNE visualization of three representations of accelerations, before pruning.

relation Coefficient (PCC) [5], this method eliminates highly linearly correlated representations with a preset absolute correlation threshold of 0.7, resulting in representations of size m_1 and 2) Principal Component Analysis (PCA) [8], it utilizes a cumulative explained variance with preset threshold of 0.95 to determine m_2, the number of components to retain. This threshold identifies the size (m_2) of the representations required to achieve it, and these representations are considered important.

4.4 Evaluation

In order to evaluate the performance of TNC4maneuvering, we evaluate three downstream tasks namely, time-series classification, clustering, and multi-linear regression across our datasets.

Classification: In this subsequent task, we employ a linear classifier due to its effectiveness in separating representations in high dimensions, assuming well-separated representations. In the TNC4maneuvering model, setting the parameter ($classify = True$) triggers the classification task. Encodings are input to a classifier comprising a dropout layer to prevent overfitting and a linear layer mapping the encodings to predefined maneuver output classes ($n_{classes}$) for classification. We evaluate using prediction accuracy and the area under the precision-recall curve (AUPRC) score, specifically suitable for imbalanced classification settings. The classification algorithm learns relationships between representations and predefined maneuver labels (defined in Sect. 5), facilitating accurate prediction and categorization of maneuvering states.

Clustering: Clustering of representations assesses their separability in the latent space using k-means [12], offering insights about resulting encoding properties with predefined maneuver labels (defined in Sect. 5). We employ two metrics for evaluation: the Silhouette score and Davies-Bouldin Index (DBI). The Silhouette score measures the similarity of an encoding within its assigned cluster

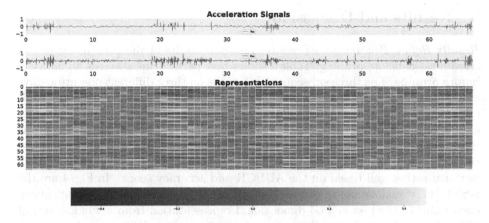

Fig. 5. One D_l accelerations in top and corresponding vector representations of size 64 encoded with a static window-size $W_t = 250$.

versus adjacent clusters, ranging from $[-1, 1]$. A higher score implies better cohesion. The DBI assesses both intra-cluster coherence and inter-cluster separation, with a lower score indicating better clusterability. Identified clusters in clustered representations are expected to reflect similar characteristics related to vehicle maneuver behavior.

Regression: In this subsequent task, peaks and valleys also known as turning points are collected. By taking consecutive differences between turning points and their square sums, quantifies their magnitudes in each window. This results to a vector $X_{man} \in \mathbb{R}^{M \times 1}$ as a summary. On the other hand, the resultant vector should offer insights into the intensity and characteristics of extrema fluctuations found in the datasets. We assume a linear mapping as a first trial where a vector $X_{man} \in \mathbb{R}^{M \times 1}$ is regressed by multivariate representations $Z \in \mathbb{R}^M$, although our perspective would be to propose a non-linear one. A train-test (70/30) data split is performed, as evaluation coefficient of determination (R^2) and learning loss are used.

Representations: Visualized representations against acceleration signals over time enhances the understanding and interpretation of extracted maneuver state and how they are modeling in the latent space ($Z \in \mathbb{R}^M$). This visual metric is crucial for comprehending vehicle maneuvering as it provides insights into maneuver behavior through visualization, facilitating the recognition of changes in maneuver states over time. Capturing these changes clearly enables deeper insights into the severity or gentleness of driver maneuvers.

5 Results and Discussion

Results Before Pruning. This section presents results of the subsequent downstream tasks before pruning. Table 2 shows task performances across all datasets before pruning. The three subsequent ML tasks on three different operation days exhibit variations. Linear regression performs the least consistently well, indicating that localized manually extracted maneuver behaviors are not linearly explained by representations. A perspective would be to resort to a non-linear mapping to better link the proposed representations with the quantity interest or further improve the quality of the representations. Overall, classification task perform rather well based on the AUPCR and accuracy scores. In Fig. 4 are the t-SNE [11] visualizations of representations of each dataset. In each visualization, each point in the plot is a 64 dimensional representation from a window-size of 250, where colors indicate different maneuvering states. With no prior domain knowledge on maneuver states, we propose a statistical approach serving as ground truth unlike in the works of authors [14]. We therefore label each dataset into four maneuvering activities, namely state 0: both a_{lat} and a_{lon} are stationary, state 1: only a_{lon} is stationary, state 2: only a_{lat} is stationary, and state 3: both a_{lat} and a_{lon} are non-stationary. Stationarity refers to cases where the ADF (p-values > 0.01) for each window-size of 250 of signals. We treat these states as a ground truth without loss of generality. In Figs. 4a and 4b, the two subgroups of states (1 and 3) and states (0 and 2) can be assumed to correspond to activities of a_{lon} and a_{lat} respectively. However, distinguishing patterns between One D_s

Table 3. Pruned representations: PCC vs. PCA on various operation days.

Operational day	Initial size (M)	PCC (m_1)	PCA (m_2)
One D_s	64	6	3
One D_l	64	4	7
Eight D	64	7	6

Table 4. Cross data performances on multi-tasks downstream after pruning.

Operational day	W_t	Classification		Clustering		Regression	
		AUPCR	Accuracy	Silhouette	DBI	R^2	Loss
TNC4maneuvering (After PCA pruning)							
One D_s	250	0.756	77.51	-	-	-0.407	2.262
One D_l	250	0.417	59.57	0.414	0.241	0.340	1.002
Eight D	250	0.450	48.15	0.497	0.202	-0.328	1.416
TNC4maneuvering (After PCC pruning)							
One D_s	250	0.956	97.02	0.404	0.758	-0.269	2.040
One D_l	250	0.936	84.86	0.247	1.211	0.330	1.017
Eight D	250	0.903	79.25	0.189	1.494	-0.313	1.400

and One D_l cluster patterns is challenging due to their difference of being short and long distance operations. In Fig. 4c, no clear-cut pattern emerges of state separability, as all states are present, reflecting that diverse driving behaviors are collected over multiple days. Figure 5 shows both accelerations and learned representations without additional for One D_l day. In this day of activity, we see that in cases where both accelerations (a_{lon}, a_{lat}) have simultaneous activity, it can also be observed with correspondence to the color code in the representation space, similar to when there is low activity. Overall, it appears that a_{lon} strongly influences the characteristics of the representations. This is due to the vehicle executing less full turns and rather accelerating and decelerating more on this particular operation day. A co-interpretation of Figs. 4b and 5 suggests that maneuverability is primarily governed by a_{lon} activity, corresponding to states 1 and 3 in Fig. 4b and the dark (negative) color codes in Fig. 5.

Results After Pruning. The two pruning methods yield different representation sizes, as shown in Table 3. These variations arise from the methods distinct selection criteria: PCC eliminates highly linear correlated representations, while PCA determines the required representation count based on cumulative explained variance. We apply pruning methods offline and subsequently evaluate model performance as post-pruned representations, as illustrated in Table 4. Our post-pruning methods further assume that representations are more disentangled since unnecessary components are removed. There was no further training to fine-tune and update model weights to recover some of the lost accuracy. Therefore, after pruning there is no major improvement on the three subsequent tasks. Linear regression performs the least further indicating that localized turning points are not linearly explained by representations. Overall, there is a decline in performance across each subsequent task. Consequently, our offline pruning methods have a reduced performance, as there is no post pruning model weights updates. In the work [16], the authors address this issue by implementing online pruning, which enhances efficiency, generalization, and interpretability without significant performance loss.

6 Conclusion

Our unsupervised representation learning method, TNC4maneuvering, effectively extracts maneuverability representations from complex MTS vehicle dataset. Its versatility is evidenced by performance in various downstream tasks, especially on a classification task. Although it allows one to capture longer temporal dependencies, scalability and speedup remain areas of challenge, which are our next points of focus. Another win to claim is that we have managed to get rid of two extra hyperparameters, the window-size and size of representations, reducing the number of hyperparameters to a bare minimum and hence reducing further complexity. TNC4maneuvering holds great promise for enhancing maneuverability analysis in smart transportation, laying a foundation for general future usage in other applications. In our future work, we will replace gradients testing with a variance due to its insensitivity to total data samples and incorporate pruning within the training framework in order to update model weights after pruning. Regarding scaling and speedup, we plan to replace the ADF test with a pre-calculated stationarity matrix.

Acknowledgement. This work has received funding from the European Union's Horizon 2020 Research and Innovation Programme, Grant Agreement n⁰ 955393. Moreover, thanks to Manufacture Française des Pneumatiques Michelin for support and car dataset provision.

References

1. Carlos, M.R., González, L.C., Wahlström, J., Ramírez, G., Martínez, F., Runger, G.: How smartphone accelerometers reveal aggressive driving behavior?-the key is the representation. IEEE Trans. Intell. Transp. Syst. **21**(8), 3377–3387 (2019)
2. Choi, H., Kang, P.: Multi-task self-supervised time-series representation learning. arXiv:2303.01034 (2023)
3. Eldele, E., et al.: Time-series representation learning via temporal and contextual contrasting. In: International Joint Conference on Artificial Intelligence, IJCAI 2021, pp. 2352–2359 (2021)
4. Eldele, E., et al.: Self-supervised contrastive representation learning for semi-supervised time-series classification. arXiv:2208.06616 (2022)
5. Eslami, T., Awan, M.G., Saeed, F.: GPU-PCC: a GPU based technique to compute pairwise Pearson's correlation coefficients for big FMRI data. In: ACM International Conference on Bioinformatics, Computational Biology, and Health Informatics, pp. 723–728 (2017)
6. Franceschi, J.Y., Dieuleveut, A., Jaggi, M.: Unsupervised scalable representation learning for multivariate time series. In: Advances in Neural Information Processing Systems, vol. 32 (2019)
7. Hyvarinen, A., Morioka, H.: Unsupervised feature extraction by time-contrastive learning and nonlinear ICA. In: Advances in Neural Information Processing Systems, vol. 29 (2016)
8. Kurita, T.: Principal component analysis (PCA). In: Computer Vision: A Reference Guide, pp. 1–4 (2019)
9. Lai, C.I.: Contrastive predictive coding based feature for automatic speaker verification. arXiv:1904.01575 (2019)
10. Lebese, T., Mattrand, C., Clair, D., Bourinet, J.M.: Unsupervised representation learning in multivariate time series with simulated data. In: 2023 Prognostics and Health Management Conference (PHM), pp. 217–225 (2023)
11. Van der Maaten, L., Hinton, G.: Visualizing data using t-SNE. J. Mach. Learn. Res. **9**(11) (2008)
12. MacQueen, J.: Classification and analysis of multivariate observations. In: 5th Berkeley Symposium on Mathematical Statistics and Probability, pp. 281–297. University of California, Los Angeles, LA, USA (1967)
13. Oord, A.V.D., Li, Y., Vinyals, O.: Representation learning with contrastive predictive coding. arXiv:1807.03748 (2018)
14. Sarker, S., Haque, M.M., Dewan, M.A.A.: Driving maneuver classification using domain specific knowledge and transfer learning. IEEE Access **9**, 86590–86606 (2021)

15. Tonekaboni, S., Eytan, D., Goldenberg, A.: Unsupervised representation learning for time series with temporal neighborhood coding. arXiv:2106.00750 (2021)
16. Weatherhead, A., et al.: Learning unsupervised representations for ICU timeseries. In: Conference on Health, Inference, and Learning, pp. 152–168. PMLR (2022)
17. Woo, G., Liu, C., Sahoo, D., Kumar, A., Hoi, S.: Cost: contrastive learning of disentangled seasonal-trend representations for time series forecasting. arXiv:2202.01575 (2022)
18. Yue, Z., et al.: Ts2vec: towards universal representation of time series. In: AAAI Conference on Artificial Intelligence, vol. 36, no. 8, pp. 8980–8987 (2022)
19. Zerveas, G., Jayaraman, S., Patel, D., Bhamidipaty, A., Eickhoff, C.: A transformer-based framework for multivariate time series representation learning. In: ACM SIGKDD Conference on Knowledge Discovery & Data Mining, pp. 2114–2124 (2021)

T-DANTE: Detecting Group Behaviour in Spatio-Temporal Trajectories Using Context Information

Maedeh Nasri[1]([⊠]) [iD], Thomas Maliappis[2] [iD], Carolien Rieffe[1,3,4] [iD],
and Mitra Baratchi[2] [iD]

[1] Department of Developmental Psychology, Leiden University,
Leiden, The Netherlands
{m.nasri,Crieffe}@fsw.leidenuniv.nl
[2] Leiden Institute of Advanced Computer Science, Leiden University,
Leiden, The Netherlands
t.maliappis@umail.leidenuniv.nl, m.baratchi@liacs.leidenuniv.nl
[3] Faculty of Electrical Engineering, Mathematics and Computer Science,
University of Twente, Enschede, The Netherlands
[4] Department of Psychology and Human Development, University College London,
London, UK

Abstract. The present study addresses the group detection problem using spatio-temporal data. This study relies on modeling contextual information embedded in the trajectories of surrounding agents as well as temporal dynamics in the trajectories of the agent of interest to determine if two agents belong to the same group. Specifically, our proposed method, called T-DANTE, builds upon the Deep Affinity Network (DANTE) [16] for Clustering Conversational Interactants using spatio-temporal data and extends it by incorporating Recurrent Neural Networks (RNN) (i.e., Long Short-term Memory (LSTM) and Gated Recurrent Unit (GRU)) to capture the temporal dynamics inherent in the trajectories of agents. Our ablation study demonstrates that including context information, combined with temporal dynamics, yields promising results for the group detection task across five real-world pedestrian and five simulation datasets using two common evaluation metrics, namely Group Correctness and Group Mitre metrics. Moreover, in the comparative study, the proposed method outperformed three state-of-the-art baselines in terms of the group correctness metric by at least 17.97% for pedestrian datasets. Although some baselines perform better in simulation datasets, the difference is not statistically significant.

Keywords: Spatio-temporal Data · Affinity Network · Group Detection

1 Introduction

Group detection using spatio-temporal trajectories has wide-ranging applications such as studying human mobility [4,10,15,16], analyzing social behavior in a community [9,11], and understanding migration patterns of animals [7,18].

I. Miliou et al. (Eds.): IDA 2024, LNCS 14642, pp. 28–39, 2024.
https://doi.org/10.1007/978-3-031-58553-1_3

Group detection research has predominantly focused on classical machine learning methods involving feature engineering [14, 19]. These approaches typically require manual feature extraction to train the model, a process that can be time-consuming and potentially introduce bias.

Recently, deep neural networks (DNN) gained popularity for modeling interactions within spatio-temporal trajectories [6, 10] due to their capability to detect complex and nonlinear relationships between variables. GD-GAN [4], NRI [6], and WavenetNRI [10] are all DNN-based approaches designed to decode the spatio-temporal patterns to identify group behavior. The preliminary limitation of this line of research is the focus of models only on the target agents, neglecting the impact of surrounding agents in detecting group behavior.

To address this issue, Swofford et al. [16] introduced DANTE, which incorporates context features representing surrounding agents using spatio-temporal trajectories, aiming to learn a graph representation for a single-frame scene via a neural network. This method considers someone's surroundings when estimating conversational group membership. The main limitation of this study is the use of multilayer perceptron (MLP), which are computationally intensive and not suitable for modeling temporally-ordered data such as spatio-temporal trajectories. Moreover, the model focuses solely on a singular frame which may overlook essential information embedded in data.

To address this gap, the present study introduces an approach, building upon DANTE [16], that represents agents and their spatio-temporal trajectories as a social graph using a DNN. The study proposes incorporating RNN layers to account for the temporal aspect of agent movements, which was overlooked in the original model and has demonstrated acceptable performance in predicting vehicle trajectories [3]. Moreover, the proposed model builds on the concept of context information that has previously shown promising results in identifying group behavior [16] and is further refined by including scenes with multiple timeframes. Subsequently, a community detection algorithm is applied to identify groups among agents. Specifically, our main contributions are:

- Introducing a novel framework for group detection from spatio-temporal trajectory datasets that extends the original DANTE by including multiple timeframes and employing RNN layers (i.e., LSTM and GRU) to capture temporal dependencies among trajectory data samples.
- Conducting extensive ablation studies to investigate the impacts of context size and different RNN layer designs on the performance of the model.
- Introducing a novel trajectory simulation framework for interaction detection by extending the spring simulation framework [6, 10] to include the concept of attraction points (i.e., points where group members often mingle around).
- Evaluating T-DANTE using five pedestrian datasets and five simulation datasets against state-of-the-art baselines (i.e., NRI [6], WavenetNRI [10], and DANTE [16]) in terms of Group Correctness and Group Mitre metrics.

The present study is organized as follows. Section 2 provides background information and reviews the related works. Section 3 presents the problem formulation and discusses the details of the proposed approach. Section 4 details the experimental setup, the datasets, the evaluation metrics, the baselines and explains the key findings from the experiments. Finally, Sect. 5 summarizes the entire study and presents potential directions for future research.

2 Background

The first models for solving the group detection task have mainly focused on employing classical machine learning approaches involving manual feature selection process [14,19]. Manual feature engineering has various drawbacks such as being time-consuming, requiring domain expertise, and having the potential to introduce bias. To tackle these issues, recent studies have integrated DNNs into their frameworks relying on their capability of automatically capturing the complex dependencies between data [6,10,12]. The present study (i) employs RNN models as a special form of DNN and (ii) incorporates context information to address group detection tasks. The following sections focus on the related works around these two main features of the proposed method.

Recurrent Neural Networks (RNNs): Many recent studies have integrated DNNs to capture the complex dependencies in spatio-temporal data [6, 10,12]. Neural Relational Inference (NRI) [6] is an RNN-based approach that models interactions between individuals. This work uses Graph Neural Networks (GNNs) and RNNs to build an auto-encoder model to learn the latent vectors that represent the interaction graph. Building upon this foundation, Nasri et al. [10] introduced WavenetNRI, a model that integrates a gated Residual Dilated Causal Convolutional Block [12] to capture both short and long-term interactions in the sequences of edge features. This approach utilizes learned interactions to effectively model group interactions among individuals. The main disadvantage of these approaches is the complete reliance on the model to understand which agents affect the trajectories of others. In contrast, the proposed method only maintains the surrounding agents as part of the same group, thus, not all agents are included in the learning process. In this way, the model focuses on the interactions between agents that are close to each other while excluding insignificant agents located at a distance from agents of interest. Excluding these agents may reduce the computational cost of the model.

Including Context Information: Recent studies have demonstrated the positive impact of modeling contextual information on the performance of the group detection model. Deep Affinity Network (DANTE) [16] utilizes a specified number of surrounding agents as context for clustering interacting agents. This approach is limited due to the use of data within a single timeframe. Similarly, Tan et al. [17] consider all agents in a given scene as contextual input and predict affinities between agents. While these methodologies demonstrate the importance of considering broader contextual cues for accurate group detection, they overlook the temporal aspects of movement.

The present study is inspired by the aforementioned approaches to tackle the group detection problem using spatio-temporal data. The proposed model uniquely captures the temporal dynamics of the data by incorporating RNN layers and further combines the temporal features with context information in a scene.

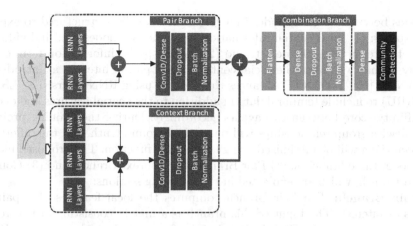

Fig. 1. Visualisation of T-DANTE framework.

3   Material and Methods

In this section, first, the problem of a group detection task is formulated. Second, the proposed framework to learn the affinities between agents in a scene is introduced. Since the proposed network is based on DANTE [16] combined with temporal features of spatio-temporal data (i.e., including RNN layers), we name it T-DANTE. Third, the Dominant Sets (DS) [5] community detection algorithm is explained, which detects groups from a given graph. Figure 1 provides a visual representation of the proposed framework.

3.1   Problem Formulation

Consider a dataset of trajectories of N agents over a scene of T consecutive time steps. X_i^t is the location and velocity of agent $i \in 1, ..., N$ in time step $t \in 1, ..., T$. The trajectory of agent i can be represented by $X_i^{1:T}$, respectively. We are interested in detecting groups $C = \{c_j | j \in [1, K]\}$ in which each agent belongs, where $1 \leq K \leq N$ is the number of groups. Assuming that agents in the same group share similar spatial behavior over a scene of T time steps, the group relationships do not change during a scene, and the duration T of the scene is fixed.

Our proposed method is based on a DNN that approximates the pairwise affinities between agents in each scene and produces the corresponding adjacency matrix. The social graph represented by each adjacency matrix will then be given to a community detection algorithm to discover groups by identifying communities in a graph. This paper uses DS [5] as the community detection algorithm.

3.2   Affinity Learning Network

This section introduces T-DANTE, a DNN that predicts the weights for each edge in an affinity graph, i.e., a graph structure that represents the relationships or affinities between different nodes in a graph. T-DANTE is structured to exploit two aspects: (1) local spatial information from the two nodes (i.e., individuals) connected to an edge of interest, and (2) global spatial information from other nearby agents, who form the social context of the pair of interest, as introduced in DANTE [16]. T-DANTE advances this idea by using RNN layers (i.e., LSTM and GRU) to include temporal data in addition to the spatial features and decide the affinity score between two agents accordingly. During the training process, the pairwise group relationships will be used as ground truth, and the difference between them will be minimized using the log loss function. The proposed model consists of three branches: (1) Pair Branch, (2) Context Branch, and (3) Combination Branch, which are explained in the following sections:

Pair Branch: The Pair Branch computes the local features of a pair of agents of interest. The input of this branch is a matrix consisting of two rows, one per agent in a pair. Each feature is transformed independently by an RNN layer. This study investigates the performance of T-DANTE by using LSTM (and conv1D) layers, so-called 'LSTM-conv', and GRU (and dense) layers, so-called 'GRU-dense', separately. The LSTM features memory cells and intricate gating mechanisms, including input, forget, and output gates, which allow them to selectively store and retrieve information over extended sequences. This capability is particularly beneficial for tasks where modeling long-term dependencies is crucial. The GRU employs simpler update and reset gates, offering computational efficiency with fewer parameters. GRUs perform well in tasks where capturing shorter-term dependencies is essential. The concatenation of the outputs of RNN layers is then managed by a series of convolutional or dense layers. In the rest of this paper, the term 'T-DANTE' is used for the model, which includes the LSTM-conv block, and the term 'T-DANTE-GD' is used for the model, which includes the GRU-dense block. Both designs are followed by a Dropout layer and a Batch Normalisation layer. The Dropout layer reduces overfitting during the training process and improves the generalization of the model. The Batch Normalisation layer is used to avoid the covariate shift that occurs when input feature distribution changes during training.

Context Branch: The Context Branch computes the global feature representations of the social context of the pair of interests as depicted in Fig. 1. The number of agents considered as social context is a hyperparameter of the model. Similar to the Pair Branch, the Context Branch first applies RNN layers to the features of each agent in the context. The convolutional/dense layer, Dropout layer, and Batch Normalisation layer sequence is repeated x times with various filters based on the defined configuration.

Combination Branch: The Pair Branch and Context Branch are followed by a concatenate layer to combine their acquired information, the so-called Combination Branch. In this branch, the tensors are flattened and used by a sequence of a dense layer, a dropout layer, and a batch normalization layer n times with different filter sizes. The number of layers and the filter size depend on the characteristics of the dataset, such as the number of frames, the maximum number of agents, and the data size. The last layer of the Combination Branch is a dense layer using a Sigmoid activation function to constrain the output to the $[0, 1]$ range. Given the specified context, this is the affinity score for a pair of agents.

3.3   Graph Community Detection

Once all the affinity values between pairs of individuals are computed within the social affinity graph, the next step is modeling the group structures in the data. To achieve this, the DS algorithm [5] is used to analyze edge-weighted graphs and identify clusters based on high relative mutual affinity. In the context of this study, the social affinity graph is used to identify groups.

4   Experiments

This section first explains the pedestrian and simulation datasets used in the experiments followed by the evaluation metrics and baselines. Lastly, the results of the experiments are reported in the ablation study and comparative study.

4.1   Datasets

Pedestrian Datasets: Five pedestrian datasets, *eth*, *hotel* [13], and *zara01*, *zara02* and *students03* [8] are used in the experiments. These datasets are available in OpenTraj repository [1][1] and are commonly used as benchmarks for group detection tasks on spatio-temporal data. They consist of the location and velocity of each agent in multiple timeframes and the ground truth of the group membership. The datasets include location data using world reference W. Each agent in the context of a sample is transformed to a local coordinate system L_{ij} that represents relative locations. This local coordinate system is defined by taking the middle point of the line connecting agents i and j. This transformation improves the learning and generalization of our proposed approach.

    Simulation Dataset: In addition to pedestrian datasets, the simulation dataset was used in the experiments, inspired by the spring simulation dataset [6,10]. The advantage of simulation data is the availability of ground truth and flexible sample size to train the model. In the original simulation dataset [6,10], particles are distributed randomly in different groups moving in a 2-D space, simulating the concept of particles moving along with each other while the particles in the same group attract each other and repel particles from other groups. The locations, velocities, and the group membership of the particles are included in this data. In the experiments, the simulation algorithm is further enhanced by including group size as a simulation parameter and attraction points to stimulate group movements towards certain points, for example, a spot in a playground where children play around. The attraction points are implemented by defining a force that points each particle toward an attraction point. All the forces have the same strength value, but their direction is based

Table 1. Characteristics of pedestrian datasets and simulation datasets regarding the duration of measurements (pedestrian dataset in seconds, and simulation dataset in timeframe), the number of agents, and the number of groups.

Pedestrian Dataset			
Dataset	Duration	Agents#	Groups#
eth	773.4	360	58
hotel	722.4	390	41
zara01	360.4	148	45
zara02	420.4	204	58
students03	215.6	428	101
Simulation dataset			
sim_1	50	8	2
sim_2	50	9	2
sim_3	50	9	3
sim_4	50	10	2
sim_5	50	10	4

on the location of the particle towards the attraction points. Table 1 presents the characteristics of pedestrian datasets and five simulation datasets.

4.2 Evaluation Metrics and Baselines

This section describes the two evaluation metrics used to assess the performance of the models in the experiments. Next, the baselines used in the experiments are explained.

Group Mitre [15]: This metric is a commonly used performance metric for group detection [2,4,10] and is built on top of a scoring scheme to measure the quality of the predicted groups. The exact procedure for calculating the Group Mitre is presented by Solera et al. [15], and the details are omitted due to the space limit. The F-1 Group Mitre score is used as an evaluation metric.

Group Correctness [5,16]: This metric considers a group as correctly estimated if at least $P * |c_d|$ of its members are correctly detected, where $P \in [0, 1]$ tunes the tolerance of the evaluation to the number of misclassified members and $|c_d|$ indicates the size of the ground truth group d. The F-1 Group Correctness metric is computed over the entire scene in the experiments.

Baselines: The performance of the proposed method, T-DANTE, is compared with three baseline methods, namely DANTE [16][2], NRI [6][3], and

---

[1] https://github.com/crowdbotp/OpenTraj.
[2] The implementation details: https://github.com/mswoff/DANTE.
[3] The implementation details: https://github.com/fatcatZF/WavenetNRI.

WavenetNRI [10] (see Footnote 3). These baseline methods are described in Sect. 2.

4.3 Results and Discussions

Settings: Experiments were designed using the Python programming language. The DNNs were implemented in Tensorflow. The detailed implementation of T-DANTE and the trajectory simulation framework is available in the GitHub repository[4]. Each pedestrian dataset has been split into 5 folds, and each method has been evaluated 5 times for each fold, leading to a total of 25 runs per method. The simulation datasets have not been split into folds, as they were generated under controlled conditions without distribution shifts that need to be addressed by cross-validation. The simulation experiments were randomly split into train, test, and validation datasets. Each method has been evaluated 25 times for each simulation dataset. The Wilcoxon signed rank test has been applied to investigate the significant differences between the top two performing models. This statistical test was selected as it is a non-parametric version of the paired T-test and provides an interpretable statistic. In the following sections, first, the results of the ablation study are demonstrated. Next, the performance of the proposed model against three state-of-the-art baseline methods is presented.

Ablation Study: The ablation study is conducted for both the pedestrian and simulation datasets. For this purpose, multiple experiments with different design layers and context sizes are performed using the Group correctness (G_C) and Group Mitre (G_M) as the evaluation metrics. The employed context sizes, includes 0, 4, and 8 context size, denoted as C_0, C_4, and C_8, respectively. In addition to various context sizes, two different DNN designs are used: (1) T-DANTE with LSTM-conv block mentioned as T-DANTE, and (2) T-DANTE-GD, which includes GRU-dense block.

Pedestrian datasets: The result of Group Correctness (G_C) and Group Mitre (G_M) of T-DANTE for different variations using the pedestrian datasets are presented in Table 2. As demonstrated in this table, including context information enhances the performance of T-DANTE, except in the *eth* dataset when using the Group Mitre as the evaluation metric. Besides, a larger context size seems to work more efficiently in both T-DANTE and T-DANTE-GD models for *zara*02 and *students*03 datasets. In summary, each dataset may benefit from using a different context size depending on the features of the dataset, such as the average number of agents per scene. Moreover, T-DANTE performs better than T-DANTE-GD in almost all of the pedestrian datasets. Thus, the LSTM-conv block better processed the spatio-temporal data compared with the GRU-dense block.

Simulation Datasets: The result of Group Correctness (G_C) and Group Mitre (G_M) evaluation metrics of T-DANTE and T-DANTE-GD using the simulation datasets are presented in Tables 2. According to this table, the context of 4 agents performed best in both evaluation metrics, G_C and G_M, either in T-DANTE or in T-DANTE-GD design. This was expected since this dataset has a maximum of

---
4 https://github.com/ADA-research/context-group-detection.

Table 2. The result of Group Correctness G_C and Group Mitre G_M for T-DANTE variations using pedestrian datasets and simulation datasets. Context sizes of C_0, C_4, and C_8 agents refer to no agent, 4 and 8 agents, respectively. The * sign shows that this result is significantly different than all the other results in the same dataset according to the Wilcoxon signed-rank test.

Pedestrian Dataset	eth		hotel		zara01		zara02		students03	
	G_C	G_M	G_C	G_M	G_C	G_M	G_C	G_M	G_C	G_M
T-DANTE C_0	0.585	**0.674**	0.523	0.602	0.810	0.825	0.849	0.848	0.542	0.713
	±0.021	**±0.015**	±0.028	±0.020	±0.018	±0.017	±0.012	±0.014	±0.096	±0.039
T-DANTE C_4	0.574	0.669	**0.534***	0.604	**0.822**	**0.838**	0.862	0.863	0.644	0.754
	±0.019	±0.012	**±0.029**	±0.021	**±0.015**	**±0.014**	±0.010	±0.011	±0.082	±0.037
T-DANTE C_8	**0.590**	0.665	0.508	0.542	0.821	0.838	**0.870***	**0.873**	**0.696**	**0.781**
	±0.030	±0.017	±0.043	±0.023	±0.015	±0.015	**±0.011**	**±0.011**	**±0.056**	**±0.028**
T-DANTE GD-C_0	0.559	0.664	0.520	0.600	0.802	0.818	0.842	0.841	0.633	0.752
	±0.027	±0.0176	±0.027	±0.019	±0.018	±0.017	±0.022	±0.015	±0.114	±0.055
T-DANTE GD-C_4	0.545	0.661	0.527	**0.612**	0.815	0.833	0.846	0.845	0.666	0.767
	±0.026	±0.017	±0.026	**±0.014**	±0.019	±0.017	±0.012	±0.011	±0.084	±0.038
T-DANTE GD-C_8	0.566	0.654	0.524	0.557	0.808	0.829	0.853	0.853	0.678	0.770
	±0.036	±0.023	±0.033	±0.032	±0.022	±0.020	±0.019	±0.015	±0.085	±0.037
Simulation Dataset	sim_1		sim_2		sim_3		sim_4		sim_5	
	G_C	G_M	G_C	G_M	G_C	G_M	G_C	G_M	G_C	G_M
T-DANTE C_0	0.965	0.979	0.954	0.970	0.947	0.975	0.948	0.979	0.932	0.974
	±0.008	±0.004	±0.009	±0.005	±0.014	±0.005	±0.0088	±0.003	±0.009	±0.003
T-DANTE C_4	0.969	0.983	**0.980***	**0.989***	**0.982***	**0.988***	0.971	0.987	0.945	0.976
	±0.002	±0.002	**±0.002**	**±0.001**	**±0.006**	**±0.003**	±0.002	±0.002	±0.011	±0.003
T-DANTE C_8	0.94	0.972	0.964	0.981	0.964	0.978	0.933	0.973	0.892	0.960
	±0.012	±0.004	±0.009	±0.004	±0.012	±0.005	±0.012	±0.005	±0.016	±0.004
T-DANTE GD-C_0	0.978	0.985	0.966	0.977	0.960	0.984	0.967	0.99	0.945	0.979
	±0.007	±0.004	±0.007	±0.004	±0.008	±0.004	±0.006	±0.002	±0.008	±0.003
T-DANTE GD-C_4	**0.981**	**0.986**	0.973	0.979	0.970	0.984	**0.975***	**0.988***	**0.960***	0.980
	±0.005	**±0.003**	±0.011	±0.006	±0.006	±0.002	**±0.004**	**±0.001**	**±0.008**	±0.003
T-DANTE GD-C_8	0.967	0.978	0.960	0.969	0.956	0.976	0.944	0.976	0.914	0.966
	±0.010	±0.005	±0.013	±0.006	±0.012	±0.004	±0.015	±0.005	±0.014	±0.004

8–10 agents across different simulations, which does not always fulfill the context size of 8 agents around the pair of agents. Thus, C_4 is the optimum number of context sizes for this dataset.

The ablation study results across both pedestrian and simulation datasets demonstrate that in almost all cases, the use of context is beneficial for the performance of the model using either of the evaluation metrics. However, the best context size is not the same for all datasets, as each dataset holds different characteristics, such as the number of agents presented in each timeframe.

Comparative Study: In this section, the performance of T-DANTE is compared with the baselines. The results of the experiments for the pedestrian datasets and simulation dataset are presented in Table 3.

Pedestrian Datasets: T-DANTE with a context size of 8 (C_8) agents is selected in this section because, on average, it performed best in the ablation study using the pedestrian datasets. According to Table 3, T-DANTE out-

Table 3. The results of Group Correctness G_C and Group Mitre G_M for T-DANTE compared with baselines using pedestrian datasets and simulation datasets. The * sign shows that this result is significantly different than all the other values in the same column according to the Wilcoxon signed-rank test.

| | Pedestrian Dataset | | | | | | | | | |
| | eth | | hotel | | zara01 | | zara02 | | students03 | |
	G_C	G_M	G_C	G_M	G_C	G_M	G_C	G_M	G_C	G_M
DANTE	0.319	0.548	0.431	**0.586**	0.731	0.793	0.633	0.705	0.024	0.502
	±0.047	±0.019	±0.043	±0.035	±0.051	±0.028	±0.038	±0.026	±0.012	±0.013
NRI	0.201	0.571	0.169	0.540	0.285	0.597	0.106	0.417	0.006	0.280
	±0.062	±0.074	±0.054	±0.097	±0.067	±0.053	±0.035	±0.019	±0.010	±0.026
WavenetNRI	0.242	0.553	0.202	0.455	0.361	0.627	0.184	0.462	0.001	0.280
	±0.059	±0.057	±0.048	±0.080	±0.091	±0.066	±0.065	±0.040	±0.004	±0.024
T-DANTE	**0.590***	**0.665**	**0.508***	0.542	**0.821***	**0.838***	**0.870***	**0.873***	**0.696***	**0.780***
	±0.030	**±0.017**	**±0.043**	**±0.023**	**±0.015**	**±0.015**	**±0.011**	**±0.011**	**±0.056**	**±0.028**
	Simulation Dataset									
	sim_1		sim_2		sim_3		sim_4		sim_5	
	G_C	G_M	G_C	G_M	G_C	G_M	G_C	G_M	G_C	G_M
DANTE	0.215	0.717	0.198	0.701	0.095	0.518	0.199	0.712	0.041	0.425
	±0.007	±0.004	±0.008	±0.003	±0.011	±0.011	±0.011	±0.005	±0.007	±0.009
NRI	0.984	0.991	0.983	0.993	**0.988***	**0.995***	0.996	0.999	**0.988***	**0.995**
	±0.004	±0.002	±0.007	±0.002	**±0.004**	**±0.002**	±0.003	±0.001	**±0.007**	**±0.003**
WavenetNRI	**0.996**	**0.998**	**0.995***	**0.998***	0.977	0.988	**0.998***	0.999	0.953	0.968
	±0.006	**±0.002**	**±0.004**	**±0.001**	±0.008	±0.004	**±0.004**	±0.001	±0.011	±0.009
T-DANTE	0.969	0.983	0.980	0.989	0.982	0.988	0.971	0.987	0.945	0.976
	±0.002	±0.002	±0.002	±0.001	±0.006	±0.003	±0.006	±0.002	±0.011	±0.003

performs all baselines, i.e., DANTE, NRI, and WavenetNRI, for all pedestrian datasets using the Group Correctness metric. These results can lead us to the conclusion that due to the superiority of T-DANTE versus DANTE, the addition of a temporal aspect using the LSTM-conv blocks enhances the performance of the model. Another notable aspect of the tables is the higher standard deviation in NRI and WavenetNRI results compared to the rest of the results. This means that these models do not consistently learn to distinguish the different classes in every experiment run, which leads to the difference between their results as represented by standard deviation.

Simulation Datasets: In the simulation dataset, the T-DANTE with a context size of 4 (C_4) was selected as it performed better in the ablation study. According to Table 3, WavenetNRI and NRI performed better than T-DANTE in all simulation datasets in both evaluation metrics, and T-DANTE performed better than DANTE. In general, the results demonstrate the positive effect of including the temporal dynamics of the data in the training process. Moreover, the results of T-DANTE and DANTE show that using data from multiple timeframes in a single sample enhances the model's performance. Another point is that T-DANTE surpasses the baselines in the pedestrian datasets, but in the simulation datasets, NRI and WavenetNRI baselines share the first- and second-best places. This behavior can be explained by the differences in the characteristics of the pedestrian and simulation datasets. Simulation datasets include a higher number of scenes with groups of over 3 members than pedestrian datasets. Besides, NRI and WavenetNRI models have the freedom to extract contextual information computationally without having a limit on the number of surrounding agents. However, this is not a realistic assumption in real-world datasets. Overall, the baselines were unable to capture the interactions of smaller groups that appear more frequently in the pedestrian datasets while being able to effectively find patterns in larger groups that are majorly included in the simulation datasets.

5   Conclusion

This study focuses on detecting group behavior using spatio-temporal trajectory data. Our proposed methodology, inspired by the neural network architecture introduced by Swofford et al. [16] and enriched with RNN layers, is effective in capturing the temporal dynamics inherent in agent movements. Our performed ablation study shows that in almost all cases, the LSTM-conv block performed better than the GRU-dense block and that the use of context is beneficial for the performance of the model. Our comparative study against state-of-the-art baselines demonstrates that T-DANTE is the superior model for the group detection task using real-world pedestrian datasets in which the sample size is limited. Whilst T-DANTE is outperformed by parameter-free models on larger simulation datasets, it still achieves competitive results.

The incorporation of dynamic context size based on the presented number of agents and dynamic group membership per scene can be explored in future research. Additionally, enhancing the generalization of the proposed approach across different datasets and its scalability to real-time applications could be another future approach. Various applications, such as analyzing students' social behavior in schoolyards, monitoring tourists' behaviors in touristic sights, and analyzing sports teams' performances, may benefit from the presented work.

Funding Information. This paper represents independent research funded by the Dutch Research Council (NWO, grant number: AUT.17.007) and Leiden-Delft-Erasmus Centre for BOLD Cities (grant number: BC2019-1).

References

1. Amirian, J., Zhang, B., Castro, F.V., Baldelomar, J.J., Hayet, J.-B., Pettré, J.: OpenTraj: assessing prediction complexity in human trajectories datasets. In: Ishikawa, H., Liu, C.-L., Pajdla, T., Shi, J. (eds.) ACCV 2020. LNCS, vol. 12627, pp. 566–582. Springer, Cham (2021). https://doi.org/10.1007/978-3-030-69544-6_34

2. Bae, I., Park, J.H., Jeon, H.G.: Learning pedestrian group representations for multi-modal trajectory prediction. In: Avidan, S., Brostow, G., Cissé, M., Farinella, G.M., Hassner, T. (eds.) Computer Vision – ECCV 2022. LNCS, vol. 13682, pp. 270–289. Springer, Cham (2022). https://doi.org/10.1007/978-3-031-20047-2_16

3. Deo, N., Trivedi, M.M.: Convolutional social pooling for vehicle trajectory prediction. In: Proceedings of the IEEE Conference on Computer Vision and Pattern Recognition (CVPR) Workshops, June 2018
4. Fernando, T., Denman, S., Sridharan, S., Fookes, C.: GD-GAN: generative adversarial networks for trajectory prediction and group detection in crowds. In: Jawahar, C.V., Li, H., Mori, G., Schindler, K. (eds.) ACCV 2018. LNCS, vol. 11361, pp. 314–330. Springer, Cham (2019). https://doi.org/10.1007/978-3-030-20887-5_20
5. Hung, H., Kröse, B.: Detecting F-formations as dominant sets. In: Proceedings of the 13th International Conference on Multimodal Interfaces, ICMI 2011, pp. 231–238. Association for Computing Machinery, New York, NY, USA (2011)
6. Kipf, T., Fetaya, E., Wang, K.C., Welling, M., Zemel, R.: Neural relational inference for interacting systems. In: Dy, J., Krause, A. (eds.) Proceedings of the 35th International Conference on Machine Learning. Proceedings of Machine Learning Research, vol. 80, pp. 2688–2697. PMLR, 10–15 July 2018
7. Lee, J.G., Han, J., Whang, K.Y.: Trajectory clustering: a partition-and-group framework. In: Proceedings of the 2007 ACM SIGMOD International Conference on Management of Data, SIGMOD 2007, pp. 593–604. Association for Computing Machinery, New York, NY, USA (2007)
8. Lerner, A., Chrysanthou, Y., Lischinski, D.: Crowds by example. Comput. Graph. Forum **26**, 655–664 (2007)
9. Nasri, M., Baratchi, M., Tsou, Y.T., Giest, S., Koutamanis, A., Rieffe, C.: A novel metric to measure spatio-temporal proximity: a case study analyzing children's social network in schoolyards. Appl. Network Sci. **8**(1), 50 (2023)
10. Nasri, M., et al.: A GNN-based architecture for group detection from spatio-temporal trajectory data. In: Crémilleux, B., Hess, S., Nijssen, S. (eds.) Advances in Intelligent Data Analysis XXI, pp. 327–339. Springer, Cham (2023). https://doi.org/10.1007/978-3-031-30047-9_26
11. Nasri, M., et al.: A novel data-driven approach to examine children's movements and social behaviour in schoolyard environments. Children **9**(8), 1177 (2022)
12. van den Oord, A., et al.: WaveNet: a generative model for raw audio. Arxiv (2016)
13. Pellegrini, S., Ess, A., Schindler, K., Van Gool, L.: You'll never walk alone: modeling social behavior for multi-target tracking. In: 2009 IEEE 12th International Conference on Computer Vision, pp. 261–268. IEEE (2009)
14. Solera, F., Calderara, S., Cucchiara, R.: Structured learning for detection of social groups in crowd. In: 2013 10th IEEE International Conference on Advanced Video and Signal Based Surveillance, pp. 7–12 (2013)
15. Solera, F., Calderara, S., Cucchiara, R.: Socially constrained structural learning for groups detection in crowd. IEEE Trans. Pattern Anal. Mach. Intell. **38**(5), 995–1008 (2016)
16. Swofford, M., et al.: Improving social awareness through DANTE: deep affinity network for clustering conversational interactants. Proc. ACM Hum.-Comput. Interact. **4**(CSCW1) (2020)
17. Tan, S., Tax, D.M., Hung, H.: Conversation group detection with spatio-temporal context. In: Proceedings of the 2022 International Conference on Multimodal Interaction, ICMI 2022, pp. 170–180. Association for Computing Machinery, New York, NY, USA (2022)
18. Tang, L.A., et al.: A framework of traveling companion discovery on trajectory data streams. ACM Trans. Intell. Syst. Technol. **5**(1) (2014)
19. Yamaguchi, K., Berg, A.C., Ortiz, L.E., Berg, T.L.: Who are you with and where are you going? In: CVPR 2011, pp. 1345–1352 (2011)

Statistical Learning

Backward Inference in Probabilistic Regressor Chains with Distributional Constraints

Ekaterina Antonenko[1,2,3,4](✉)(iD), Michael Mechenich[5](iD), Rita Beigaitė[5](iD),
Indrė Žliobaitė[5](iD), and Jesse Read[1](iD)

[1] LIX, Ecole Polytechnique, Institut Polytechnique de Paris, Palaiseau, France
ekaterina.antonenko@minesparis.psl.eu
[2] Mines Paris, PSL Research University, CBIO-Centre for Computational Biology,
75006 Paris, France
[3] Institut Curie, PSL Research University, 75005 Paris, France
[4] INSERM, U900, 75005 Paris, France
[5] University of Helsinki, Helsinki, Finland

Abstract. State-of-the-art approaches for multi-target prediction, such as Regressor Chains, can exploit interdependencies among the targets and model the outputs jointly, by flowing predictions from the first output to the last. While these models are very useful in applications where targets are highly interdependent and should be modeled jointly, they are however unable to answer queries in situations when targets are not only mutually dependent but also have joint constraints over the output. In addition, existing models are unsuitable when certain target values are fixed or manually imputed prior to inference, and as a result, the flow of predictions cannot cascade backward from an already-imputed output. Here we present a solution to the aforementioned problem as a backward inference algorithm for Regressor Chains via Metropolis-Hastings sampling. We evaluate the proposed approach via different metrics using both synthetic and real-world data. We show that our approach notably reduces errors when compared to traditional marginal inference methods that overlook joint modeling. Furthermore, we show that the proposed method can provide useful insights into a problem in conservation science in predicting the distribution of potential natural vegetation.

Keywords: Regressor Chains · Probabilistic inference · Multi-output modeling · Potential natural vegetation

1 Introduction

In Regressor Chains, increasingly used for multi-output prediction, predictions are cascaded across the outputs and are used as input features to model the subsequent targets. Figure 1a illustrates the standard setting of Regressor Chains for three targets. The same approach has been widely used in the context of multi-label classification as Classifier Chains for binary outputs [15]. There are recent successes also in the multi-target regression context with continuous outputs:

I. Miliou et al. (Eds.): IDA 2024, LNCS 14642, pp. 43–55, 2024.
https://doi.org/10.1007/978-3-031-58553-1_4

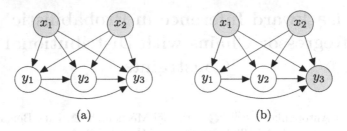

(a)   (b)

Fig. 1. Illustration of a Regressor Chain depicted as a Bayesian network (shaded nodes indicate fixed observations) for inputs $\boldsymbol{x} = \{x_1, x_2\}$ and outputs $\boldsymbol{y} = \{y_1, y_2, y_3\}$. (a) illustrates the standard setting with forward inference; (b) demonstrates challenges we address, i.e. backward propagation of imputed output information (label y_3) while maintaining a joint constraint and without training a new structure.

for example, Regressor Stacking [17], Ensembles of Regressor Chains [2,18], and probabilistic frameworks [14].

However, the above-cited works make some standard assumptions: all outputs are to be predicted in a pre-determined order, individually, and new models can be structured and trained with relative ease. We consider a new computational task setting bringing in the following constraints (c.):

1. Any output may be pre-imputed/fixed *prior to* prediction;
2. Base regression models cannot be retrained;
3. Outputs satisfy a joint constraint.

We aim at inferring a joint posterior distribution over labels, i.e. probabilistic Regressor Chains, under these constraints.

Such constraints are realistic in many settings, e.g. if the data is not accessible after training the model due to ethical or privacy concerns or if computational resources and human resources required to form a model are tight.

Indeed, in this work, we consider the following motivating example: the estimation of hypothetical *distribution* (i.e., constraints 3) of vegetation and landcover types ('potential vegetation') based on climatic conditions supposing that no urban or agricultural activities are present (i.e., constraints 1) while only data with urban activity observed was available for the model training (i.e., constraints 2).

Generally, let $\boldsymbol{y} = [y_1, \ldots, y_L]$ be the targets. Since the outputs represent compositional data, i.e. sum up to 1, $\sum_{l=1}^{L} y_l = 1$ and $y_l \geq 0$. Our goal is to answer queries of the form

$$p(\boldsymbol{y}_{\neg F} \mid \boldsymbol{x}, \boldsymbol{y}_F), \tag{1}$$

where $F \subset \{1, \ldots, L\}$ is a set of fixed/observed outputs, and $\neg F = \{1, \ldots, L\} \backslash F$ are the remaining outputs to predict; e.g. $\boldsymbol{y}_{\neg F} = [y_1, y_2]$ and $\boldsymbol{y}_F = [y_3]$ in Fig. 1b.

A Regressor Chain $\mathcal{H} = [h_1, \ldots, h_L]$ involves a model (regressor) h_l for each of the outputs $y_1, \ldots, y_L$ providing prediction $\hat{y}_l = h_l(\boldsymbol{x}, y_1, \ldots, y_{l-1})$ which is, typically, a function of probability density function (pdf) $p(y_l \mid \boldsymbol{x}, y_1, \ldots, y_{l-1})$, e.g. with the expected value $\hat{y}_i = E_{y_i \sim p(y_i \mid \boldsymbol{x}, y_1, \ldots, y_{i-1})}[y_i]$. This allows us to provide a prediction for all outputs,

$$\hat{\boldsymbol{y}} = [\hat{y}_1, \ldots, \hat{y}_L] = [h_1(\boldsymbol{x}), h_2(\boldsymbol{x}, \hat{y}_1), \ldots, h_L(\boldsymbol{x}, \hat{y}_1, \ldots, \hat{y}_{L-1})].$$

Recall that each prediction becomes a feature for the following model in the chain. By this mechanism, Regressor Chains aim to model the outputs jointly. If the pdf is explicitly modeled (as in the case of Probabilistic Regressor Chains [14]), Regressor Chains also provide the joint posterior distribution (e.g., by forward or ancestral sampling):

$$p(\boldsymbol{y} \mid \boldsymbol{x}) = \prod_{l=1}^{L} p(y_l \mid \boldsymbol{x}, y_1, \ldots, y_{l-1}).$$

Thus, Fig. 1a implies $p(\boldsymbol{y} \mid \boldsymbol{x}) = p(y_1 \mid \boldsymbol{x}) \cdot p(y_2 \mid \boldsymbol{x}, y_1) \cdot p(y_3 \mid \boldsymbol{x}, y_1, y_2)$. However, here we face the challenge posed by the interaction of constraints 1. (fixed output) and 3. (joint constraint): if y_3 is a fixed observation, forward inference along the chain cannot be completed while respecting the other constraints, specifically the term $p(y_3 \mid \boldsymbol{x}, y_1, y_2)$. A naive approach of simply predicting $\hat{y}_1$ and $\hat{y}_2$ and then normalizing them to meet the constraint $\sum_{l=1}^{L} \hat{y}_l = 1$ is not valid, because it answers the query $p(\boldsymbol{y}_{\neg F} \mid \boldsymbol{x})$ but not the target query $p(\boldsymbol{y}_{\neg F} \mid \boldsymbol{x}, \boldsymbol{y}_F)$.

We propose Metropolis-Hastings sampled Regressor Chains (mhsRC) and Ensembles of Regressor Chains (mhsERC) which can provide a solution to the aforementioned problem by combining Regressor Chains and Metropolis-Hastings sampling for backward inference in the prediction step. We apply our approach, mhsERC, on synthetic and real-world datasets and find that in the case of synthetic data, the resulting distribution provided by mhsERC is very close to the ground-truth, for given fixed values. The model naturally provides a distribution for each instance in addition to a predicted mean value. In three multi-target regression datasets, we provide values for one target explicitly and compare predictions of the other targets, where mhsERC successfully infers unknown targets and reaches significantly better performance than the base-line methods. Finally, in the real-world climate and land-cover data, we extract insights into the potential distribution of vegetation in the absence of urban cover.

2   Related Work

Although there exist a variety of methods for multi-target regression, e.g. Pre-dictive Clustering Trees [11], Regressor Chains (RC) and Ensembles of Regressor Chains (ERC) [18], Regressor Stacking [17], these approaches are typically used in a standard predictive setting and do not directly target joint prediction of tar-gets when some of the output values are provided before the prediction. In [14], Regressor Chains were further developed into a probabilistic framework, however only ancestral, or forward, inference along the chain is available which does not respond to the set constraints 1–3. Also, the authors did not propose how to use non-probabilistic base classifiers such as Decision Trees.

The given problem (recall example in Fig. 1b) can be represented as a Bayesian Network, specifically a Hybrid Bayesian Network [16] which can handle continuous variables during inference as opposed to classic Bayesian Networks. However, learning the structure of a Bayesian Network is extremely costly and inference options are limited, normally corresponding to linear-Gaussian models or approximate methodologies based on sampling, and variational inference.

The problem setting involving the distributional constraint, with missing value, has been called 'structurally incomplete' by [3]; but authors here use a neural network approach that can be built arbitrarily. Oppositely, we develop a probabilistic inference approach under Regressor Chains. The general setting for predicting a composition of outputs is known in the statistics literature as 'compositional data analysis' [1].

As an example of a real-world problem involving constraints 1–3, we consider the prediction of potential vegetation distribution in the absence of urban activity. A similar setting was considered in [3]. In our work, we also face the issue of the evaluation of ground-truth distribution for comparison, since the goal is to explore alternative hypotheses. Our solution is to study the probabilistic challenge of deriving a joint distribution directly; whereas the authors of [3] focus on the accuracy of predicting dominant vegetation types. Another important difference is that we consider the additional constraint of tackling the problem at *inference* time, rather than selecting different training regimes.

In ecology and biogeography, a related research question concerns the inference of potential natural vegetation, i.e. the anticipated state of vegetation under specific environmental conditions, without a notable human intervention [4]. In recent years, statistical and machine-learning techniques have gained popularity for their application in constructing such models [9,10] but, in these works, the focus is on exploring the relationship from climatic input observations to the targets, rather than the probabilistic relationship among the targets, as we do.

3   Our Method: Metropolis-Hasting Regressor Chains

As mentioned in Sect. 1, we target inferring the probability defined by Eq. (1). In other words, we want to evaluate the probability

$$\pi(\hat{\boldsymbol{y}}) = p(\hat{\boldsymbol{y}}_{\neg F} \mid \hat{\boldsymbol{x}}, \hat{\boldsymbol{y}}_F)$$

for any particular $\hat{\boldsymbol{y}}$ and $\hat{\boldsymbol{x}}$, where $\hat{\boldsymbol{y}}_F$ are fixed and $\hat{\boldsymbol{y}}_{\neg F}$ are unknown. By the definition of condition probability,

$$\pi(\hat{\boldsymbol{y}}) = \frac{p(\hat{\boldsymbol{x}}, \hat{\boldsymbol{y}}_{F \cup \neg F})}{p(\hat{\boldsymbol{x}}, \hat{\boldsymbol{y}}_F)} = \frac{\Pi_{l=1}^{L} p(\hat{y}_l \mid \hat{\boldsymbol{x}}, \hat{y}_1, ..., \hat{y}_{l-1})}{p(\hat{\boldsymbol{x}}, \hat{\boldsymbol{y}}_F)} \propto \Pi_{l=1}^{L} p(\hat{y}_l \mid \hat{\boldsymbol{x}}, \hat{y}_1, ..., \hat{y}_{l-1}).$$

To evaluate the probabilities $p(\hat{y}_l \mid \hat{\boldsymbol{x}}, \hat{y}_1, ..., \hat{y}_{l-1})$, we assume that for each base estimator h_l, the corresponding distribution may be presented as a normal distribution, and it is possible to obtain its parameters, the mean μ and standard

---

Algorithm 1: Evaluate the probability of the proposed estimate

---

    Input : proposed estimate y', Regressor Chain $\mathcal{H} = [h_1, ..., h_L]$
    Output: probability π of the proposed estimate
1: **procedure** π
2:    **for** l in $1, ..., L$ **do**
3:        $\mu, \sigma \leftarrow$ mean and std of $h_l(\hat{x}, \hat{y}_1, ..., \hat{y}_{l-1})$
4:        $p_l \leftarrow pdf(y'_l)$ for $\mathcal{N}(\mu, \sigma)$
5:    $\pi = \Pi_{l=1}^{L} p_l$

Algorithm 2: Metropolis-Hastings sampled Regressor Chains

    Input : number of iterations T, probability π estimated in Algorithm 1
    Output: T samples from distribution defined by probabilities π
1 **procedure** MHSRC
2    $y^{[0]} \leftarrow initialization$ ▷ First step of random walk
3   **for** $0 < t < T$ **do**
4        $y' \leftarrow y^{[t]} + noise$ ▷ Propose new y'
5   **if** y' *satisfies evaluation criteria based on* π **then**
6            $y^{[t+1]} \leftarrow y'$ ▷ Accept proposed point
7   **else**
8            $y^{[t+1]} \leftarrow y^{[t]}$ ▷ Refuse proposed, keep previous

---

deviation σ (see Algorithm 1). In our work, we use Random Forests and calculate the mean and the standard deviation for the predictions of individual trees from the ensemble. Other possibilities to query these parameters include, for example, direct inference from the model for Bayesian regression models, Monte Carlo Dropout [7], or input perturbation (shallow Monte Carlo Dropout).

We propose to use the Metropolis-Hastings sampling [8,13] (though other sampling approaches may be applied), basically a random walk where each new proposed estimate y' is evaluated by the distribution probability function $\pi(y')$ and is accepted as a sample if it is likely to be found in the desired distribution, summarized in Algorithm 2. Our method, Metropolis-Hastings sampled Regressor Chains (mhsRC), is not specific to any particular chain order and can be applied to an RC of any order with any set of fixed outputs. If an ERC was given as a prior trained model, then the procedure is performed for all individual chains in the ensemble, and their predictions are averaged, resulting in Metropolis-Hastings sampled Ensembles of Regressor Chains (mhsERC). The code is available on https://github.com/ekaantonenko/mhsERC.

4 Experiments

We remind the reader that our goal is to estimate Eq. (1). To evaluate possible solutions to this problem, we perform the following experiments. First, we generate synthetic data where joint and marginal distributions including Eq. (1) are

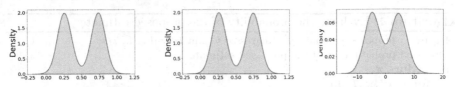

(a) Ground-truth distributions of snow $P(y_1 \mid y_3 = 0)$ and grass $P(y_2 \mid y_3 = 0)$ types while urban $y_3 = 0$; temperature $P(x)$.

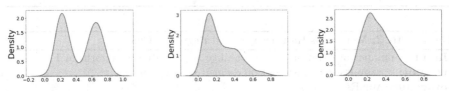

(b) Observed distributions of snow $P(y_1)$, grass $P(y_2)$, and urban $P(y_3)$.

Fig. 2. Synthetic dataset.

fully known, and we can compare the distributions directly. Then, we perform experiments on real-world data where we provide the values of fixed targets $\boldsymbol{y}_F$ and evaluate predictions for the other targets $\boldsymbol{y}_{\neg F}$. Finally, we use real-world data and expert intuition to make conclusions with regard to Eq. (1), given hypothetical (not observed in the data) values of $\boldsymbol{y}_F$.

Synthetic Dataset. In this work, we present a synthetic dataset with one feature x and targets $\boldsymbol{y} = \{y_1, y_2, y_3\}$, where $\boldsymbol{y}_F = \{y_3\}$ and the ground-truth distribution $p(\boldsymbol{y}_{\neg F} \mid x, \boldsymbol{y}_F)$ is known. It is based on our motivating example of potential vegetation, distribution of land-cover y_1, y_2, and y_3, with climatic features x. We might suppose, e.g., snow, grass, and urban land cover $\boldsymbol{y}$, and temperature x. We designed this dataset such that a hypothetical urban activity is more likely to be settled in the grass type, and snow and grass are not equally masked by the presence of human activities. In this setting, we observe y_1 and y_2 in the presence of urban and are interested in evaluating $P(y_1 \mid y_3 = 0)$ and $P(y_2 \mid y_3 = 0)$, i.e. vegetation distribution in the absence of urban; see Fig. 2a.

This can be represented as follows: first sample three target variables of interest, $y_1^{y_3=0} \sim \alpha$, $y_2^{y_3=0} \sim 1 - \alpha$, and $y_3 = 0$, where α is a bi-modal mixture of normal distributions, and y_1, y_2 are normalized afterwards so that $0 \le y_1^{y_3=0}, y_2^{y_3=0} \le 1$ and $y_1^{y_3=0} + y_2^{y_3=0} = 1$. This gives us $P(y_1 \mid y_3 = 0)$ and $P(y_2 \mid y_3 = 0)$. After that, the joint distribution is generated: $y_1 = y_1^{y_3=0} \cdot (1-p)$, $y_2 = y_2^{y_3=0} \cdot (1 - q)$, and $y_3 = y_1^{y_3=0} \cdot p + y_2^{y_3=0} \cdot q$, where $p \sim \mathcal{N}(0.1, 0.1)$ and $q \sim \mathcal{N}(0.5, 0.2)$, respectively (taking the absolute value if negative is generated). This significantly shifts the distribution of the y_2 variable when compared to $y_2^{y_3=0}$, see Fig. 2b. The x feature is generated by adding noise to the parameter α and further linear transformation: $x = -20 \cdot (\alpha + \varepsilon) + 10$, $\varepsilon \sim \mathcal{N}(0, 0.1)$.

Real-World Benchmark Data. Subsequently, we use three real-world multi-target regression datasets. The compositional Arctic lake dataset [1]

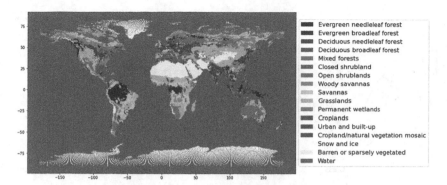

Fig. 3. Grid cells (learning instances) represented by a categorical distribution over 17 vegetation types; the map shows the dominant type within each grid cell.

describes the distribution of sand, clay, and silt (3 targets) in 39 water samples, at different depths (1 feature). The Slump dataset [20] describes the concrete properties (3 targets) and ingredients (7 features) in 103 samples. The Energy Building dataset (Enb) [19] describes the heating load and cooling load requirements of buildings (2 targets) and building parameters (8 features) for 768 instances. For evaluation, we split the data into train and test subsets (80:20, 5-fold cross-validation), and in the prediction phase provide explicitly the values of the first target $y_F = [y_1]$. The metrics are calculated for the predicted targets $y_{\neg F} = [y_2, \ldots]$.

Vegetation Data. Finally, we apply our method to a dataset describing the distribution of land cover globally to infer a possible vegetation distribution in the absence of urban activity (i.e. force the corresponding classes to 0 explicitly). We aim to predict the set of 17 land cover classes, including water [6]. We use 19 bioclimatic variables [5] as predictive features, representing ecologically relevant means, minima, and maxima in temperature and precipitation, averaged for the period 1970–2000. Both land cover targets and bioclimatic features were obtained from the Eco-ISEA3H database [12]. The dataset consists of 56,821 instances which correspond to terrestrial cells measuring approximately 2600 km² each.

We are interested in inferring the fractional distribution of natural land cover classes in the absence of three cover classes associated with human activity, namely croplands, urban and built-up lands, and cropland/natural vegetation mosaics (mapped together in red in Fig. 3).

Evaluation. In the synthetic dataset both ground-truth distributions $P(y_1 \mid y_3 = 0)$ and $P(y_2 \mid y_3 = 0)$ and observed distributions $P(y_1)$, $P(y_2)$, and $P(y_3)$ are known, the goal is to reconstruct the ground-truth distributions from the observed distributions. As objective evaluation metrics, we use Mean Squared Error (MSE), Uniform Cost Function (UCF), and Wasserstein Distance (WD). SE-based metrics are standard for point-wise evaluation, and WD is suited to comparing distributions. UCF is motivated in [2]; here we use $\delta = 0.5$.

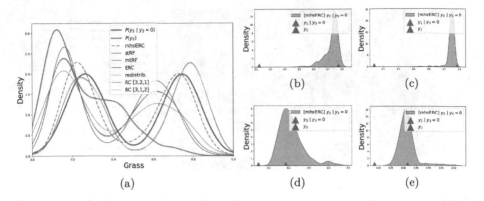

Fig. 4. (a) Re-discovering of ground truth $P(y_2 \mid y_3 = 0)$ distribution in synthetic data. Note, that $y_1 \mid y_3 = 0$ is equal to $1 - (y_2 \mid y_3 = 0)$ (by nature of compositional data) so technically we are evaluating distributions of both targets. (b–e) Predicted per-instance distributions for four individual instances.

We use Random Forests as base estimators for all chain-based methods. Namely, we compare our proposed method to Regressor Chain (RC) with direct order $[y_1, ..., y_L]$, when the target with fixed values (y_1) comes first in the chain, and thus the fixed values are cascaded directly, without backward inference. Second, we evaluate several marginal models that do not take the joint constraint into account. Straightforwardly, we may set fixed targets to corresponding values and re-normalize remaining targets without training a predictive model on features (redistrib.), this is applicable only for the synthetic dataset. Otherwise, we may predict with trained models, Ensembles of Regressor Chains (ERC), multi-target Random Forests (mtRF), and individual single-target Random Forests (stRF), plug in the fixed values *after* prediction and re-normalize the targets so that their sum is equal to one.

5   Results and Discussion

Synthetic Data. Table 1 shows the comparison of different methods for the synthetic data, where a model is expected to uncover the ground-truth distribution without urban activity. First, to support the choice of Regressor Chains for a predictive task, we evaluate the performance of all methods in a standard setting when prediction from x to $y = \{y_1, y_2, y_3\}$ is required. To this end, we perform 5-fold cross-validation and observe that Regressor Chains and Ensembles of Regressor Chains outperform single- and multi-target Random Forests.

Second, we compare empirically the predictions of $\hat{y}_1, \hat{y}_2$ when $y_3 = 0$ by the models listed above and ground-truth y_1, y_2 when $y_3 = 0$. We observe that the metrics values differ significantly for different chain orders and mhsRC with the order $[3, 1, 2]$ shows the best result: this is unsurprising as we plug in directly the constraint to the first regressor of the chain and further propagate its inference.

Table 1. Performance on synthetic data: classic cross-validated regression ($x \to y$) and prediction with $y_3 = 0$ vs. ground truth $y_1, y_2 \,|\, y_3 = 0$ (a smaller value is better). Best values are in **bold**, second best values are underlined. Orders of Regressor Chains (permutations of $1, 2, 3$) are given in square brackets.

Model	$x \to y$	$x \to y_1, y_2 \,|\, y_3 = 0$		
	MSE	MSE	WD	UCF
mhsRC [1, 2, 3]	0.016	0.019	0.052	0.099
mhsRC [1, 3, 2]	**0.015**	0.011	0.038	0.073
mhsRC [2, 1, 3]	0.017	0.169	0.132	0.281
mhsRC [2, 3, 1]	0.017	0.029	0.114	0.272
mhsRC [3, 1, 2]	0.017	**0.007**	**0.016**	0.054
mhsRC [3, 2, 1]	0.017	0.011	0.039	0.085
mhsERC	**0.015**	0.010	0.027	**0.037**
stRF	0.018	0.018	0.115	0.172
mtRF	0.018	0.019	0.115	0.176
ERC	0.016	0.016	0.108	0.117
redistrib.	–	0.024	0.122	0.213
RC [3, 1, 2]	0.017	0.009	0.038	0.053
RC [3, 2, 1]	0.017	0.009	0.036	0.055

For some of the orders ([2, 1, 3], [2, 3, 1]) the task is more difficult. However, mhsERC consisting of all possible 6 chain orders shows high-performing results when compared to 'naive' models without joint inference: mhsERC runs first best w.r.t. UCF and second best w.r.t. MSE and WD. Figure 4a also demonstrates graphically the resulting distributions. The inference of the mhsERC model is very close to the original bi-modal symmetric distribution of grass and snow. Examples of individual per-instance distributions of $\hat{y}_2 \,|\, y_3 = 0$ (for given x) are presented in Fig. 4b–4e. Again, we see that the predicted distributions tend to center around the desired value of ground truth $y_2 \,|\, y_3 = 0$.

Real-World Benchmark Data. For three multi-target regression datasets, we provide the values of the target y_1 explicitly, and other targets are to be predicted. Table 2 shows the comparison of mhsERC, an RC with direct order $[y_1, ..., y_L]$ (fixed values of y_1 are simply propagated via chain), and three marginal methods (ERC, mtRF, stRF). We observe that mhsERC obtains significantly better results than the marginal methods and close to the ones of RCs with direct orders concerning all three metrics. The statistical significance is illustrated by the Friedman-Nemenyi diagrams in Fig. 5 for all three metrics, the mhsERC method ranked along with RCs with direct order and significantly higher than other methods.

Vegetation Data. First, we point out that ERC is a well-performing model for the prediction of vegetation types from climate, see Table 3. The experiments are done under 10-fold cross-validation with splits designed to account for spatial

Table 2. Multi-target regression data, predictions when the first target is provided explicitly; a smaller value is better. Best value in **bold**, second best value underlined.

Model	Arctic Lake			Slump			Enb		
	MSE	WD	UCF	MSE	WD	UCF	MSE	WD	UCF
mhsERC	**0.002**	0.030	**0.000**	0.177	0.205	**0.806**	0.016	0.065	0.152
RC direct	**0.002**	**0.029**	**0.000**	**0.173**	**0.188**	0.825	0.015	0.068	**0.132**
mtRF	0.008	0.039	0.150	0.373	0.446	0.951	0.025	**0.063**	0.187
stRF	0.008	0.041	0.150	0.322	0.415	0.951	0.020	0.069	0.174
ERC	0.008	0.036	0.125	0.333	0.408	0.971	0.027	0.069	0.193

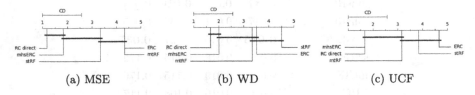

(a) MSE   (b) WD   (c) UCF

Fig. 5. Friedman-Nemenyi diagrams for the rankings of the empirically tested methods for multi-target regression data (the y_1 value is given explicitly). Lower rank is better, statistically indistinguishable methods are connected by a horizontal line.

Table 3. Performance of multi-label models for vegetation prediction, 10-fold cross-validation, best values in **bold**.

Model	MAE	std
stRF	0.047	5.07e−05
mtRF	**0.045**	**4.55e−05**
RC (with RFs)	0.061	6.12e−05
ERC (with RFs)	0.050	4.87e−05

correlations between neighboring grid cells to avoid information leakage between train and test partitions. While stRF and mtRF show the best predictive performance, we are not aware if it is possible to force these models to modify particular targets in the prediction phase. ERC runs only slightly worse, and we propose a powerful mechanism to impute fixed targets for any chain in the ensemble, while other targets take this value into account. We set the values of three variables (croplands, urban and built-up, cropland/natural vegetation mosaic) to zero and apply the proposed method, mhsERC. Figure 6 demonstrates the predicted vegetation distribution in the absence of human activity in Europe. Subjectively, the results are visually plausible, with no noticeable anomalies. This adds support to our claim that our method can be used flexibly for real-world tasks. Although, inherently, there can be no objective ground-truth evaluation for such a task,

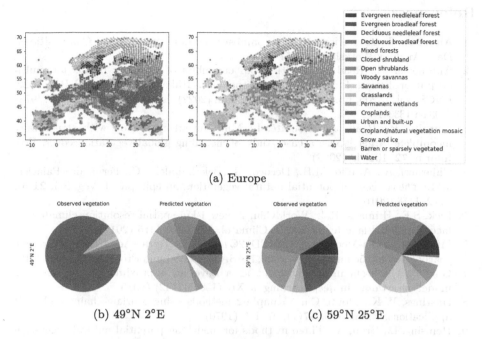

(a) Europe

(b) 49°N 2°E   (c) 59°N 25°E

Fig. 6. Vegetation distribution per grid cell (a) in Europe and (b-c) at specific locations, with (observed) and without (predicted) human activity.

we can take confidence in the relatively high performance on the synthetic and real-world tasks investigated earlier.

6   Conclusion

We tackled a new challenging multi-output prediction setting with continuous target variables: target prediction under a joint (distributional) constraint with some targets fixed prior to prediction, without retraining base models. In the context of Regressor Chains, we proposed a novel approach employing Metropolis-Hastings backward inference and providing a posterior distribution for each target while comprising an estimated distribution and meeting all constraints. We evaluated it on synthetic data and several use cases. It performed competitively in all cases. We conclude that the proposed method successfully solves the task, allowing flexibility and applicability of Regressor Chains beyond their predictive performance in standard multi-target regression settings. Our method may significantly contribute to practical applications, as demonstrated in its ability to predict the distribution of potential natural vegetation, highlighting its effectiveness in scenarios involving complex and interrelated data.

Acknowledgements. Research leading to these results was supported by Research Council of Finland (grants no 314803 and 341623 to IŽ).

References

1. Aitchison, J.: A concise guide to compositional data analysis. In: Compositional Data Analysis Workshop (2005)
2. Antonenko, E., Read, J.: Multi-modal ensembles of regressor chains for multi-output prediction. In: Bouadi, T., Fromont, E., Hüllermeier, E. (eds.) IDA 2022. LNCS, vol. 13205, pp. 1–13. Springer, Cham (2022). https://doi.org/10.1007/978-3-031-01333-1_1
3. Beigaitė, R., Read, J., Žliobaitė, I.: Multi-output regression with structurally incomplete target labels: a case study of modelling global vegetation cover. Eco. Inform. **72**, 101849 (2022)
4. Chiarucci, A., Araújo, M.B., Decocq, G., Beierkuhnlein, C., Fernández-Palacios, J.M.: The concept of potential natural vegetation: an epitaph? J. Veg. Sci. **21**(6), 1172–1178 (2010)
5. Fick, S.E., Hijmans, R.J.: WorldClim 2: new 1-km spatial resolution climate surfaces for global land areas. Int. J. Climatol. **37**, 4302–4315 (2017)
6. Friedl, M., Sulla-Menashe, D.: MCD12Q1 MODIS/Terra+Aqua land cover type yearly L3 global 500m SIN grid v006, NASA EOSDIS Land Processes DAAC (2019)
7. Gal, Y., Ghahramani, Z.: Dropout as a Bayesian approximation: representing model uncertainty in deep learning. arXiv (1506.02142) (2015)
8. Hastings, W.K.: Monte Carlo sampling methods using Markov chains and their applications. Biometrika **57**(1), 97–109 (1970)
9. Hemsing, L., Bryn, A.: Three methods for modelling potential natural vegetation (PNV) compared. Nor. Geogr. Tidsskr. **66**(1), 11–29 (2012)
10. Hengl, T., Walsh, M.G., Sanderman, J., Wheeler, I., Harrison, S.P., Prentice, I.C.: Global mapping of potential natural vegetation: an assessment of machine learning algorithms for estimating land potential. PeerJ **6**, e5457 (2018)
11. Kocev, D., Vens, C., Struyf, J., Džeroski, S.: Tree ensembles for predicting structured outputs. Pattern Recogn. **46**(3), 817–833 (2013)
12. Mechenich, M.F., Žliobaitė, I.: Eco-ISEA3H, a machine learning ready spatial database for ecometric and species distribution modeling. Sci. Data **10**, 77 (2023)
13. Metropolis, N., Rosenbluth, A., Rosenbluth, M., Teller, A., Teller, E.: Equation of state calculations by fast computing machines. J. Chem. Phys. **21**(6), 1087–1092 (1953)
14. Read, J., Martino, L.: Probabilistic regressor chains with Monte-Carlo methods. Neurocomputing **413**, 471–486 (2020)
15. Read, J., Pfahringer, B., Holmes, G., Frank, E.: Classifier chains: a review and perspectives. J. Artif. Intell. Res. (JAIR) **70**, 683–718 (2021)
16. Salmerón, A., Rumí, R., Langseth, H., Nielsen, T., Madsen, A.: A review of inference algorithms for hybrid Bayesian networks. J. Artif. Intell. Res. **62**, 799–828 (2018)
17. Santana, E., Mastelini, S., Barbon, S.: Deep regressor stacking for air ticket prices prediction. In: Anais do XIII Simpósio Brasileiro de Sistemas de Informação, pp. 25–31. SBC (2017)
18. Spyromitros-Xioufis, E., Tsoumakas, G., Groves, W., Vlahavas, I.: Multi-target regression via input space expansion: treating targets as inputs. Mach. Learn. **104**(1), 55–98 (2016)

19. Tsanas, A., Xifara, A.: Accurate quantitative estimation of energy performance of residential buildings using statistical machine learning tools. Energy Build. **49**, 560–567 (2012)
20. Yeh, I.C.: Modeling slump flow of concrete using second-order regressions and artificial neural networks. Cement Concr. Compos. **29**(6), 474–480 (2007)

Empirical Comparison Between Cross-Validation and Mutation-Validation in Model Selection

Jinyang Yu[1], Sami Hamdan[1,2], Leonard Sasse[1,2,5], Abigail Morrison[3,4], and Kaustubh R. Patil[1,2(✉)]

[1] Institute of Neuroscience and Medicine, Brain and Behaviour (INM-7), Research Center Jülich, Jülich, Germany
[2] Institute of Systems Neuroscience, Medical Faculty, Heinrich Heine University Düsseldorf, Düsseldorf, Germany
k.patil@fz-juelich.de
[3] Institute for Neurosciennce and Medicine (INM-6) and Institute for Advanced Simulation (IAS-6), Research Center Jülich, Jülich, Germany
[4] Department of Computer Science 3 - Software Engineering, RWTH Aachen University, Aachen, Germany
[5] Max Planck School of Cognition, Stephanstrasse 1a, Leipzig, Germany

Abstract. Mutation validation (MV) is a recently proposed approach for model selection, garnering significant interest due to its unique characteristics and potential benefits compared to the widely used cross-validation (CV) method. In this study, we empirically compared MV and k-fold CV using benchmark and real-world datasets. By employing Bayesian tests, we compared generalization estimates yielding three posterior probabilities: practical equivalence, CV superiority, and MV superiority. We also evaluated the differences in the capacity of the selected models and computational efficiency. We found that both MV and CV select models with practically equivalent generalization performance across various machine learning algorithms and the majority of benchmark datasets. MV exhibited advantages in terms of selecting simpler models and lower computational costs. However, in some cases MV selected overly simplistic models leading to underfitting and showed instability in hyperparameter selection. These limitations of MV became more evident in the evaluation of a real-world neuroscientific task of predicting sex at birth using brain functional connectivity.

Keywords: model selection · mutation validation · cross-validation

1 Introduction and Related Work

The model selection process aims to find a model from a pool of candidate models, taking into account a variety of performance criteria encompassing predictive accuracy and computational efficiency [9,12]. The estimated model generalization error, which represents the expected error on unseen data, is a commonly

I. Miliou et al. (Eds.): IDA 2024, LNCS 14642, pp. 56–67, 2024.
https://doi.org/10.1007/978-3-031-58553-1_5

used criterion for model selection. Generalization error can be empirically estimated using resampling techniques like cross-validation (CV) [9]. Notably, while holdout validation is particularly effective when a wealth of data is available, CV emerges as the preferred method when dealing with limited data [10]. Nonetheless, it is imperative to acknowledge that CV, especially the commonly used k-fold CV, can be computationally intensive due to the necessity of fitting multiple models. Additionally, research has demonstrated that CV, which relies on excessively reusing the validation set, might lead to overfitting [7].

To address these challenges, Zhang et al. (2023) introduced a novel model selection approach that does not partition the dataset. Unlike CV, this method uses the entire dataset for training and validation while injecting noise into the fitting process. This noise is generated by mutating the sample labels, hence the name mutation validation (MV). While an over-complex classifier with a large capacity can generate a flexible decision boundary resulting in high accuracies on both original and mutated data, an over-simple classifier shows poor performance on both. A good classifier, on the other hand, can detect the ground-truth pattern despite the noise in the training labels and thus should show good performance on the original data and perform worse on the mutated data. This behavior is the intuition behind MV [18].

Zhang et al. [18] conducted extensive experiments on a broad range of datasets. They demonstrated that MV consistently and effectively captured underlying data patterns, thereby offering successful recommendations for the most suitable machine learning algorithm. In contrast, CV occasionally struggled to deliver comparable results. Further experiments showed MV's consistent preference for less complex models, when the algorithms were configured with specific capacity-related hyperparameters. While Zhang et al. provided valuable insights into the MV method, they did not assess the generalization performance of models selected by MV using nested-CV, which is commonly used in many application domains. Our study aimed to conduct a comprehensive comparison of the generalization estimates of the selected models, seeking to fully evaluate and understand MV's performance in contrast to the more commonly used CV method on further benchmark and real-world datasets. It is worth noting that none of our benchmark datasets were previously utilized by Zhang et al. Additionally, we employed Bayesian inference to support our analysis, a step not taken by Zhang et al. Furthermore, our study delved into differences in runtime (energy consumption), expanding the evaluation beyond merely generalization performance.

In summary, our study provides insights into strengths and limitations of MV with a basis on generalization performance, which can aid machine learning researchers and practitioners in making informed decisions regarding the trade-offs associated with those model selection approaches.

2 Methods and Experimental Setup

CV and MV. We implemented CV in a standard way such that the samples were randomly split in equally sized folds and each fold was used as the test set once.

As MV is a relatively new method, we provide an overview here for completeness. MV injects noise in the labels to generate mutated data. The generation of noise is a crucial aspect of MV. To illustrate the mechanism of label mutation, let's consider a binary classification problem. First, a class list is created, encompassing all unique class labels, i.e., 0, 1. Then, η proportion of the original labels are randomly selected and swapped, resulting in mutated labels. That is, labels of class 0 are replaced with 1, class 1 is replaced with 0. We used the recommended value of $\eta = 0.2$ [18].

Two models f and f_η are then trained, on the original training data S and on the mutated training data S_η, respectively. The performance difference between the models provides information regarding model complexity which is then used for model selection. We use $\hat{T}_S(f)$ to refer to the accuracy of model f on the original training data S, $\hat{T}_{S_\eta}(f_\eta)$ for the accuracy of model f_η on the mutated training data S_η, and $\hat{T}_S(f\eta)$ for the accuracy of model f_η on the original training data S. An empirical scoring metric m is used to assess model performance based on the theoretical metamorphic relation in MV [18]:

$$m = (1 - 2\eta)\hat{T}_S(f_\eta) + \hat{T}_S(f) - \hat{T}_{S_\eta}(f_\eta) + \eta$$

The score m aims to capture the changes in training accuracies before and after mutation, forming the foundation of MV model selection. For a good classifier, the performance difference $(\hat{T}_S(f) - \hat{T}_{S_\eta}(f_\eta))$ is expected to be substantial, and the accuracy $\hat{T}_S(f_\eta)$ to be high, resulting in a high score m. Conversely, an over-complex classifier, characterized by a small performance difference and a low accuracy $\hat{T}_S(f_\eta)$, the score m is expected to be low. An over-simple classifier is also anticipated to yield a low m score as both models will perform poorly. Finally, the model with the highest m is selected.

Datasets. In this investigation, our emphasis was on binary classification problems. We utilized 12 benchmark datasets sourced from the OpenML platform and the UCI repository [5,16] (Table 1).

The brain functional magnetic resonance imaging (fMRI) datasets were taken from the Amsterdam Open MRI Collection (AOMIC) [15]. The collection comprises three datasets: ID1000, PIOP1, and PIOP2 (Table 2). We used the functional MRI data from all three datasets: PIOP1 and PIOP2 obtained from resting-state task, while ID1000 based on a movie-watching task.

For each of the fMRI datasets, the functional connectivity (FC) representing synchrony between brain regions across time was extracted. We employed standard preprocessing steps, including motion correction and registration to Montreal Neurological Institute (MNI) space with the fMRIPrep pipeline [6], denoising and feature extraction with xcpEngine [1]. The parcellation of the processed fMRI images was carried out using the Schaefer 100 parcellation scheme

Table 1. Overview of the benchmark datasets. The datasets obtained from UCI are marked with an asterisk (*).

Index	Dataset name	Number of instances	Number of instances with label 0	Number of instances with label 1	Number of features
1	mfeat-fourier	2000	200	1800	76
2*	autism-screening	609	180	429	92
3	mfeat-karhunen	2000	200	1800	64
4	mammography	11183	260	10923	6
5	letter	20000	813	19187	16
6	satellite	5100	75	5025	36
7	fri-c2-1000-10	1000	420	580	10
8	segment	2310	330	1980	19
9*	sonar	208	97	111	60
10	qsar-biodeg	1055	356	699	41
11*	early-stage-diabetes-risk	520	200	320	16
12	ozone-level-8hr	2534	160	2374	72

which partitions the whole brain into one hundred non-overlapping parcels [14], resulting in 100 time series (1 per brain region). Finally, the FC was calculated as Pearson's correlation coefficients between the time series of all pairs of brain regions. The lower triangle of this symmetrical matrix was vectorized resulting in 4950 features. This process was done for all participants (i.e. samples) resulting in 2-dimensional tabular data.

All benchmark datasets and FC datasets in our study are presented in a tabular format, with features and a target associated with each sample. Specifically, the target variable of FC datasets was binary labels with female (F) and male (M) according to the participant's sex assigned at birth, an important task for basic and applied neuroscience [17].

Table 2. Overview of the FC datasets.

Dataset	ID1000	PIOP1	PIOP2
Subjects	764	158	186
Target	382 (F) / 382(M)	79 (F) / 79 (M)	93 (F) / 93 (M)
Features	4950	4950	4950
Age min	19	18.25	18.25
Age max	26	26.25	25.75
Age mean	22.862	22.081	21.958
Age std	1.713	1.809	1.787

Machine Learning Algorithms. Machine learning algorithms often have a set of hyperparameters which need to be adjusted during the learning process. In many cases this results in a set of candidate models with different capacity. The capacity of a model can be seen as a measure of its ability to capture the complex relationships present in the underlying data pattern [2]. Following the principle of Occam's razor, when dealing with multiple models exhibiting similar performance, a preference is given to simpler models, namely those with smaller capacities [13]. Hence, comparing model capacity can provide crucial information regarding the behavior of the model selection methods.

Specific algorithms, such as decision trees (DT), are associated with capacity-related hyperparameters, meaning that the resulting model's capacity heavily relies on the specific hyperparameter configuration. For instance, trees with higher depth can be considered more complex. Similarly, support vector machines (SVM) and Kernel Ridge Classifiers (KRC), when configured with the polynomial kernel, namely the polynomial SVM and polynomial KRC, also exemplify this characteristic. Additionally, multi-layer perceptrons (MLP) with various dropout rates are well suited to investigate model capacity. Considering these factors and the findings of Zhang et al., we designed experiments involving the above algorithms to examine different capacity-related hyperparameters. Specifically, DT involves the hyperparameter of maximum depth, with a range from 1 to 30. For MLP, the dropout rate serves as the hyperparameter, varying between 0.2 and 0.8. In the case of Polynomial SVM and Polynomial KRC, the hyperparameter is the polynomial degree, spanning from 1 to 15.

Bayesian Analysis. Considering the importance of properly comparing generalization performance in model selection, the comparison of model capacity and computational efficiency were based on Bayesian probabilistic analysis. Bayesian analysis provides a valuable alternative to traditional Null Hypothesis Significance Testing (NHST), offering potentially richer and more informative insights [11]. The Bayesian framework involves modeling the posterior probability distribution across the parameter space based on the observed data.

To illustrate, in situations involving two groups of data with equal length, a difference vector is computed, and the posterior probability distribution of the parameter, represented as the mean difference and denoted as μ, is subsequently formulated. Crucially, in this approach it is feasible to accept the null value of the evaluated parameter by setting the Region of Practical Equivalence (ROPE). This region encompasses parameter values that are considered practically indistinguishable from the null value. The size of the ROPE is determined according to the characteristics of the applications. For our empirical comparison of the two validation methods, the ROPE was set to [-0.025, 0.025], corresponding to a 5% difference.

In our study, the comparison framework was applied to each dataset, resulting in a difference vector derived from the two sets of scores (CV and MV). Two variations of the Bayesian paradigm were employed to evaluate the performance of CV and MV. The Bayesian correlated t-test for a single dataset [3] offered a probabilistic comparison on each dataset using the difference vector. Additionally, the Bayesian hierarchical test for multiple datasets [4] provided a final estimation based on the concatenated difference vectors from all datasets. The Bayesian analysis yields three posterior probabilities:

1. P_{CV}: the probability that the model selected by CV outperforms the model selected by MV;
2. $P_{P.E.}$: the probability that both models selected by CV and MV perform practically equivalent;
3. P_{MV}: the probability that the model selected by MV outperforms the model selected by CV.

Comparison Framework. It is important to define an appropriate scheme to allow for one-to-one comparison between CV and MV. We devised a framework based on nested cross-validation, which is particularly suitable for real-world datasets with a limited sample size. The inner loop is utilized for model selection, e.g. by selecting particular hyperparameters, while the outer loop is employed to evaluate the generalization performance of the selected model. We used a ten-times repeated 10-fold CV following previous recommendation to obtain reliable and robust estimates [3]. This generated three types of results:

1. 100 validation scores for the chosen models from the outer iterations;
2. 100 hyperparameter settings of the corresponding selected models;
3. the final model with the best hyperparameter setting, linked to the highest validation score, trained on the entire dataset.

In the assessment of nested cross-validation results, we considered two primary aspects. First, we compared the generalization estimates through the application of Bayesian tests. Second, guided by the probabilities obtained from these tests, we compared two quantities; (1) model performance and capacity, as signified by the selected hyperparameter configurations, and (2) evaluation of computational efficiency, encompassing runtime and CO_2 emissions.

3   Results

Comparison of Model Performance and Capacity. The comparative analysis of generalization performance, conducted through Bayesian correlated t-tests, showed that the models selected by both CV and MV exhibit practically equivalent performance (Fig. 1a, highlighted by inner cells in bright yellow), which is also confirmed by the results from the Bayesian hierarchical test (Fig. 1b).

This observed pattern holds across all three machine learning algorithms: polynomial KRC, polynomial SVM, and DT. On the majority of benchmark datasets examined, probabilities exceeding 90% affirm the practical equivalence between the two methods. In the case of MLP, it's worth noting that while certain specific datasets exhibit lower posterior probabilities regarding practical equivalence compared to the other three algorithms, the overarching trend still leans toward practical equivalence, remaining notably above chance level.

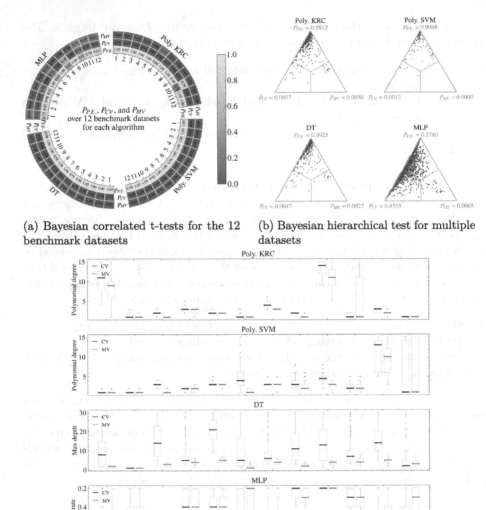

(a) Bayesian correlated t-tests for the 12 benchmark datasets

(b) Bayesian hierarchical test for multiple datasets

(c) Hyperparameter selection

Fig. 1. An overview of four algorithms evaluated on 12 benchmark datasets. Each subfigure consists of four sectors, one for each algorithm, with dataset indices cross-referenced in Table 1. (a) Each sector displays three tracks representing the posterior probabilities $P_{\text{P.E.}}$, P_{CV}, and P_{MV} for each case. These probabilities are presented as a heat-map. (b) The points represent samples drawn from the posterior probability distribution of 4000 samplings (default setting [4]). The final posterior probabilities $P_{\text{P.E.}}$, P_{CV}, and P_{MV} are located in the corners of each sector. (c) For each algorithm, boxplots indicate results obtained from the top 100 hyperparameter values generated by the comparison framework. Note that the dropout rate in the last sector is displayed on an inverted vertical axis, inline with the interpretation of capacity across all four sectors.

We compared the model capacities selected by both methods, as shown in Fig. 1c. A lower median line of the boxplots was interpreted as smaller model capacity. For KRC and SVM, MV-selected models exhibited similar or slightly lower median polynomial degrees. The interquartile range was small and comparable for both methods. In the case of DT, MV consistently chose models with notably lower maximum depths compared to CV, resulting in a more uniform and less varied selection. For MLP, the dropout rates chosen by MV were largely equivalent to those by CV. In summary, MV consistently favored models with lower capacity. Given the practical equivalence in performance between models selected by both methods, this suggests a preference for models determined by MV.

Comparison of Computational Efficiency. To achieve a balance between bias and variance, Kohavi [10] suggests using $k = 10$ folds for CV. However, selecting an appropriate value for k is not trivial. For instance, starting from version 0.22, the default value used by the scikit-learn[1] library was changed to $k = 5$ from previous $k = 3$. Hence, to analyse the effect of different values of k on CV and MV, we compared them using $k = 3, 5, 10$.

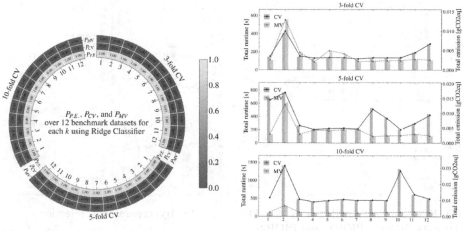

(a) Bayesian correlated t-tests for the 12 benchmark datasets

(b) Runtime and emission analysis

Fig. 2. The three sectors of each subfigure correspond to $k = 3$-, $k = 5$-, and $k = 10$-CV. (a) This subfigure contains results obtained from Bayesian correlated t-test across the 12 benchmark datasets. The indices of the datasets are listed in Table 1. In each sector, there are three tracks, representing the three posterior probabilities $P_{P.E.}$, P_{CV}, and P_{MV}. (b) The results from CV are shown in black and those from MV are shown in red. In each sector, the horizontal axis lists the indices of the benchmark datasets. The left vertical axis shows the total runtime of the procedure, and the right vertical axis shows the equivalent CO_2 emission.

---

[1] https://scikit-learn.org/stable/modules/generated/sklearn.model_selection.KFold.html.

Here we applied the algorithms to the benchmark datasets, KRC is presented as an example (Fig. 2a and Fig. 2b). The probability of $P_{\text{P.E.}}$ approaches 1, indicating practical equivalence in generalization performance between the two selection methods (Fig. 2a). The findings lead to further comparisons in computational efficiency. Overall, the computational efficiency of CV with varying k compared to MV yielded a consistent pattern (Fig. 2b). With $k = 3$, the performance of both methods is comparable. However, as k increases to 5, MV shows higher efficiency than CV. Finally, at $k = 10$, MV demonstrated a noticeable advantage over CV in terms of efficiency and carbon emission.

Comparison on Brain FC Datasets. The brain FC datasets, like many other real-world data, contain numerous features (in our case 4950 Pearson's correlation coefficients per sample). In this context, feature selection can be a desirable preprocessing step [8]. Besides, the three FC datasets used differ largely in sample size (Table 2). In particular, ID1000 has over four times more samples than PIOP1 and PIOP2.

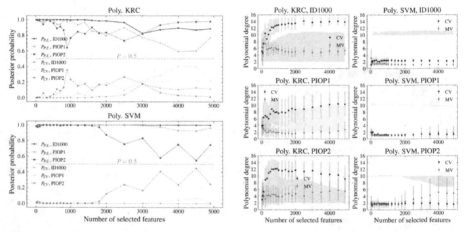

(a) Bayesian correlated t-tests for FC datasets (ID1000, PIOP1, and PIOP2)

(b) Hyperparameter selection

Fig. 3. (a) The Bayesian correlated t-test was used to calculate $P_{\text{P.E.}}$ and P_{CV} across subsets of the FC domain with varying numbers of selected best features. The above sector shows the results obtained from polynomial KRC, while the below sector displays those obtained from polynomial SVM. The probability curves in purple, yellow, and blue correspond to the datasets ID1000, PIOP1, and PIOP2, respectively. (b) The mean of the 100 best polynomial degrees across subsets of the FC domain ID1000, PIOP1, and PIOP2 for the polynomial KRC and SVM algorithms. Each point in the plot represents the mean of the polynomial degrees and the error bars demonstrate the standard deviation. The shaded areas in each sector shows the difference between the mean polynomial degrees from CV and MV. (Color figure online)

A commonly used feature selection method is calculation of F-scores using an ANOVA, as provided by the `SelectKBest` in `scikit-learn`[2] which selects the top K informative features corresponding to K highest F-scores. We generated several subsets of the FC datasets, each by selecting a different number of important features. This allowed us to compare CV and MV in a controlled manner on a large number of datasets each with different number of informative features. Each FC dataset was analyzed separately.

Two popular kernelized algorithms in neuroscience were investigated in this experiment. The probability values of polynomial KRC and polynomial SVM across the range of K (from 0 to 4950) are illustrated in Fig. 3a, respectively. Notably, the trend reflected a consistent linear decline in $P_{\text{P.E.}}$ in four cases, presenting a diminishing level of confidence in the practical equivalence between the two methods. This decline is mirrored by the probability P_{CV}, indicating a shift towards better performance of CV associated with generalization estimates as the number of features increased.

Exploring the relative relationship between the hyperparameters selected by CV and MV can aid in a deeper interpretation of these results. On the ID1000 dataset, the mean polynomial degree selected by CV was higher than that of MV for both cases (Fig. 3b, Poly. KRC, ID1000 and Poly. SVM, ID1000). The variance of the polynomial degrees selected by MV was generally lower or similar to that of CV, indicating more stable model selection by MV. Overall, MV selected models with lower complexity, consistent with the observations and results obtained on the benchmark datasets.

The results on the PIOP1 dataset were similar to those on ID1000 (Fig. 3b, Poly. KRC, PIOP1). Generally, the models selected by MV were less complex than those selected by CV. The relative performance of MV worsened as indicated by declining $P_{\text{P.E.}}$ (Fig. 3a, Poly. KRC) which might suggest that MV might tend to penalize complex models excessively and may encounter some level of underfitting. The mean hyperparameter values selected by both methods were similar (Fig. 3b, Poly. SVM, PIOP1). However, here MV displayed a more substantial variance, implying potential instability in the tuning process.

The PIOP1 dataset is challenging due to its limited sample size, a problem that PIOP2 also shares. However, the results on the PIOP2 dataset exposed further inconsistencies. As the number of features increased, MV selected models with higher capacity and $P_{\text{P.E.}}$ decreased (Fig. 3b, Poly. SVM, PIOP2). Furthermore, the variance of MV also showed an increase. In this specific instance, MV appears to have forfeited all of its advantages and performed worse than CV.

4   Discussion and Conclusion

Our systematic evaluation of generalization estimates involved the use of Bayesian correlated t-tests and Bayesian hierarchical tests, revealing that CV and MV exhibited practical equivalence in performance across the benchmark

---

[2] https://scikit-learn.org/stable/modules/feature_selection.html.

datasets. Building upon this observation, further experiments unveiled distinctions in terms of both model capacity and computational efficiency. First, comparison of hyperparameters indicating model capacity revealed that MV generally tended to select lower complexity models compared to CV, which is desirable in the light of Occam's razor. Second, MV consistently demonstrated advantages in runtime and a reduction in carbon emissions, particularly when the number of CV folds k exceeded 3. Comparing the two methods on the neuroscientific FC datasets, with number of selected most informative features ranging from low to high, the results on the largest dataset (ID1000), revealed that MV remained a practical alternative to CV.

Collectively, the results suggest that MV could function as a valuable complement to CV. In particular, MV can be leveraged as a preliminary tool to augment the efficiency of hyperparameter tuning, particularly in resource-constrained environments. This would facilitate a more thorough exploration of the hyperparameter space while maintaining an acceptable runtime and a lower carbon footprint. Nevertheless, it is important to acknowledge limitations of MV. As our experiments on the FC data showed, MV may be susceptible to underfitting and instability leading to suboptimal model selection.

There remain opportunities for further enhancements that warrant exploration in future research endeavors. While this study primarily focused on binary classification problems, future investigations could extend this comparative analysis to encompass multiclass classification and regression tasks. Furthermore, considering the widespread prevalence of neural networks in diverse domains, it would be intriguing to examine how MV fares in comparison to CV within the realm of deep learning architectures. Finally, examining MV's behavior with respect to data characteristics could provide further insights.

Acknowledgements. This work was partly supported by the Helmholtz Portfolio Theme "Supercomputing and Modelling for the Human Brain" and by the Max Planck School of Cognition supported by the Federal Ministry of Education and Research (BMBF) and the Max Planck Society (MPG).

References

1. xcpengine-container 1.0.1. https://pypi.org/project/xcpengine-container/
2. Barbiero, P., Squillero, G., Tonda, A.P.: Modeling generalization in machine learning: a methodological and computational study. CoRR abs/2006.15680 (2020)
3. Corani, G., Benavoli, A.: A Bayesian approach for comparing cross-validated algorithms on multiple data sets. Mach. Learn. **100**, 285–304 (2015)
4. Corani, G., Benavoli, A., Demšar, J., Mangili, F., Zaffalon, M.: Statistical comparison of classifiers through Bayesian hierarchical modelling. Mach. Learn. **106**(11), 1817–1837 (2017)
5. Dua, D., Graff, C.: UCI machine learning repository (2017). http://archive.ics.uci.edu/ml
6. Esteban, O., et al.: fMRIPrep: a robust preprocessing pipeline for functional MRI (2022)

7. Feldman, V., Frostig, R., Hardt, M.: The advantages of multiple classes for reducing overfitting from test set reuse. CoRR abs/1905.10360 (2019)
8. Guyon, I., Elisseeff, A.: An introduction to variable and feature selection. J. Mach. Learn. Res. **3**, 1157–1182 (2003)
9. Hastie, T., Tibshirani, R., Friedman, J.: The Elements of Statistical Learning: Data Mining, Inference, and Prediction. Springer Series in Statistics, Springer, New York (2009). https://doi.org/10.1007/978-0-387-21606-5
10. Kohavi, R.: A Study of Cross-validation and Bootstrap for Accuracy Estimation and Model Selection, pp. 1137–1143. IJCAI 1995, Morgan Kaufmann Publishers Inc., San Francisco, CA, USA (1995)
11. Kruschke, J.: Bayesian estimation supersedes the t test. J. Exp. Psychol. General **142**, 573–603 (2012)
12. Mitchell, T.M.: Machine Learning. McGraw-hill, New York (1997)
13. Raschka, S.: Model evaluation, model selection, and algorithm selection in machine learning. CoRR abs/1811.12808 (2018)
14. Schaefer, A., et al.: Local-Global Parcellation of the Human Cerebral Cortex from Intrinsic Functional Connectivity MRI. Cerebral Cortex (2018)
15. Snoek, L., et al.: The Amsterdam open MRI collection, a set of multimodal MRI datasets for individual difference analyses. Sci. Data **8**(1), 85 (2021)
16. Vanschoren, J., van Rijn, J.N., Bischl, B., Torgo, L.: OpenML: networked science in machine learning. SIGKDD Explor. **15**(2), 49–60 (2013)
17. Weis, S., Patil, K.R., Hoffstaedter, F., Nostro, A., Yeo, B.T.T., Eickhoff, S.B.: Sex classification by resting state brain connectivity. Cereb. Cortex **30**(2), 824–835 (2019)
18. Zhang, J.M., Harman, M., Guedj, B., Barr, E.T., Shawe-Taylor, J.: Model validation using mutated training labels: an exploratory study. Neurocomput. **539**(C), 126116 (2023). https://doi.org/10.1016/j.neucom.2023.02.042

Amplified Contribution Analysis
for Federated Learning

Maciej Krzysztof Zuziak$^{(\boxtimes)}$ (iD) and Salvatore Rinzivillo (iD)

Consiglio Nazionale Delle Ricerche, ISTI-KDD, 56124 Pisa, Italy
{maciejkrzysztof.zuziak,rinzivillo}@isti.cnr.it

Abstract. The problem of establishing the client's marginal contribution is essential to any decentralised machine-learning process that relies on the participation of remote agents. The ability to detect harmful participants on an ongoing basis can constitute a significant challenge as one can obtain only a very limited amount of information from the external environment in order not to break the privacy assumption that underlies the federated learning paradigm. In this work, we present an Amplified Contribution Function - a set of aggregation operations performed on gradients received by the central orchestrator that allows to non-intrusively investigate the risk of accepting a certain set of gradients dispatched from a remote agent. Our proposed method is distinguished by a high degree of interpretability and interoperability as it supports the gross majority of the currently available federated techniques and algorithms. It is also characterised by a space and time complexity similar to that of the leave-one-out method - a common baseline for all deletion and sensitivity analytics tools.

Keywords: Federated Learning · Contribution Metrics · Sensitivity Analysis

1 Introduction

Contribution analysis is based on the necessity to quantify the potential impact of including (or excluding) a certain client from a learning cohort. A potential solution to such a problem may be applicable in a range of scenarios. In the cross-silo scenario, it may be that a number of enterprises want to indirectly use their data to train one model. However, being competitors, they want to ensure that no agent serves as a *free rider* - either noising its data completely or sending a random noise as parameters. In a multi-device scenario, we may have remote access to a number of devices without the possibility to assess directly whether some of them are malfunctioning. We accept the risk of incorporating unreliable data to some extent, but we still need to blacklist or recover those devices that are negatively impacting the training. In both cases, a dynamic evaluation method is crucial to the success of the federated process.

© The Author(s), under exclusive license to Springer Nature Switzerland AG 2024
I. Miliou et al. (Eds.): IDA 2024, LNCS 14642, pp. 68–79, 2024.
https://doi.org/10.1007/978-3-031-58553-1_6

In the current literature, this challenge is being approached from various angles. A suitable solution should be comparatively straightforward to implement and employ - so as to interpolate it in the existing learning frameworks smoothly. Our main motivation consists of giving the central orchestrator an efficient and interpretable tool for dynamically validating clients' contributions to either filter out those that make training more difficult (to the point of jeopardizing the whole process) or do not contribute in the long run anything to the data discovery process. Our overall contribution is as follows:

1. We formalize the marginal contribution evaluation problem in a theoretical framework that is suitable for federated learning. In connection with that, we introduce the notion of **Aggregation Masks** and **Collaborative Contribution Function**. This set of definitions can help with describing and evaluating solutions similar to the one presented in this paper.
2. We introduce the **Alpha-Amplified Function** that can amplify and capture the marginal impact of a selected client on a global model's performance. The proposed method is characterized by a time and space complexity similar to the baseline leave-one-out while visibly enhancing the detection capabilities of the orchestrator. Moreover, by introducing a modifiable parameter α, it is possible to adjust the sensitivity of the analysis while still retaining the same execution time.
3. We perform a simulation on three different datasets, exploring the possibilities and limitations of the presented solution. Different task complexity allows us to explore how the particular learning setting is influencing the behaviour of our method. Additionally, we compare the behaviour of the method in two different settings: one that envisages the independence of samples distributed across the nodes and another that covers a case of non-independent and identically distributed data.
4. We introduce an experimental library that allows us to perform the federated learning process and an amplified contribution analysis in a simulated environment. The code is available under the link: https://github.com/MKZuziak/IDA_2024_demo.

2   Prerequisites and Related Work

2.1   Federated Learning

Federated Learning was first introduced in [10] as a method of decentralised learning that allows for training a one global model using weights provided by the nodes (often referred to as clients) that perform local training. Formally, the federated learning objective function is formulated as follows:

$$min_{W \in R^{n \times d}} F(W), \textbf{ where } F(W) = \frac{1}{|S|} \sum_{j \in S} f_j(w_j) \tag{1}$$

where W is the $n \times d$-dimensional matrix containing all the weights of the model, $F(\cdot)$ is the global objective function, and $f_j(\cdot)$ is the local objective function, which is generally the same on all respective clients. Each client receives a matrix of weights W^t, performs at most τ training epochs using local data, and then broadcasts the weights w_j^t back to the orchestrator. The orchestrator performs an aggregation operation yielding a new set of weights W^{t+1}. This process continues until convergences. In each round, the broadcasting and aggregation are performed with a set of sampled clients S, which may not necessarily equal the whole sampled population.

This vanilla method, called Federated Averaging (FedAvg) can be rewritten in terms of pseudo-gradients that depict the *instantaneous* change in the learnable parameters of each local model. Formally:

$$W^{t+1} = \frac{1}{|S|} \sum_{j \in S} w_j^t = W^t - \frac{1}{|S|} \sum_{j \in S} (W^t - w_j^t) = W_t - \frac{1}{|S|} \sum_{j \in S} \Delta_j \qquad (2)$$

where $\Delta_j^t \in \mathbb{R}^{n \times d}$. It is easy to see that $W^t - w_j^t = \Delta_j^t$ is a type of *pseudo-gradient* that reflects the change of parameters before and after the local training. This observation was originally derived by [11], and it allows us to introduce server-side optimization and momentum to the learning algorithm. It is also surprisingly convenient for contribution analysis, as it allows us to think about the process in terms of *instantaneous* changes[1] in the direction of the global or local optimum of the objective function, which we will use extensively throughout our contribution. On the final note, Eq. 2 can be generalised as:

$$W^{t+1} = \sum_{j \in S} \phi_j w_j^t = W^t - \sum_{j \in S} \phi_j (W^t - w_j^t) \qquad (3)$$

where ϕ_j for $j = 1, 2, \cdots, |S|$ are weights associated with each client s.t. $\sum_j^{|S|} \phi_j = 1$. If the weights are distributed uniformly, i.e. $\phi_j = \frac{1}{|S|}$, then the Eq. 3 is equivalent to Eq. 2 describing the basic Federated Averaging.

2.2 Contribution Measures for Federated Learning

The issue of quantifying the marginal impact of the particular subset of the dataset on the general model has been under examination for several years up to this date. In the current literature, the majority of available solutions are based on the Shapley Value [4]. In 2017, [3] introduced the idea of using cooperative games theory to evaluate personal data in networks. Similarly, [1] introduced Data Shapley - a method for assessing the impact of each observation on the final model's performance. This was later expanded by [2]. Works of [5,6] employed this approach to the federated learning scenario. By labelling each client as a player and using the loss function as a value function, the whole learning process is structured as a cooperative game and the client's marginal contribution is

---

[1] We acknowledge the term *instantaneous* is an abuse of a concept here, hence the evident usage of the italics. The changes are, in fact, made over the training rounds. However, this metaphor is still handy for illustrative purposes.

embedded in the final pay-off vector. This approach comes with several issues, such as relatively high computational complexity that can act as a prohibitive factor, even if a sampling schema is employed to decrease the number of subsets that must be considered (baseline Shapley Value requires calculating all the $\mathcal{O}(2^N)$ subsets).

Some authors seek solutions outside the realms of cooperative game theory. In the [8], authors propose using pairwise correlation for detecting malicious clients. This approach can work efficiently, given that the group of unwanted models will always share some degree of similarity in parameters, which was not proven to be always guaranteed in the case of deep learning models. A data-driven approach was presented in [9], where authors propose constructing a separate model for grading the marginal contributions of clients. The most similar method to our work was presented in [12], where authors also observe the usefulness of employing the cosine similarity between different vectors of parameters of local clients. However, they employ a different approach to this problem, as the contribution is calculated at the end of the whole training cycle, comparing the route of optimal gradient descent to a proposed one. Conversely, we present a dynamic schema that can produce contribution vectors each round without considerably slowing the training process. Another work that bears a strong connection to ours was presented in [7], where the observation that gradients received from clients can be (temporarily) preserved to assemble and compare the efficiency of different models. However, this technique was limited solely to Shapley Values. In the course of our work, we have decided to create and present a uniform notation that scales to every method of calculating the marginal differences between the assembled clients.

3 Amplified Deletion Analytics

3.1 Notation and Aggregation Masks

The most important element that we define to simplify the further analysis is the Aggregation Mask. The term aggregation mask is used to describe a process of manipulating the client's weights before the actual aggregation in order to detect the difference in performance between the global and marginal coalition of clients. Hence, the sole usage of masks neither influences the actual aggregation of the clients nor impacts the sampling schema. However, results of comparing different coalitions can be used to either adjust the weighting or sampling schema or simply blacklist some clients from attending the training.

Definition 1 (Aggregation Mask). *Given the aggregation procedure*

$$W_{(m)}^{t+1} = \sum_{j \in S} \phi_j w_j^t = W^t - \sum_{j \in S} \phi_j (W^t - w_j^t) = W_t - \frac{1}{|S|} \sum_{j \in S} \phi_j \Delta_j \quad (4)$$

aggregation mask is a set of weights $\{\phi_j | \forall j \in S\}$ such that it fulfils $\sum_j^{|S|} \phi_j = 1$ and $W_{(m)}^{t+1} \neq W^{t+1}$.

In other words, a mask is an auxiliary weights aggregation that does not serve as a direct update. Using the concept of aggregation mask, we will define the two most important masks used throughout this work. Namely, the Leave-one-out Mask and the Alpha-Amplification Mask, denoted respectively with $W_{(L,i)}^{t+1}$ and $W_{(\alpha,i)}^{t+1}$ (Fig. 1).

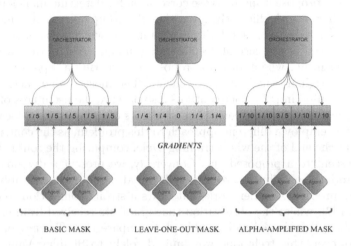

Fig. 1. Visual representation of the mask concept. Each time a set of agents (small rectangles at the bottom of the infographic) sends gradients to an orchestrator (large rectangle at the top of the infographic), the orchestrator can prepare a corresponding mask that will be used for evaluation, irrespective of the weighting schema used for performing an actual update. Leave-one-out mask zeros the contribution of the i-th client, while the alpha-amplification reinforces it by alpha factor. Those masked gradients may then be used in specific evaluation functions.

Definition 2 (Leave-one-out Mask). *The Leave-one-out mask denoted by* $W_{(L,i)}$ *is an aggregation mask that fully conceals the contribution of client* i. *Formally:*

$$W_{(L,i)}^{t+1} = W^t - \sum_{j \in S} \phi_j \Delta_j \tag{5}$$

$$\textbf{where } \phi_i = 0 \textbf{ and } \sum_{j \in S} \phi_j = 1 \tag{6}$$

Definition 3 (The Alpha-Amplification Mask). *The Alpha-Amplification Mask denoted* $W_{(\alpha,i)}$ *is an aggregation mask that amplifies the contribution of client* i *by the parameter* α. *Formally:*

$$W_{(\alpha,i)}^{t+1} = W^t - \sum_{j \in S} \phi_j \Delta_j \tag{7}$$

$$\textit{where } \phi_j = \alpha \textit{ and } \alpha + \sum_{j \in S} \phi_j = 1 \tag{8}$$

Remark 1. The L (e.g. $W_{(L,i)}$) is a constant signalizing use of the leave-one-out mask. However, $\alpha \in \mathbb{R}$ is a parameter that controls the amplification. If α is less or equal to one, it should be treated as a weight multiplier. If the α is larger than one, then it should be treated as preserving α copies of the client's i gradient during the weight's aggregation. This provides an intuitive interpretation of the mask and allows us to convert between those two interpretations. Hence, if $\alpha > 1$, $\phi_i = \frac{\alpha}{n}$ and $\phi_j = \frac{1-\frac{\alpha}{n}}{n-1}$.

Remark 2. The parameter α allows us to control the sensitivity of the analysis. More precisely, it allows us to decide to what extent we want to amplify the contribution of the evaluated client. Setting the parameter α to 0 reduces the alpha-amplification mask to a leave-one-out mask, i.e. $W_{(\alpha=0,i)} = W_{(L,i)}$. On the other hand, an increase in the alpha parameter will result in the marginalisation of other clients. At the limit, the aggregation mask will provide weights that are identical to those of an evaluated client.

In addition to what was presented above, we denote a loss function with $\mathcal{L}$: $W \longrightarrow \mathbb{R}$. To simplify the notation, we use a shorthand notation that omits the concept of the hypothesis function while defining the loss function, substituting just the matrix of weights in that place. We often omit the dependence on data matrix X and the corresponding label vector $\overline{y}$.

3.2 Leave-one-out and Alpha-Amplified Function

Definition 4 (Collaborative Contribution Function). *The Collaborative Contribution Function Ψ, given the previous matrix of weights W^t, set of gradients $\{\Delta^t | \forall \Delta_j^t \in S\}$ and a loss function $\mathcal{L}(\cdot)$ returns an n-dimensional vector $\overline{\psi} \in \mathbb{R}^d$ that maps every client to its respective marginal contribution calculated by using a selected mask (m). Formally:*

$$\Psi_{(m)} : \mathcal{L}, W^t, \Delta^t \longrightarrow \overline{\psi} \in R^d \tag{9}$$

$$\textit{where } \forall i \in S : \overline{\psi}_i \approx \mathcal{L}(W^{t+1}) - \mathcal{L}(W^{t+1}_{(m,i)}) \tag{10}$$

*is called a **Contribution Index** of a client i.*

This definition implies that the contribution index of each client should always be bounded by a co-domain of a loss function. Hence, two collaborative contribution functions are always expressed in the same units as long as they are calculated using the same loss function.

Definition 5 (The Leave One Out Function). *The Leave-one-out Function is a Collaborative Contribution Function Ψ that calculates the contribution of the client i using the leave-one-out mask:*

$$\Psi_{(L)} = \mathcal{L}(W^{t+1}) - \mathcal{L}(W^{t+1}_{(L,i)}) \quad \forall i \in S \tag{11}$$

Definition 6 (The Alpha-Amplified Function). *The Alpha-Amplified Function is a Collaborative Contribution Function Ψ that amplifies the possible contribution of the client i by parameter α using an amplification mask and is defined as:*

$$\Psi_{(\alpha)} = \mathcal{L}(W^{t+1} - \mathcal{L}(W_{(\alpha,i)}^{t+1}) \quad \forall i \in S \tag{12}$$

Remark 3. Both functions calculate client contribution, but the yielded contribution indices will have an opposite sign. For the leave-one-out function, if the client has a positive impact on the training, masking its presence will result in a lower score. Because of that, positive scores returned by the leave-one-out method are evidence of a positive contribution, while negative scores are on the contrary. For the alpha-amplification function, the difference will be positive if the client is deemed to be harmful to the training while remaining negative if the client is positively contributing. For this reason, alpha-amplification can be interpreted as a *threat score*.

Remark 4. We must highlight the importance of noticing the difference between the score returned by the collaborative contribution functions and the score of the model that was updated using masked weights. The score of a model is simply a loss value returned when evaluating this model against a given dataset, i.e. $\mathcal{L}(W_{m,i}^t)$. The contribution function is always a difference between the loss of the baseline model and the masked model, i.e. $\mathcal{L}(W^t) - \mathcal{L}(W_{(m,i)}^t)$.

A remark must be made about the time complexity. Without the loss of generality, we assume that forward propagation through the neural network (without backpropagation) constitutes one computational unit. The complexity of the baseline leave-one-out is $\mathcal{O}(N)$, as it is required to perform exactly $N+1$ forward propagations through the network. Similarly, alpha-amplification requires testing only $N+1$ combinations, so its time complexity is $\mathcal{O}(N)$. However, as evidenced by the next section, it tends to capture better the behaviour of the malfunctioning sensors (clients).

4 Experiments

4.1 Description

In the experimental section, we were mainly interested in testing two hypotheses. The first hypothesis concerns the behaviour of the alpha-amplification function. It states that the function, in relation to the leave-one-out baseline, can better detect a set of weights that are harmful to the general training, i.e. the average scores of such clients exceed one standard deviation of the sample. The second hypothesis is connected to non-heterogeneous data splits. It states that when the local splits are non-IID, the collaborative contribution function (be it LOO or alpha-amplification) exhibits lower confidence in clients with original data, which was not subject to any transformation or dilatations.

In order to test formulated hypotheses, we perform six different simulations using three different datasets, namely: MNIST [13], FMNIST [14], and CIFAR10

[15]. The datasets are split using two different methods to obtain heterogenous and homogenous label distribution across the clients. Homogeneous distribution assumes that the labels are distributed uniformly across all the clients. Heterogenous distribution is generated using Dirichlet Distribution, as this method was employed before in some papers concerning Federated Learning, e.g. [11] uses Dirichlet Distribution to create highly heterogeneous splits. While we test homogeneous and heterogeneous distributions, we assume that the total number of data samples at each node is the same. Also, we assume that each client shares the same neural network architecture. For solving the MNIST classification problem, we employ a convolutional neural network with three convolutional layers and another four fully-connected layers. Both types of layers use Rectified Linear Units. Additionally, the fully connected layers employ neuron dropout at a rate of 20%. FMNIST classification task is solved by a modified version of ResNet34 with a changed classification head. The CIFAR10 problem is solved with a modified version of ResNet101. For all three problems, we resort to a generalized version of FedAvg - the FedOpt [11] with a learning rate $\eta = 1.0$ and a number of global epochs equal to 80. During the local training, we use Stochastic Gradient Descent, with a learning rate equal to $\hat{\eta} = 0.01$ and batch size equal to 32:

To simulate clients whose participation is detrimental to the training, we perform transformation and dilatations on the first five clients in each split. We distinguish three types of data modification: Gaussian noise addition, blur transformation and rotation. All clients affected by those modifications are labelled as *malfunctioning sensors*. The remaining five clients are labelled as *valid sensors*. Noise addition is performed by firstly sampling a noise matrix from a Gaussian Distribution $\mathcal{N}(0, 1)$ that is multiplied by a scaling constant c and then adding it to the original image. For client number 0, the scaling constant equals 0.4, while for client numbers 1 and 2, the constant equals 0.15 and 0.10, respectively. Client number 3 is blurred by using a 3×3 Gaussian kernel, while the rotation on client number 4 is performed by first selecting a random angle from the interval o to 45 and then applying the rotation.

4.2 Results

In Fig. 2, we represent average values obtained in the six simulations run on MNIST, FMNIST and CIFAR10 datasets. - one in homogeneous and one in heterogeneous setting for each dataset. The leave-one-out and alpha-amplification values were calculated simultaneously in the same simulation-run to keep all the constant fixed. While the leave-one-out is impossible to tune by default, the alpha parameter in alpha-amplification was set to $\alpha = 5$. The left-hand side of Fig. 2 contains six plots representing leave-one-out and alpha-amplification values obtained during MNIST, FMNIST and CIFAR 10 simulations run in an IID (homogeneous) setting. The right-hand side of this figure contains six plots representing the leave-one-out and alpha-amplification values obtained during the simulations run in a Non-IID (heterogeneous) setting. The identification numbers of nodes are plotted on the x-axis, while the average score is plotted on the y-axis. The *malfunctioning sensors* are coloured red, while the *valid sensors* are coloured green. The solid black line represents the baseline value at point 0.0 (no visible contribution), and the dotted black line represents the *cut-off* value at the level of one standard deviation of a sample. Note that the leave-one-out and alpha amplification scores are sign-flipped by definition: a negative leave-one-out score indicates a negative impact on the trained model, while the positive alpha-amplification value represents the same situation.

4.3 Commentary

Examination of the presented results allows us to draw several conclusions about the behaviour of the alpha-amplification function and, as a consequence, allows us to formulate a preliminary answer to our hypothesis. Firstly, the Alpha-Amplification serves well when used as a threat detection tool. In almost every of the simulation runs (except the non-IID CIFAR10 dataset), the tested function can detect all three clients with excessive noise, often also detecting clients with datasets transformed by blur or rotation. On the other hand, the leave-one-out fails in the majority of the cases, often detecting no more than one *noised* client. Hence, it seems that the alpha-amplification can handle better the task of *malfunctioning clients* detection than the leave-one-out baseline. This option provides us with an answer to the first hypothesis. The empirical evidence suggests that the alpha-amplification score of *malfunctioning sensors* exceeds one standard deviation of the sample, while the leave-one-out score of the same sensors often fails to do so.[2] Secondly, in the case of alpha-amplification, there is no observable detrimental effect of the data heterogeneity on the *malfunctioning sensor detection*. Although alpha-amplification works well for threat detection, it generally does not penalize the clients with healthy data, even in the case of a heterogeneous environment. It can be explained by the fact that while increasing the impact of a *malfunctioning sensor* is detrimental to training, the amplification of weights provided by a *valid sensor* does not affect the training to that extent in the homogeneous scenario or may cause drift of weights towards values that better fit the local (client's) distribution in the heterogeneous case. This also implies, that while alpha-amplification is a promising threat-detection tool, it may now be always advisable to use it as a method of universal contribution quantification. However, this depends also on the chosen sensitivity. As the alpha parameter was set to $\alpha = 5$ during the experiments, the alpha-amplification function tends to heavily amplify the signal from the tested client. A lower value of alpha would allow us to build a more universal contribution index (capturing also the positive behaviour of valid sensors), but would not yield such conclusive results for the *malfunctioning* batch. This also portrays well the flexibility of the presented method, which is not present in the leave-one-out baseline.

---

[2] The placed threshold was defined in relation to the standard deviation of the sample, but it is possible to test both functions against a different detection threshold. We have chosen standard deviation, as it is fairly straightforward to interpret and present.

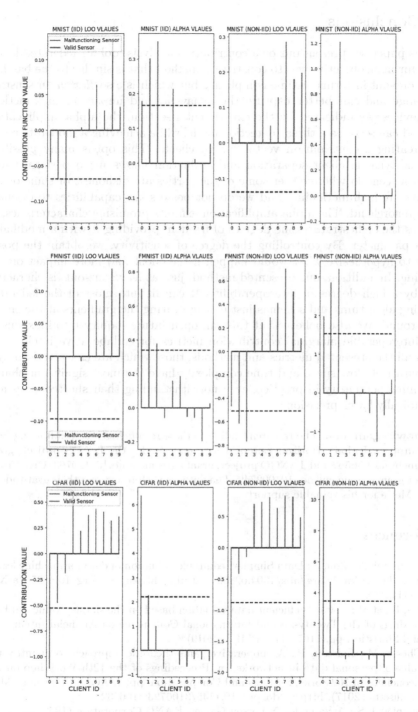

Fig. 2. The average value of leave-one-out and alpha-amplification scores obtained in each of the six simulations.

5   Conclusions

In this paper, we present our own contribution analysis tool for federated learning. Our main objective was to construct a method that is similar to the baseline leave-one-out in terms of time complexity but which is also *flexible* in sensitivity range and can better capture the behaviour and impact of malfunctioning devices. As indicated by the experimental section, the alpha-amplification method can serve as a threat detection model while achieving a moderate result for creating a contribution vector for all clients. This opens many possibilities, as dynamic client separation and evaluation allows us to perform training in a scenario where either some of the parties are dishonest or some of the sensors are malfunctioning, and we do not possess the capabilities to evaluate them beforehand. The alpha-amplification exhibits promising characteristics, as it gives the orchestrator some degrees of freedom providing it with a modifiable alpha parameter. By controlling the degree of sensitivity, we obtain the possibility to hypothesize what would happen if the client had more impact on the training. In addition, our presented method, just as leave-one-out, is characterized by a high degree of interoperability. It can fit into most of the federated learning algorithms and schemas just by aggregating the gradients at the end of each round. We also believe that this can open future inquiry into inexpensive and interoperable marginal contribution metrics for collaborative intelligence. From all the presented metrics and methods, those which are characterized by a comparatively low space and time complexity have the most significant chance of gaining popularity in practice, even notwithstanding their shortcomings and sporadically lower precision.

Acknowledgements. The research leading to these results has received funding from the European Union's Horizon Europe Programme under the LeADS project, grant agreement no. 956562 and TANGO project, grant agreement no.101120763, CREXData project grant agreement no. 101092749. The authors wish to express their gratitude to Carlo Metta for his valuable support.

References

1. Ghorbani, A., Zou, J.: Data Shapley: equitable valuation of data for machine learning. http://arxiv.org/abs/1904.02868 (2019). https://doi.org/10.48550/arXiv.1904.02868
2. Jia, R., et al.: Towards efficient data valuation based on the shapley value. In: Proceedings of the Twenty-Second International Conference on Artificial Intelligence and Statistics, pp. 1167-1176. PMLR (2019)
3. Chessa, M., Loiseau, P.: A cooperative game-theoretic approach to quantify the value of personal data in networks. In: Proceedings of the 12th Workshop on the Economics of Networks, Systems and Computation, pp. 1. ACM, Cambridge Massachusetts (2017). https://doi.org/10.1145/3106723.3106732
4. Shapley, L.S.: A Value for N-Person Games. RAND Corporation (1952)
5. Wang, T., Rausch, J., Zhang, C., Jia, R., Song, D.: A principled approach to data valuation for federated learning. In: Yang, Q., Fan, L., Yu, H. (eds.) Federated

Learning: Privacy and Incentive, pp. 153–167. Springer, Cham (2020). https://doi. org/10.1007/978-3-030-63076-811

6. Liu, Z., Chen, Y., Yu, H., Liu, Y., Cui, L.: GTG-shapley: efficient and accurate participant contribution evaluation in federated learning. ACM Trans. Intell. Syst. Technol. **13**, 60:1-60:21 (2022). https://doi.org/10.1145/3501811

7. Song, T., Tong, Y., Wei, S.: Profit allocation for federated learning. In: 2019 IEEE International Conference on Big Data (Big Data), pp. 2577-2586 (2019). https:// doi.org/10.1109/BigData47090.2019.9006327

8. Lv, H., et al.: Data-free evaluation of user contributions in federated learning. http://arxiv.org/abs/2108.10623 (2021)

9. Shyn, S.K., Kim, D., Kim, K.: FedCCEA: a practical approach of client contribution evaluation for federated learning. http://arxiv.org/abs/2106.02310, https:// doi.org/10.48550/arXiv.2106.02310 (2021)

10. McMahan, B., Moore, E., Ramage, D., Hampson, S., Arcas, B.A. y: Communication-efficient learning of deep networks from decentralized data. In: Proceedings of the 20th International Conference on Artificial Intelligence and Statistics, pp. 1273-1282. PMLR (2017)

11. Reddi, S.J., et al.: Adaptive federated optimization. In: Presented at the International Conference on Learning Representations, 26 March 2022

12. Zhang, J., Wu, Y., Pan, R.: Incentive mechanism for horizontal federated learning based on reputation and reverse auction. In: Proceedings of the Web Conference 2021, pp. 947-956. ACM, Ljubljana Slovenia (2021). https://doi.org/10.1145/ 3442381.3449888

13. MNIST handwritten digit database, Yann LeCun, Corinna Cortes and Chris Burges. http://yann.lecun.com/exdb/mnist/, Accessed 02 Nov 2023

14. Xiao, H., Rasul, K., Vollgraf, R.: Fashion-MNIST: a novel image dataset for benchmarking machine learning algorithms. http://arxiv.org/abs/1708.07747, https:// doi.org/10.48550/arXiv.1708.07747 (2017)

15. CIFAR-10 and CIFAR-100 datasets. https://www.cs.toronto.edu/~kriz/cifar.html, Accessed 02 Nov 2023

Data Mining

Monitoring Concept Drift in Continuous Federated Learning Platforms

Christoph Düsing(✉) and Philipp Cimiano

CITEC, Bielefeld University, Bielefeld, Germany
{cduesing,cimiano}@techfak.uni-bielefeld.de

Abstract. Continuous federated learning (CFL), a recently emerging learning paradigm that facilitates collaborative, yet privacy-preserving machine learning (ML), bears the potential to shape the future of distributed ML. In spite of its great potential, it is - similar to continuous ML - prone to suffer from concept drift (a change in data properties over time). In turn, CFL can greatly benefit from employing drift detection to react adequately to emerging drifts. Although various such approaches exist, respective research lacks application of drift detection to CFL with dynamic client participation as well as detailed analysis of the advantages of different drift detection approaches such as error-based or data-based drift detection. To this end, we apply these drift detection approaches to a CFL platform that allows new clients to join even after the training has started and measure the negative impact of concept drift on model performance. Moreover, we uncover distinct differences between the error- and data-based drift detection. In particular, we find the former ones to be more suitable to detect the point in time where the joint models stops benefiting from concept drift whereas the latter allows for a more precise detection of the first occurrence of concept drift.

Keywords: Concept Drift · Federated Learning · Continuous Learning

1 Introduction

Federated learning (FL) has become an increasingly popular, privacy-preserving approach towards distributed machine learning (ML) [13,20]. As such, it enables mutually distrustful or legally separated participants (referred to as *clients* in FL literature) to jointly train a ML model while maintaining full autonomy over their respective data [13,19]. To do so, a central server orchestrates several rounds of joint training, each consisting of (1) the central server sending out the globally shared model to participating clients, (2) each active client locally fitting the received model to their private data and afterwards sending model updates back to the central server, and (3) the central server aggregating all model updates received to a new shared model [13,19].

As of recently, traditional FL algorithms such as *FedAvg* [19] and *FedProx* [15] have been extended to facilitate continuous learning [4,5,23]. This opens

© The Author(s), under exclusive license to Springer Nature Switzerland AG 2024
I. Miliou et al. (Eds.): IDA 2024, LNCS 14642, pp. 83–94, 2024.
https://doi.org/10.1007/978-3-031-58553-1_7

up various areas of research and application, including FL platforms open to any client willing to participate [8]. Unlike previous applications of FL, such platforms allow new clients to join even after the training has started (we refer to this as *dynamic client participation* in the following) and therefore potentially improving the performance of the joint model further. Likewise, dynamic client participation bears the potential to introduce concept drift to the continuous FL (CFL) platform that compromises the co-created value of CFL, i.e., the improvements in terms of performance. In turn, concept drift - a change in data properties over time - constitutes a major issue in CFL [11] and has therefore recently been subject to a vast variety of research [3,6,11]. The goal of existing methods is usually either to minimize the time it takes to recognize concept drift [7] or maximize the performance of the shared model in face of concept drift [6].

In order to detect concept drift among clients in CFL, existing methods can be divided into error-based drift detection or data-based detection [4,5,8]. Here, the former methods detect concept drift by monitoring either each client's local loss [6] or the loss of the global model [5] during each round of training. The latter type of approaches relies on federated analytics [21] or secure aggregation [2] to monitor data properties such as label distributions and detect concept drift accordingly [4]. However, existing approaches to detect concept drift and adapt to it only consider FL platforms with static client participating (e.g., [3,25]). Hence, they might not suite FL platforms allowing clients to join the federation once the CFL procedure has started. Unfortunately, this significantly hinders their real-world applicability. Furthermore, to the best of our knowledge, no previous work compared error- and data-based drift detection approaches with respect to their time required to identify concept drift as well as the performance of the model trained using CFL. In this work, we therefore address these evident gaps in existing literature by applying and evaluating both error- and data-based drift detection for CFL platforms with dynamic client participation. More precisely, the contributions of our work are as follows:

1. We apply drift detection to CFL platforms with dynamic client participation and showcase the need for closely monitoring concept drift to sustain the value of CFL for its clients;
2. We identify that error- and data-based approaches differ in terms of the point in time when concept drift is detected. More specifically, error-based approaches exceed their data-based alternatives in terms of the performance of the jointly trained model at the time of intervention. This is due to the fact that they can leverage the short period of *beneficial drift* during which it actually improves model performance instead of harming it;
3. In contrast, data-based approaches excel in a timely detection of concept drift after its first occurrence. Accordingly, subsequent drift adaptation can intervene sooner, avoiding potentially harmful concept drift;
4. In a broader sense, our work necessitates a more nuanced discussion of concept drift and fine-grained evaluation of drift detection approaches, introduces the notions of *beneficial* and *harmful drift*, and encourages future research on the subject.

In turn, our findings are of great values for researchers and practitioners concerned with the improvement and operation of FL platforms under concept drift. Moreover, they complement existing work that neither addressed dynamic client participation in CFL nor investigated the unique advantages that different types of drift detection approaches come along with. Finally, our results suggest that future work on concept drift for CFL should consider the inherent differences of the aforementioned types of drift detection approaches and that their selection should be done according to the task at hand.

To do so, we elaborate on related work in the following section. Afterwards, we outline the proposed methodology in Sect. 3 before presenting the preliminary results of our analyses in Sect. 4. Finally, we conclude the paper and discuss limitations of our study and future research directions.

2 Related Work

FL is a novel distributed ML paradigm that, by design, maintains data privacy of all involved clients. In turn, it is a perfect fit for a wide range of domains with particularly sensitive data [13,22]. The most common approach towards FL, namely *FedAvg*, relies on a central server orchestrating several rounds of training among participating clients [19]. During each round, a global model held by the central server is sent to the clients which fit it to their respective local data [13,19]. Afterwards, all local model updates are sent back to the server in a privacy-preserving manner [2,22]. Lastly, the central server aggregates all received model updates to compute a new global model [19].

More recently, FL has been extended to CFL (e.g., [23]), allowing the jointly trained model to smoothly integrate available continuous data streams [4] and retain useful knowledge from environments that evolve over time [5]. Instead of training a federated model once and deploying it afterwards, CFL fosters life-long learning, continuously fitting the model to incoming data [4]. Despite its unique advantages, the performance of CFL suffers from concept drift, posing the threat of catastrophic forgetting and performance deterioration [8]. As concept drift is a major cause of poor CFL performance [8], recent studies aim towards making CFL more resilient against it [23]. Here, concept drift refers to changes in various statistical properties such as feature and label distribution over time [17]. To properly adapt to concept drift, it is crucial to reliably detect it in the first place [17]. Respective concept drift detection approaches can be categorized into error-based and data-based approaches [4,5], which we elaborate upon in more detail in the following.

2.1 Error-Based Drift Detection

Error-based drift detection detects concept drift by constantly monitoring the error of the model during training [4,8]. Although the most intuitive choice of error measurement is the model's loss (e.g., cross-entropy-loss), many extensions and alternatives have been proposed in the past [24]. The intuition of such approaches is to identify concept drift through rapid changes in the model error, as new concepts cannot (yet) be adequately predicted by the continuous model [8]. As relying on a single increment in model error to detect concept drift is vulnerable to reporting false-positives, sliding windows can increase their reliability significantly [18]. These approaches define a sliding window size and intervene only if the increased error persists throughout the window size.

2.2 Data-Based Drift Detection

Unlike previous approaches, data-based detection relies on clients' input data or label distributions to detect concept drift [8]. Typically however, such approaches require access to the respective client's data (e.g., [10]), which would violate data privacy when applied to FL environments. To avoid this, approaches such as secure aggregation [2] allow the central server to collect data- and client-specific characteristics as aggregates. Hence, they facilitate privacy-preserving data analysis, also referred to as federated analytics [21]. Subsequently, these data-specific properties allow to detect significant changes in feature and label distributions that indicate the presence of concept drift [4].

3 Methodology

In order to apply concept drift detection to FL platforms with dynamic client participation as well as to investigate advantages and disadvantages of different drift detection approaches, we first set up a respective FL platform. Once the platform is established, we apply different types of concept drift to the federation, namely *sudden, gradual, incremental,* and *reoccurring* drift. Finally, we apply both error- and data-based drift detection and compare their point in time of intervention regarding the proximity to the first occurrence of concept drift as well as the best performance of the joint model. Accordingly, we aim to unravel their distinct advantages over another.

3.1 Concept Drift in CFL Platforms

CFL platforms extend existing FL strategies to leverage continuous streams of input data [4,5]. Unlike traditional continuous learning, where a single client constantly provides additional train data [17], in FL, this continuous growth in data can be attributed to two different sources: either, participating clients within a cohort generate new train data that is made available to the federation [3], or new clients join the federation and serve as additional source of data [8]. Existing literature on CFL only considers the former type of FL platform, i.e., platforms that allow clients to continuously provide new data but do not allow new clients to join the federation throughout the training (e.g., [3–5]). In this work, we argue that this disregards an essential part of CFL platform applications and consider such CFL platforms with dynamic client participation.

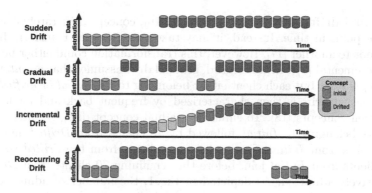

Fig. 1. Types of concept drift applied to FL platforms [17] with dynamic client participation, where each data silo represents 5 clients joining the federation.

To adequately simulate such platforms, we apply CFL using *FedAvg* [19] as aggregation strategy. More precisely, we initiate the FL procedure among 5 simulated clients and subsequently perform 50 rounds of joint training. Afterwards, we admit 5 new clients to the federation and repeat training for 50 rounds. Through several such repetitions, each adding 5 new clients to the federation and triggering continuous learning for another 50 rounds, the federation as well as the amount of data contained in it grows step-wise.

In order to now introduce concept drift to the federation, our work relies on simulating four different notions of concept drift, each of which being previously considered in various studies on concept drift [4,9,18]. Therefore, we first sort the dataset D used during evaluation (for more details, see Sect. 4) according to the regression target value. Afterwards, the first half of data points (i.e., those with target values smaller than the median) is assigned to a dataset named $D^{Initial}$, whereas the second half belongs to the dataset $D^{Drifted}$. Then, we split $D^{Initial}$ into 80% train ($D^{Initial}_{Train}$) and 20% test set ($D^{Initial}_{Test}$) before distributing it among 40 simulated clients that belong to the cohort labeled *Initial*. Finally, we do the same with the cohort named *Drifted* that consists of 80 clients and that will introduce concept drift to the platform. The choice of cohort sizes and data distribution is motivated by the subsequent simulation of different concept drifts, which we elaborate upon in more detail hereafter. Given these 120 clients belonging to one of the two cohorts, we provide details on the simulation of different concept drifts in the following and visualize them in Fig. 1:

1. **Sudden drift.** Sudden drift is among the most frequently applied forms of concept drift (e.g., [9,17]). In short, it assumes an immediate drift of concepts in the data. Applied to CFL, sudden drift is simulated by first allowing all clients from the *initial* cohort to join the federation and then admitting all *drifted* clients. Complying with our previously outlined methodology, we admit 5 clients at a time before continuing model training. Accordingly, 8 rounds of admission for *initial* clients take place, followed by 16 rounds of admitting the remaining 80 *drifted* clients.

2. **Gradual drift.** As stated by Liu et al. [16], concept drift rarely occurs at a single point in time. Instead, it may take some time to transform from one concept to another [17]. However, this transformation could either be gradual or incremental [17]. In this regard, gradual drift assumes that no *intermediate concepts* exist but each client either belonging to the *initial* or *drifted* cohort. The actual drift is then characterized by frequent back and forth changes between concepts [18]. To simulate gradual concept drift, we first admit 20 clients belonging to *Initial*, followed by 10 clients from *Drifted* and another 10 clients from *Initial*. Finally, 20 more clients from the *drifted* cohort and 10 clients from *Initial* join, before the remaining 70 clients from *Drifted* are iteratively admitted. As depicted in Fig. 1, the number of admitted clients holding the *drifted* concept increases gradually.

3. **Incremental drift.** To better understand how incremental drift differs from gradual drift, Gama et al. [9] introduce the term *intermediate concept*, referring to the slow transformation from one concept to another [17]. During incremental drift, some clients introduce these intermediate concepts and do not strictly belong to the *initial* or *drifted* concepts. We simulate incremental drift by controlling how the data contained within $D_{Train}^{Drifted}$ is allocated to the 80 simulated clients. Instead of randomly distributing it, as we do for the other three notions of concept drift, we sort $D_{Train}^{Drifted}$ according to the regression target again. The lower half of samples is split among 40 clients in accordance with the ordering we received previously. The other half is again randomly distributed among the remaining 40 clients. Finally, we first admit the 40 clients holding the *initial* concept interatively, followed by the 40 intermediate clients (maintaining their order) and the remaining 40 clients.

4. **Reoccurring drift.** Reoccurring drift is rather similar to sudden drift. But for sudden drift, the shift from one concept to another is final, whereas for reoccurring drifts, the shift might reverse at a later point in time. A typical example for this would be seasonal changes in data that might reoccur several times during the CFL lifetime [9]. Finally, we simulate this type of concept drift by starting with 20 *initial* clients and then admitting all 80 *drifted* clients. At the end, 20 clients from *Initial* are admitted to simulate shifting back to the initial concept.

3.2   Drift Detection

"Drift detection refers to the techniques and mechanisms that characterize and quantify concept drift via identifying change points" [17]. Usually, these change points, i.e., occurrences of concept drift, can be detected by employing error-based or data-based drift detection [4,5,8]. As our study aims to analyse and compare both techniques, we follow a general approach towards implementing them. As proposed by Mahgoub and colleagues [18] as well as Casado et al. [4], we utilize window-based detection for drift detection, where a sliding-window is used to detect concept drift based on measurements of model loss or data distribution. In what follows, we present details on both techniques applied to identify concept drift in CFL systems.

1. **Error-based detection.** To apply error-based drift detection to CFL with dynamic client participation, our work follows previously outlined approaches such as discussed by Criado et al. [8]. Thus, our implementation of error-based drift detection monitors the model error carefully and stores it persistently throughout the CFL platform's lifetime. Whenever new clients join the federation, the error reported by the model is compared to previous errors. If it is further off the previous mean error than standard deviation, drift is detected internally. However, to increase its reliability, we employ a sliding widow of size 3 and drift detection only intervenes once three rounds of clients joining subsequently fulfill the previous requirement. Here, a window size of 3 offers the best trade-off between false-positives and false-negatives for the given task.

2. **Data-based detection.** On the other hand, our implementation of data-based detection is inspired by existing approaches tracking feature and label distributions of training data [10]. To maintain data-privacy, we facilitate tools from federated analytics [21] throughout the process. More precisely, our implementation of data-based drift detection requires clients to self-report on their individual label distribution. Here, each client is tasked to compute their mean target value. Then, secure aggregation is applied to collect the information about the average client labels in a privacy-preserving manner. Whenever a new client joins the cohort, we again compare it to the existing cohort's mean labels and decide if the new client is within standard deviation of the previous mean. If not, the client is subsequently marked as potentially introducing concept drift. Similarly to our implementation of error-based detection, we deploy a sliding window of size 3 to reliably detect concept drift and avoid false positives.

4 Preliminary Evaluation

4.1 Experimental Setup

Our preliminary evaluation relies on the public *Uber Fares Dataset* [1] containing information and prices of nearly 200.000 Uber rides. The rationale behind our choice of dataset is that its use-case fits to CFL platforms. Think of different taxi companies that seek to provide their customers precise estimates of the cost of transport from one place to another. In order to increase the accuracy of these estimates, companies may join forces and utilize FL to jointly train a ML model for fare prediction. However, the constant change of prices over time (e.g., due to inflation) necessitates the use of CFL to properly adapt to them. As companies start benefiting from their platform participation, additional taxi providers might get attracted to join as well. These companies may either further improve the shared model's performance, or could introduce concept drift. The latter could for example be the case if the companies are from other geographical areas significantly different in terms of the fares per ride.

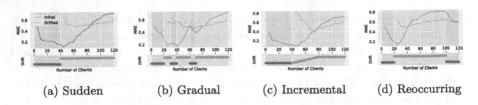

(a) Sudden   (b) Gradual   (c) Incremental   (d) Reoccurring

Fig. 2. Model performance under different concept drifts. Blue and orange areas indicate if clients of the *Initial* or *Drifted* cohort join the FL platform. (Color figure online)

Next, we prepare the datasets in accordance with the methodology presented in Sect. 3. Hence, we define the two cohorts named *Initial* and *Drifted*, allocate respective train and test datasets, and assign samples to clients.

Next, we define model architecture and FL setup for the CFL platform. The model is a rather simple, 5-layered neural network, that is trained to minimize the mean *mean square error* (MSE) defined in Eq. 1. Its input layer is of size 11, the hidden layers are of sizes 50, 30, and 10 with ReLU-activation and dropout of 0.2. As the target is a regression value, the output layer is a single neuron. In terms of FL aggregation strategy, we apply *FedAvg* [19], the de-facto standard due to its popularity and widespread application [14]. Training is applied for 100 rounds, during each of which all clients perform 3 epochs of local training. To evaluate the performance of the model, we apply 5-fold cross-validation throughout all of our analyses and adopt the universally applicable, scale-dependent *mean absolute error* (MAE) [12]. We chose MAE over alternatives such as *root mean square error*, as it is less prone to high sample variance that would possibly affect evaluation results. MAE is computed as follows:

$$MSE = \frac{1}{n} \sum_{i=1}^{n} (y_i - \hat{y}_i)^2 \quad \text{and} \quad MAE = \frac{1}{n} \sum_{i=1}^{n} |y_i - \hat{y}_i|, \tag{1}$$

where $\hat{y}_i$ is the predicted and y_i the actual fare for datapoint i.

4.2 CFL Performance Under Concept Drift

First, we apply the four previously introduced types of concept drift to our CFL platform. During application, we monitor the model performance in terms of MAE on $D_{Test}^{Initial}$ and $D_{Test}^{Drifted}$ and plot it in Fig. 2. This aims to assess the impact of concept drift on federated model performance.

The results depicted in Fig. 2 indicate the detrimental effect of all types of drift on model performance for the non-drifted, i.e., *initial* cohort. Figures 2 (a) to (d) show that the MAE of *initial* clients starts to worsen shortly after the first occurrence of concept drift. Surprisingly however, for a short period of time (usually between 1 to 2 rounds of admitting 5 drifted clients), they benefit from the presence of concept drift. This is likely due to the added information on such rides that were previously on the higher end of fares. In contrast to

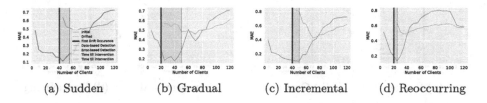

(a) Sudden	(b) Gradual	(c) Incremental	(d) Reoccurring

Fig. 3. Distance to first occurrence of different concept drifts

the subsequent phase of *harmful drift*, we refer to this type of concept drift as *beneficial drift*. Moreover, Figs. 2 (b) and (d) show that whenever clients from the *initial* cohort join the federation again, the respective MAE start to recover immediately. Finally, it shows that clients from the *drifted* cohort experience high MAE with small improvements throughout the platform's lifetime. This indicates that concept drift does not only affect clients that participated prior to the concept drift, but also those clients introducing the drift to the federation.

Overall, our findings are mostly in line with previous work addressing concept drift in CFL with static client participation. Beyond existing work however, we identify the notion *beneficial drift*, which constitutes the small period of time in which concept drift in FL platforms with dynamic client participation can actually improve the predictive performance of the shared model.

4.3 Detecting First Occurrence of Concept Drift

Next, we compare drift detection with respect to the proximity of their intervention to the first occurrence of concept drift. Accordingly, we determine the type of detection with lowest time until intervention after concept drift first appears.

Figure 3 visualizes the points in time when concept drift first occurs during our experiments as well as when error-based and data-based detection intervene. Figures 3 (a) to (d) show that data-based approaches reliably detect concept drift at the time of its first occurrence. Accordingly, it is capable of precisely detecting concept drift as soon as possible and outperforms error-based approaches in this regard. Error-based approaches tend to intervene at a later point in time and therefore only provide detection with a delay in time. The magnitude of the delay depends, among other things, on the ratio of non-drifted and drifted clients. In (a) and (c), for example, detection is delayed by 10 to 15 clients. Here, however, concept drift only occurs after the initial cohort already consists of 40 clients. In (d), on the other hand, drift already appears after the initial cohort consists of only 20 clients. In turn, the error-based detection recognizes the emerging drift sooner. In (b), the constant change between drifted and non-drifted clients joining the federation seems to delay the detection of error-rate based concept drift, as this is where the intervention is delayed the most. This is likely due to our choice of setting the window size to 3, negatively affecting the sensitivity towards gradual drift but increasing its overall robustness.

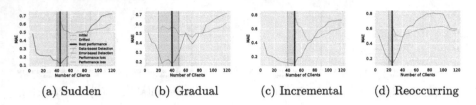

Fig. 4. Distance to best performing model under concept drift

4.4 Detecting Best Performance Under Concept Drift

Previous analyses investigated how well drift detection identifies the first occurrence of concept drift. However, this disregards our previous findings regarding *beneficial drift*. In Fig. 4, we therefore monitor the distance of drift detection intervention to the optimal performance of the joint model for the *initial* cohort. To do so, we first identify the time at which the MAE is smallest for the *initial* cohort and relate it to the interventions of the respective drift detection approaches. Thereby, we aim to identify the type of drift detection more suitable to differentiate between *beneficial* and *harmful drift*.

Figure 4 shows that while data-based detection has previously been found to identify the first occurrence of data drift more precisely, their error-based alternative excels when it comes to identifying best model performance. More precisely, among all types of concept drift, error-based detection is off the point in time providing best performance by not more than 10 clients (i.e., 2 rounds of admitting 5 clients each). For (c) and (d), they even precisely identify the point in time without delay. In turn, they outperform data-based detection for all but *sudden* concept drifts. For Fig. 4 (b), we previously argued that error-based detection is the furthest off the first occurrence of drift. Our findings regarding the best model performance however show that *gradual* concept drift comes along with the longest period of *beneficial drift* which is only taken into account by this type of drift detection.

5   Conclusion, Limitation, and Future Work

In this paper, we apply concept drift detection to CFL platforms with dynamic client participation and compare error- to data-based drift detection. In particular, we demonstrate that - similar to CFL with a static participation - concept drift harms the performance of CFL platforms in terms of MAE and should be taken care of. To do so, both type of drift detection reliably detect sudden, gradual, incremental, and reoccurring concept drift. However, we find error-based drift detection to better approximate the point in time where the jointly trained model performs best. Accordingly, they are capable of facilitating *beneficial drift*, which referrs to the small period of time where concept drift positively affects model performance before eventually harming it. Data-based drift detection, on the other hand, intervenes closer to the first occurrence of concept drift and is therefore to be preferred when fast drift adaptation is the main objective.

From our findings we argue that the novelty of our work does not lie within the methods we apply to detect concept drift in CFL platforms but in our choice to (1) consider dynamic client participation throughout the platform's lifetime, and (2) compare existing drift detection with respect to their point in time of intervention. Accordingly, our work suggests that there is no approach towards drift detection that performs better per se. Instead, the choice of approach can greatly benefit from our presented findings.

Limitations. To start with the obvious: our findings are yet derived from a single dataset only. While the results among the different types of concept drift show the validity of our previous conclusions, applications on different datasets might differ from those presented in this work. Moreover, due to the limited scope of this work, we only consider a single drift detection approach for error- and data-based detection, respectively.

Future Work. Our results indicate the differences between error- and data-based drift detection. In future work, we will significantly extend our preliminary evaluation to cover different model architectures, data types, and drift detection approaches. Finally, some follow-up work will aim to develop a drift detection framework leveraging both types of approaches to dynamically adjust drift detection to the current environment.

Acknowledgements. This research was partially funded by the German Federal Ministry of Health as part of the KINBIOTICS project. The research of Philipp Cimiano is partially funded by the Ministry of Culture and Science of North Rhine-Westphalia under the grant no NW21-059A SAIL.

References

1. Uber Fares. https://www.kaggle.com/datasets/yasserh/uber-fares-dataset
2. Bonawitz, K., et al.: Practical secure aggregation for privacy-preserving machine learning. In: proceedings of the 2017 ACM SIGSAC Conference on Computer and Communications Security, pp. 1175–1191 (2017)
3. Canonaco, G., Bergamasco, A., Mongelluzzo, A., Roveri, M.: Adaptive federated learning in presence of concept drift. In: 2021 International Joint Conference on Neural Networks (IJCNN), pp. 1–7. IEEE (2021)
4. Casado, F.E., Lema, D., Criado, M.F., Iglesias, R., Regueiro, C.V., Barro, S.: Concept drift detection and adaptation for federated and continual learning. Multimed. Tools App. **81**, 1–23 (2022)
5. Casado, F.E., Lema, D., Iglesias, R., Regueiro, C.V., Barro, S.: Ensemble and continual federated learning for classification tasks. Mach. Learn. **112**, 3413–3453 (2023)
6. Chen, Y., Chai, Z., Cheng, Y., Rangwala, H.: Asynchronous federated learning for sensor data with concept drift. In: 2021 IEEE International Conference on Big Data (Big Data), pp. 4822–4831. IEEE (2021)
7. Chow, T., Raza, U., Mavromatis, I., Khan, A.: Flare: detection and mitigation of concept drift for federated learning based IoT deployments. arXiv preprint arXiv:2305.08504 (2023)

8. Criado, M.F., Casado, F.E., Iglesias, R., Regueiro, C.V., Barro, S.: Non-IID data and continual learning processes in federated learning: a long road ahead. Inf. Fusion **88**, 263–280 (2022)

9. Gama, J., Žliobaitė, I., Bifet, A., Pechenizkiy, M., Bouchachia, A.: A survey on concept drift adaptation. ACM Comput. Surv. (CSUR) **46**(4), 1–37 (2014)

10. Hammoodi, M.S., Stahl, F., Badii, A.: Real-time feature selection technique with concept drift detection using adaptive micro-clusters for data stream mining. Knowl. Based Syst. **161**, 205–239 (2018)

11. Hinder, F., Vaquet, V., Hammer, B.: Suitability of different metric choices for concept drift detection. In: Bouadi, T., Fromont, E., Hüllermeier, E. (eds.) IDA 2022. LNCS, vol. 13205, pp. 157–170. Springer, Cham (2022). https://doi.org/10.1007/978-3-031-01333-1_13

12. Hyndman, R.J., Koehler, A.B.: Another look at measures of forecast accuracy. Int. J. Forecast. **22**(4), 679–688 (2006)

13. Kairouz, P., et al.: Advances and open problems in federated learning. Found. Trends® Mach. Learn. **14**(1–2), 1–210 (2021)

14. Li, Q., Diao, Y., Chen, Q., He, B.: Federated learning on non-IID data silos: an experimental study. In: 2022 IEEE 38th International Conference on Data Engineering (ICDE), pp. 965–978. IEEE (2022)

15. Li, T., Sahu, A.K., Zaheer, M., Sanjabi, M., Talwalkar, A., Smith, V.: Federated optimization in heterogeneous networks. Proc. Mach. Learn. Syst. **2**, 429–450 (2020)

16. Liu, A., Song, Y., Zhang, G., Lu, J.: Regional concept drift detection and density synchronized drift adaptation. In: IJCAI International Joint Conference on Artificial Intelligence (2017)

17. Lu, J., Liu, A., Dong, F., Gu, F., Gama, J., Zhang, G.: Learning under concept drift: a review. IEEE TKDE **31**(12), 2346–2363 (2018)

18. Mahgoub, M., Moharram, H., Elkafrawy, P., Awad, A.: Benchmarking concept drift detectors for online machine learning. In: Fournier-Viger, P., Hassan, A., Bellatreche, L. (eds.) Model and Data Engineering. MEDI 2022. LNCS, vol. 13761, pp. 43–57. Springer, Cham (2023). https://doi.org/10.1007/978-3-031-21595-7_4

19. McMahan, B., Moore, E., Ramage, D., Hampson, S., y Arcas, B.A.: Communication-efficient learning of deep networks from decentralized data. In: Artificial Intelligence and Statistics, pp. 1273–1282. PMLR (2017)

20. Truex, S., et al.: A hybrid approach to privacy-preserving federated learning. In: Proceedings of the 12th ACM Workshop on Artificial Intelligence and Security, pp. 1–11 (2019)

21. Wang, D., Shi, S., Zhu, Y., Han, Z.: Federated analytics: opportunities and challenges. IEEE Netw. **36**(1), 151–158 (2021)

22. Yang, Q., Liu, Y., Chen, T., Tong, Y.: Federated machine learning: concept and applications. ACM TIST **10**(2), 1–19 (2019)

23. Yoon, J., Jeong, W., Lee, G., Yang, E., Hwang, S.J.: Federated continual learning with weighted inter-client transfer. In: International Conference on Machine Learning, pp. 12073–12086. PMLR (2021)

24. Zhang, Z., Sabuncu, M.: Generalized cross entropy loss for training deep neural networks with noisy labels. In: Proceedings of NIPS (2018)

25. Zhu, J., Ma, X., Blaschko, M.B.: Confidence-aware personalized federated learning via variational expectation maximization. In: Proceedings of the IEEE/CVF Conference on Computer Vision and Pattern Recognition, pp. 24542–24551 (2023)

S+t-SNE - Bringing Dimensionality Reduction to Data Streams

Pedro C. Vieira[1]([⊠]) [ID], João P. Montrezol[1] [ID], João T. Vieira[1] [ID],
and João Gama[2,3] [ID]

[1] Department of Computer Science, Faculty of Sciences, University of Porto, Porto,
Portugal
{pedrocvieira,joao.antunes,up201905419}@fc.up.pt
[2] Faculty of Economics, University of Porto, Porto, Portugal
jgama@fep.up.pt
[3] INESC TEC, Porto, Portugal

Abstract. We present S+t-SNE, an adaptation of the t-SNE algorithm
designed to handle infinite data streams. The core idea behind S+t-SNE
is to update the t-SNE embedding incrementally as new data arrives,
ensuring scalability and adaptability to handle streaming scenarios. By
selecting the most important points at each step, the algorithm ensures
scalability while keeping informative visualisations. By employing a blind
method for drift management, the algorithm adjusts the embedding
space, which facilitates the visualisation of evolving data dynamics. Our
experimental evaluations demonstrate the effectiveness and efficiency of
S+t-SNE, whilst highlighting its ability to capture patterns in a streaming scenario. We hope our approach offers researchers and practitioners a
real-time tool for understanding and interpreting high-dimensional data.

Keywords: dimensionality reduction · data streams · algorithm

1 Introduction

Dimensionality reduction techniques are an object of great interest in applications such as image or natural language processing. Dimensionality reduction
techniques simplify complex data, ensuring better interpretability of such data
and helping its visualisation. Furthermore, they enhance model performance by
employing feature selection and thus improving computation speed Constructing efficient algorithms for dimensionality reduction in a streaming context opens
the possibility of working with potentially infinite datasets. Hence, such algorithms could be used with arbitrary-size datasets, offline or online. For example,
it would be possible to visualise static, very large datasets typically used in deep
learning. Another example application would be to use this algorithm to improve
computation speed and provide a human-readable visualisation of data streams
in services like electricity, gas, and water maintenance.

P. C. Vieira and J. P. Montrezol—Equal contribution, order defined by coin flip.

© The Author(s), under exclusive license to Springer Nature Switzerland AG 2024
I. Miliou et al. (Eds.): IDA 2024, LNCS 14642, pp. 95–106, 2024.
https://doi.org/10.1007/978-3-031-58553-1_8

In this paper, we explain the limitations of existing dimensionality reduction techniques when applied to data streams and propose a new method named S+t-SNE to solve some limitations.

In Sect. 3, we explain how our approach works and handles common challenges, such as the continuously increasing volume of historic data and the constant flow of new data entries, together with the possibility of concept change. In Sect. 4, we delve into the tests performed to evaluate the performance of the proposed algorithm. The version used for the tests is available in a code repository[1] and follows implementation specifics to be integrated with the River[2] framework, as described in the aforementioned framework's documentation.

2   Related Work

The t-distributed stochastic neighbour embedding (t-SNE) [1] specialises in transforming a high-dimensional dataset into a two or three-dimensional dataset. It does so by using a t-distribution as a basis for calculating the similarity between points in the projected space and the original space while using the KL divergence to guide the point positioning.

Dimensionality reduction techniques are classified into "out-of-sample" and "in-sample" categories [5]. Out-of-sample techniques start with a small data subset and map the rest accordingly, making them more scalable but less accurate when the subset does not accurately represent the full dataset. In-sample techniques, like t-SNE, classical Multidimensional Scaling (MDS), and UMAP [12], process the entire dataset at once, resulting in more accurate results but with higher computational costs. The algorithm discussed in this paper falls into the out-of-sample category.

Developments in out-of-sample techniques focus on incorporating user knowledge into the projection process. Examples include Piecewise-Laplacian Projection (PLP) [14], Least Squares Projection (LSP) [13], and Local Affine Multidimensional Projection (LAMP) [7]. While these techniques can handle larger datasets, they are often unsuitable for data streams, as they rely on the quality of the initial data subset. Some online strategies have been developed to mitigate this problem. Basalaj [4] introduces an online version of classical Multidimensional Scaling. In his work, when a new data entry is received, MDS is applied considering both the new entry and the already processed ones to create a new full pairwise distance matrix. Alsakran et al. [3] apply a force-based approach, updating to consider new instances and recomputing the full pairwise distance matrix in memory. Jenkins et al. [6] and Law et al. [10] introduce online versions of the ISOMAP in-sample technique. Upon receiving a new data entry, every previous entry is also processed, and a full pairwise distance matrix must be computed. Law et al. [9] make this process faster by sidestepping the requirement of a full pairwise matrix in their introduction of an online version of LMDS. In this technique revision, the only distances the algorithm needs to compute are

---

[1] github.com/PedrV/S–t-SNE.
[2] riverml.xyz.

the ones between the new data entry and the pre-existing entries. Kouropteva et al. [8] and Schuon et al. [17] present online revisions of the in-sample LLE technique through the definition of strategies that update neighbourhood relationships when a new entry is introduced. Rauber et al. [16] introduce a Dynamic t-SNE. This technique facilitates the projection of windows of datasets that depend on time while maintaining spatial consistency in the positions of points in projections. Although this has interesting results, Dynamic t-SNE must reproject *all* received data to create an entirely new projection to compute the most recent data. Consequently, it is necessary to keep the whole dataset in memory, which is inadequate for streaming scenarios in which the data continues to grow. This is something that the method proposed in this paper attempts to address. Furthermore, there is, to the best of our knowledge as of writing, no attempt made by any of the methods described to deal with concept drift. Our approach will address this problem.

3   Streaming T-SNE (S+t-SNE)

In a streaming scenario, data arrives continuously. This assumption hinders the application of the traditional t-SNE. The two inherent challenges to this application are (1) the duration or termination point of the data stream is often unknown, rendering it impossible to determine when to halt the accumulation of points and start the application of t-SNE; (2) the potential accumulation of points due to an extensive or even infinite data stream poses obstacles on the computation time and the computational resources.

3.1   Problem One - When to Start

One possible approach to using t-SNE in a streaming scenario involves accumulating all encountered points until a change in the data stream is detected, at which point the accumulated points are projected. However, for this technique to work, the data would need to exhibit drift and we would have to establish a threshold of "how much drift is enough drift". An alternative is to adopt a fixed batch-wise approach to mitigate these challenges in our work. Points are accumulated until a predetermined batch size, B, is attained. Subsequently, t-SNE is applied to project the accumulated points. This approach offers a swift and agnostic solution that does not rely on specific data patterns, enabling its off-the-shelf application.

One iteration of S+t-SNE consists of accumulating a new data point, checking if the total accumulated is equal to B, and if so, applying t-SNE. The first projection is made by applying t-SNE to the batch of data. Since there are no points in the projection space, normal t-SNE will be applied in the first projection. However, after the first iteration of S+t-SNE, subsequent iterations will project in a space where points already exist.

In our approach, incorporating new data points from iteration $t + 1$ into the projection space involves considering conditional probabilities between new and previously embedded points from iteration t. To achieve the intended outcome, our approach is grounded in the openTSNE framework [15], with a primary reliance on the technique of partial embedding. The concept of partial embedding facilitates the incorporation of new points into an established embedding space by considering only the conditional probabilities between points in t and $t+1$. However, one limitation of this approach is that focusing solely on these conditional probabilities may omit the natural inclusion of conditional probabilities between new data points. Hence, groups with similar conditional probabilities to already embedded points yet exhibiting low conditional probabilities between themselves may converge into the same area of the lower-dimensional space. This constraint arises from relying solely on information from old points for new point embeddings. The method from Subsect. 3.3 will mitigate this effect by removing unnecessary points.

3.2 Problem Two - Reduction of Space Fingerprint

Section 3.1, overcomes the issue of determining when to apply t-SNE in the streaming context (Problem 1). However, the concern of accumulating points in the projected space persists.

To address the accumulation of points in the lower-dimensional space, we propose retaining the shape of groups of points by using a clustering algorithm applied to t-SNE projections. Each group's shape will be represented by convex hulls (the calculation of a convex hull is done in $O(n \log n)$, where n is the size of a cluster), minimising the number of retained points. However, convex hull points are not informative enough for new point incorporations. We introduce "*PEDRUL*" (Points Expected to Define Regions of Unambiguous Location) to represent important points.

The PEDRUL within each 2D group is determined by their density in the original D-dimensional space. Hence, for each group of points in the embedding space, their PEDRUL is defined as the points in the original space with the largest number of neighbours within a search radius, denoted as R. To control the transitivity of the neighbourhood of dense points, the search identifies points as candidates for being PEDRUL only if they are not in a neighbourhood of an already defined PEDRUL point. This methodology may sacrifice the selection of *de facto* densest points but maintains the integrity and unambiguity of the dense regions that each PEDRUL spans within each group.

To efficiently obtain the PEDRUL, the KDTree [11] data structure is used for nearest neighbour searches in multi-dimensional spaces. KDTree has a construction time complexity of $O(n \log n)$ and a balanced structure (the height of the tree does not exceed $O(\log n)$). Once the KDTree structure is constructed, it can be queried to retrieve all points within a specified distance from a given point. Each point in the current batch's neighbourhood limited on R is estimated and sorted by size. The resulting neighbours are stored in a hashable set, enabling efficient intersection calculations and allowing the identification and selection of sparse density points as required.

In summary, PEDRUL reduces the number of points in the low-dimensional space by preserving only points with maximal information (maximal points around them) that will be used for subsequent embeddings. To aid visualisation and limit regions, we store the points delimiting the convex hull of each group. Accumulating too many PEDRUL will not be a problem since for an entire application of our algorithm, the number of PEDRUL remains constant. If the initial indication is to hold 100 points, there will be 100 points in the projection at all times. However, the points can change across iterations.

3.3 Handling Drift

Data streams often have non-stationary distributions, with drift taking the form of sudden or gradual changes. Sudden drift is an abrupt shift in data distribution without temporal overlap between the pre and post-change distributions. Gradual drift involves a slower, incremental overlapping transition [2]. Adapting t-SNE for online use requires addressing these drift types.

In the following paragraphs, we propose a method that can be coupled with incremental t-SNE proposed earlier and with any method using convex regions. The proposed method is fit for online scenarios and handles drift by updating the projections in the space of interest - the low-dimensional space. Furthermore, it mitigates the possible artefacts referred to in Sect. 3.2.

When sudden drift occurs, data embeddings are likely to experience steep changes. Hence, the old embeddings correspond to old views of the data and must be immediately removed. Gradual drift necessitates gradually removing examples as they become less relevant. Adjusting for different drifts is connected with the concept of forgetting different points at different speeds.

Our solution involves using the convex hulls obtained from clustering and dividing them into parts. Each partition will employ blind drift detection by exponential decay based on the number of iterations in S+t-SNE. This allows parts without new points during a period (given by exponential decay) to disappear, ensuring consistency. We parameterise the exponential decay with three parameters, $\alpha = 0.88, \beta = 1.6$ and $\eta = 0.01$, yielding $N(t) = \alpha e^{-t\eta+\beta}$ where t is the number of iterations. This expression encapsulates our definition of drift. In this configuration, a polygon in iteration 200 will have section x cut if said section does not receive points for more than $N(200)$ iterations.

The selected partitions are along the medians of the polygon (Fig. 1 - B) and its concentric regions (Fig. 1 - A). These partitions allow us to monitor any translation with arbitrary precision as long as enough iterations are completed. Furthermore, they allow for arbitrary deletion while ensuring the result is always a convex shape. Since shapes are always convex, maintaining the algorithm is efficient.

Let n denote the maximum number of vertices of a polygon from the embedding space, k the number of polygons, p the maximum number of PEDRUL registered in a polygon, and m the number of concentric regions. The temporal complexity to determine the median regions is $\Theta(n)$, testing ownership of a point to a region is $\Theta(1)$. Performing a cut takes $O(n)$, where n is the number of

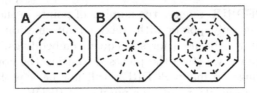

Fig. 1. In shape A) we see the concentric cuts over a convex hull; In B) the median cuts; In C) all cuts together, forming a structure similar to a cobweb.

vertices of the polygon. A cut is done by deleting points counter-clockwise from one end of a point of a median region to another. Maintaining all median regions of all polygons takes $O(2\,p\,k\,n)$. An important fact to speed up this procedure is that all operations can be efficiently parallelisable polygon-wise. This means that all polygons can run in parallel, reducing the total time to independent of the number of polygons and linear on the number of PEDRUL.

Calculating a concentric region for a polygon has a $\Theta(n)$ complexity. As for testing for point membership, it takes $O(log\,n)$ using a Delaunay tessellation of triangles. Constructing such a tessellation takes $O(n\,log\,n)$, but this is only done once per region. The total iteration complexity, including the construction of the tesselation, takes $O(m\,k\,p\,log\,n) + O(m\,k\,n\,log\,n)$. The factor m is generally small (≤ 5). Hence, we will disregard it in asymptotic analysis. Like the median regions, the concentric regions have highly independent operations, making them highly parallelisable. A parallel implementation reduces the complexity to $O(p\,log\,n) + O(n\,log\,n)$, which is effectively $O(p\,log\,n)$.

From this point onwards, we will refer to the method described above as Exponential Cobweb Slicing (ECS). The interplay between S+t-SNE and ECS is displayed in Algorithm 1.

Points that suffer from the problem described in Subsect. 3.1 are projected onto the wrong region A of the space. If new points that truly belong to A appear, the algorithm self-corrects. If no points appear mapped to A then ECS will trigger and cut A out. Hence, the visual artefacts are reduced.

4   Experiments

Our methodology does not compare itself with the alternatives delineated in Sect. 2 due to those lacking an available implementation by the authors, blocking, from our point of view, a fair comparison between algorithms.

However, we evaluate and compare S+t-SNE against t-SNE. With these tests, we aim to understand the strengths and weaknesses of the proposed method in comparison to its original variant. Hopefully, the comparison against a strong baseline demonstrates the strength of S+t-SNE.

All tests used a system running Windows 10 Pro 22H2, with an AMD Ryzen 7 5800 × 3.8 GHz (boosting to 4.5 GHz) processor and 32 GB (3200 MHz) RAM.

Algorithm 1: S+t-SNE

    Require: *conn* is alive
    Ensure : dataStorage = ∅
1: proj ← **EmptyProjection**(NULL)
2: **while** *True*
3:     newPoint ← **ReceiveData**(conn)
4:     dataStorage ← **StoreData**(newPoint)
5:     **if** *dataStorage.length* = $\mathcal{B}$ **then**
6:         proj ← **ProjectPoints**(proj, dataStorage)
7:         proj.PEDRUL ← **CalculatePEDRUL**(proj)   ▷ For new projections
8:         proj.hulls ← **CalculateConvexHulls**(proj)   ▷ For visualization
9:         dataStorage ← ∅
10:        proj.iterations ← proj.iterations + 1
11:        proj.hulls ← **ECS**(proj.hulls, proj.iterations)
12:        **for** *PEDRUL in proj.PEDRUL*
13:            **if** *PEDRUL* **not in** *proj.hulls* **then**
14:                **RemovePoint**(proj.PEDRUL, PEDRUL)
15:            **end if**
16:        **end for**
17:     **end if**
18: **end while**

Datasets. We use two datasets: MNIST, used in [8,9,16], and a synthetic dataset to evaluate drift. The latter dataset was created by randomly selecting points from three different spaces (structures) of dynamic 3D distributions. Each distribution is configured with spatial movement, adjusting both mean and covariance parameters at each time step emulated by a tick count and consist of 525000 points, or 175000 per structure. Two of the structures used will suffer translation, contractions, and dilations in space, and the other will only experience the last two effects. The structures overlap in the high-dimensional space, increasing the difficulty of the dataset. The purpose of this dataset is to assess the performance of S+t-SNE in a scenario closer to a real online one.

Configurations. All datasets will be streamed, emulating an online paradigm for data acquisition. The parameters for t-SNE were used empirically, in a way that resulted in the best-looking projections. For t-SNE to work, we adopt batch projections akin to those utilized in S+t-SNE. Specifically, upon the accumulation of $\mathbb{B}$ data points, t-SNE reprojects the entire dataset. Consequently, the last reprojected iteration encapsulates a conventional application of t-SNE to the entire dataset, similar to an offline approach. The parameters for the internal t-SNE used in S+t-SNE were the same as the ones used in the pure t-SNE. Said parameters can be viewed in the code repository. The number of iterations for t-SNE is 500 rounds of optimisation and 250 of early exaggeration. These settings allow for the overall best results. Regarding the number of iterations, factoring the time for ECS, S+t-SNE using 400 optimisation iterations and 250

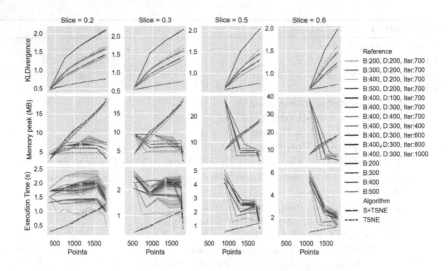

Fig. 2. Break-down of KLD, Peak Memory and Time for the MNIST dataset.

early exaggeration iterations is the comparable setting to t-SNE. Furthermore, the parameters regarding the ECS and the selection of density points can be revised in the repository. Since using different parameters yields cuts at different iterations for the same dataset, hence a different "definition" of drift, we opted to keep the parameters from Sect. (3.3) as they present good results visually.

4.1 Results

MNIST. Figure 2 shows the curves for comparing the algorithms on the MNIST dataset. The S+t-SNE algorithm is represented by the continuous lines and the t-SNE by the dashed lines. Each point in a line represents an action by the algorithm, typically at the end of the batch. The main point of comparison is the initial slice of the data used to define the opening space. That is the quantity of data accumulated before projecting the first time. The slice is represented fractionally regarding the total amount of data from 0 to 1. The lines representing S+t-SNE represent the variation in the size of batch (B), the number of PEDRUL (or density points) considered (D), and the number of iterations allowed for the algorithm to run (Iter).

KLD. We used the Kullback-Leibler divergence (KLD) to measure the entropy between the projection and the points in the original space. Looking at Fig. 2, we see that the KLD for t-SNE increases as the number of points increases, meaning that having more points deteriorates the t-SNE performance. This result is not unexpected since more points can cause more entropy. As for S+t-SNE, all configurations present a slow increase in the divergence as the number of points increases. Based on our analyses, the global rate of change manifests, in the

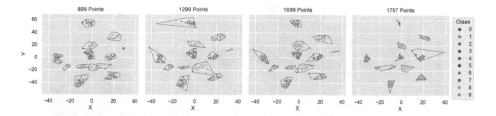

Fig. 3. Projections by S+t-SNE for MNIST. B:400 D:400 IT:700 Slice:0.2

least favourable scenario, as a progressively slow function relative to the number of batches processed by the S+t-SNE algorithm (figures in the repository). We see that larger batches and more PEDRUL cause a smaller KLD. A larger number of PEDRUL has more influence than larger batches. This may happen because more PEDRUL points mean more anchors for incoming points to have as a reference, meaning better projections. Technically, t-SNE uses all points of the dataset as PEDRUL, serving as a benchmark for performance.

Memory. As expected, the peak memory used by t-SNE (Fig. 2) aggressively increases as the number of points increases, achieving the same memory peak in all slices in the last reprojection. As for S+t-SNE, the peak memory use is achieved when getting the opening projection. The initial S+t-SNE iteration (getting the opening projections) for slice x uses the same process of a t-SNE application for the same slice x. However, the peak memory of S+t-SNE is higher because it has to account for the search of the PEDRUL points. We notice that larger slices further increase the initial memory peak gap between algorithms because searching for PEDRUL points within more data is more costly. After the first iteration, the peak memory reduces drastically and remains constant.

Time. The results for execution time (Fig. 2) were as expected. Larger batches, number of points and iterations increase the computational time. Furthermore, a large slice increases the initial time of S+t-SNE but contributes to a faster decrease in the time consumed.

Knowing the final drop in memory and time for S+t-SNE is due to reaching the end of the dataset, for the general case, based on this dataset, we would want to have the initial slice as large as possible. We should hold as many points as possible before the initial projection, increasing the quality of the representation by trying to obtain a stable enough opening projection. However, we must consider the larger memory footprint and initial computation time for larger slices. The number of PEDRULs and batches should be as large as possible until the limit of memory and time is available. However, more PEDRUL is highly preferred to a larger batch. We believe the number of iterations has a negligible influence. Hence, we advocate for selecting the minimum iteration count necessary for convergence.

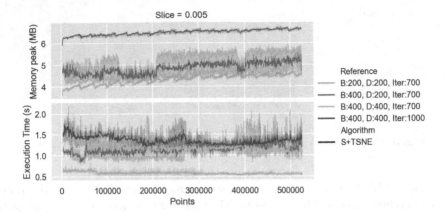

Fig. 4. Break-down of KLD, Peak Memory and Time for the dataset with drift.

Figure 3 shows the evolution of the S+t-SNE embeddings for MNIST for B:400 D:400 IT:700 Slice:0.2. Besides the points delimiting the hulls, the PEDRUL are also visible. All ten classes are well distinguishable. Moreover, ECS eliminates inaccurately projected points, as posited in Sect. 3.1. One example is in the transition from the top-right figure to the bottom-left, where the blue cluster had its stretched side removed. A typical ECS cut.

Synthetic Data. Figure 5 shows the results obtained for the synthetic dataset. We were not able to run t-SNE past the initial slice. Hence, we ran S+t-SNE only with a slice size of 0.005. Furthermore, it was impossible to calculate the KLD due to the size of the matrix.

Memory. From Fig. 4, we can see that increasing the batch size and number of points increases the peak memory consumption. However, the increase is very small (in the order of single bytes), even though we are doubling batch size and number of points. This result happens due to the capacity of larger batches and the larger number of points, allowing us to pick points with better representational capabilities, lasting more iterations and triggering fewer operations and consequently less accumulation of objects in memory. The results confirm the ones obtained with MNIST: the memory used is near constant after reaching a stable point where the representation is stable (initial peak removed in image).

Time. Figure 4 corroborates the findings derived from the MNIST dataset. Notably, the quasi-constant computational time after the first iteration (initial peak removed in image).

We believe the considerations taken from MNIST regarding the value to use for each parameter hold in this case. Figure 5 shows the projections for the synthetic data. The bottom-right plot shows the embeddings if no ECS was used and all points were considered as PEDRUL. The plots show that our techniques allow us to focus on the points of interest. Hence, we can see the most

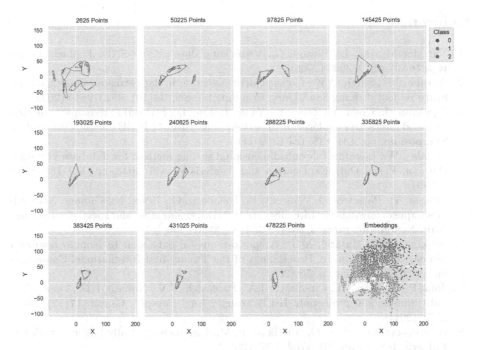

Fig. 5. S+t-SNE for synthetic data and total accumulated points (bottom-right).

recent movements of the dataset, these being the gradual shrinking of the green structure and the orange and blue ones slowly drifting closer to each other until they overlap.

The effect of ECS is not noticeable in the computational time due to its small runtime footprint. Even without a parallel implementation, the main difference between configurations is the number of hulls/clusters found at each iteration. However, this number is similar to all configurations because the clustering parameters and dataset are the same for all configurations. Furthermore, the number of PEDRUL has a constant effect in the time taken by ECS, and since the gap between configurations in this regard is not very large the effect is negligible. The drift has virtually no impact on memory consumption because of its gradual nature.

5   Conclusion

In this paper, we developed an efficient adaptation of t-SNE called S+t-SNE to work with data streams. Our version supports dimensionality reduction of online data and can adapt to data drift. A possible direction for future work is to test different mechanisms for obtaining PEDRUL and test the robustness of the method to different types of drift. To allow for a uniform comparison across online methods, it would be interesting to develop a metric for the comparison of projections of arbitrary size.

References

1. Van der Maaten, L., Hinton, G.: Visualizing data using t-SNE. J. Mach. Learn. Res. **9**(86), 2579–2605 (2008)
2. Agrahari, S., Singh, A.K.: Concept drift detection in data stream mining?: a literature review. J. King Saud Univ. Comput. Inf. Sci. **34**(10), 9523–9540 (2022)
3. Alsakran, J., Chen, Y., Zhao, Y., Yang, J., Luo, D.: StreamIT: dynamic visualization and interactive exploration of text streams. In: 2011 IEEE Pacific Visualization Symposium, pp. 131–138. IEEE (2011)
4. Basalaj, W.: Incremental multidimensional scaling method for database visualization. In: Visual Data Exploration and Analysis VI, vol. 3643, pp. 149–158. SPIE (1999)
5. Bengio, Y., Paiement, J., Vincent, P., Delalleau, O., Roux, N., Ouimet, M.: Out-of-sample extensions for LLE, ISOMAP, MDS, eigenmaps, and spectral clustering. Adv. Neural. Inf. Process. Syst. **16**, 1–8 (2003)
6. Jenkins, O.C., Matarić, M.J.: A spatio-temporal extension to Isomap nonlinear dimension reduction. In: Proceedings of the Twenty-first International Conference on Machine Learning, p. 56 (2004)
7. Joia, P., Coimbra, D., Cuminato, J.A., Paulovich, F.V., Nonato, L.G.: Local affine multidimensional projection. IEEE Trans. Visual Comput. Graph. **17**(12), 2563–2571 (2011)
8. Kouropteva, O., Okun, O., Pietikäinen, M.: Incremental locally linear embedding. Pattern Recogn. **38**(10), 1764–1767 (2005)
9. Law, M.H., Jain, A.K.: Incremental nonlinear dimensionality reduction by manifold learning. IEEE Trans. Pattern Anal. Mach. Intell. **28**(3), 377–391 (2006)
10. Law, M.H., Zhang, N., Jain, A.K.: Nonlinear manifold learning for data stream. In: Proceedings of the 2004 SIAM International Conference on Data Mining, pp. 33–44. SIAM (2004)
11. Maneewongvatana, S., Mount, D.M.: Analysis of approximate nearest neighbor searching with clustered point sets. CoRR cs.CG/9901013 (1999)
12. McInnes, L., Healy, J., Melville, J.: Umap: Uniform manifold approximation and projection for dimension reduction (2018)
13. Paulovich, F.V., Nonato, L.G., Minghim, R., Levkowitz, H.: Least square projection: a fast high-precision multidimensional projection technique and its application to document mapping. IEEE Trans. Visual Comput. Graph. **14**(3), 564–575 (2008)
14. Paulovich, F.V., Eler, D.M., Poco, J., Botha, C.P., Minghim, R., Nonato, L.G.: Piece wise Laplacian-based projection for interactive data exploration and organization. In: Computer Graphics Forum, vol. 30, pp. 1091–1100. Wiley Online Library (2011)
15. Poličar, P.G., Stražar, M., Zupan, B.: openTSNE: a modular python library for t-SNE dimensionality reduction and embedding, August 2019
16. Rauber, P.E., Falcao, A.X., Telea, A.C., et al.: Visualizing time-dependent data using dynamic t-SNE (2016)
17. Schuon, S., Durković, M., Diepold, K., Scheuerle, J., Markward, S.: Truly incremental locally linear embedding. In: CoTeSys 1st International Workshop on Cognition for Technical Systems (2008)

λ-DBSCAN: Augmenting DBSCAN with Prior Knowledge

Joel Dierkes, Daniel Stelter, and Christian Braune(✉)

Otto von Guericke University, Institute for Simulation and Graphics,
Magdeburg, Germany
{joel.dierkes,daniel.stelter,christian.braune}@ovgu.de

Abstract. State-of-the-art density based cluster algorithms offer remarkable speed and robustness. However, they do not allow the user to make local changes without affecting the global outcome. The user thus has to choose between clustering a local region well or keeping the global result.

We present a new approach, λ-DBSCAN, which augments the DBSCAN algorithm to include local a priori knowledge. The parameters can be specified per observation, rather than globally, which enables the user to include local knowledge about the data without modifying other regions. Furthermore, we define regions in the data that should be affected by certain parameter choices, to reduce the workload for a user.

Keywords: prior knowledge · density-based clustering · DBSCAN

1 Introduction

Finding relations in data is one of the most common problems in data analysis. For the unsupervised task of clustering the goal is to combine similar observations to clusters, whereas dissimilar observations should be in different clusters. However, the solution to this task can often be ambiguous, since in general there is no *perfect* clustering, i.e. a given ground truth, but rather multiple good ones. This is especially true for real life datasets, since normally no ground truth exists at all and all clusterings are open for interpretation by the viewer.

There exists a variety of cluster models, such as connectivity-, centroid-, and density-based models. Density models assume a cluster to be a dense region of observations bounded by less dense regions. The number of clusters does not have to be known *a priori*, and neither are any assumption made about their possible shapes. In contrast centroid-based clustering algorithms (like k-means [6]) implicitly assume that all points within a cluster can be properly represented by a single point. However, trivial density-based clustering algorithms, like DBSCAN [3], encounter difficulties if the non-dense regions of observations that divide dense regions do not clearly separate clusters. Several state-of-the-art density based cluster models, like OPTICS [1] or HDBSCAN [2], can successfully find clusters with varying densities. They do however tend to perform poor on datasets with close high density clusters in combination with distant low density ones [7].

© The Author(s), under exclusive license to Springer Nature Switzerland AG 2024
I. Miliou et al. (Eds.): IDA 2024, LNCS 14642, pp. 107–118, 2024.
https://doi.org/10.1007/978-3-031-58553-1_9

Each approach to find clusters in a given data set must make assumptions about what a cluster actually is and how clusters should be separated. These assumptions are usually made implicitly by the choice of the clustering algorithm (e.g. k-means finds hyperspherical clusters, or single linkage clustering is suitable to follow lines in the data set) or explicitly by the choice of parameters for a clustering algorithm (e.g. the choice of the metric used to calculate (dis)similarites, or the choice of linkage method for hierarchical clustering). Another way to direct the clustering process would be to incorporate *a priori* knowledge into the clustering process.

However, even with *a priori* knowledge there might not be a sufficiently good choice of a clustering method and its parameters. This can be the case for datasets which show too large variety in different regions of the domain. Although there are algorithms which try to adapt to local properties of the data, they still can fail as the selected global parameters might only deliver good results for some (or even only for single) regions of the whole domain.

In this work we present λ-DBSCAN which is a modification of the DBSCAN algorithm presented in [3]. Our idea is to extend the algorithm slightly to allow the integration of prior knowledge of the data to the parameters. This enables us to change the way clusters are formed depending on the specific location of a data point, i.e., besides *global* prior knowledge the user is able to apply *local* prior knowledge as well.

2   Related Work

DBSCAN is a well studied density based cluster algorithm. Each observation is classified as either a core point, a border point, or noise. The algorithm defines a local neighborhood as a hypersphere of radius ε around each point and counts the number of other data points within this neighborhood to decide which class (core, border, or noise) a point belongs to.

$$N_\varepsilon(\vec{p}) = \{q \in D \mid dist(\vec{p}, \vec{q}) \leq \varepsilon\}. \tag{1}$$

A point $\vec{p}$ is considered to be a *core* point if and only if the size of its ε-neighborhood is at least $minPts$ or μ:

$$cores_{\varepsilon,\mu} = \{\vec{p} \in D \mid \mu \leq |N_\varepsilon(\vec{p})|\} \tag{2}$$

Thus, ε and μ are the two major parameters of DBSCAN.

Furthermore, an observation is a border point if at least one and less than μ observations are contained within its ε-neighborhood, otherwise it is noise

$$borderPoints_{\varepsilon,\mu} = \{\vec{p} \in D \mid 1 \leq |N_\varepsilon(\vec{p})| < \mu\} \tag{3}$$

$$noise_{\varepsilon,\mu} = \{\vec{p} \in D \mid 0 = |N_\varepsilon(\vec{p})|\}$$
$$= D \setminus (cores_{\varepsilon,\mu} \cup borderPoints_{\varepsilon,\mu}) \tag{4}$$

To form clusters DBSCAN needs to decide which of the core and border points belong into the same cluster. For this [3] defines a set of relations to describe the interaction between different points. The first is *directly density reachable*. Two points are directly density-reachable if their distance is less than ε and at least one of them is a core point. Secondly, two points $\vec{p}$ and $\vec{q}$ are *density reachable* if there exists a sequence of mutually directly density reachable points which starts with $\vec{p}$ and ends with $\vec{q}$. Clusters are then formed by the transitive closure of the density reachability relation.

Note here that $\vec{p}$ is in the ε-neighborhood of a point $\vec{q}$ if and only if $\vec{q}$ is in the ε-neighborhood of $\vec{p}$:

$$\vec{p} \in N_\varepsilon(\vec{q}) \Leftrightarrow \vec{q} \in N_\varepsilon(\vec{p}) \tag{5}$$

While DBSCAN enables the user to find clusters with similar densities, it struggles to produce good results if the densities within different clusters vary vastly. Other versions of the DBSCAN algorithms have been invented to cope with its limitations, such as OPTICS [1] or HDBSCAN [2].

OPTICS [1] (Ordering Points To Identify the Clustering Structure) is a density-based clustering algorithm designed to find patterns in spatial datasets. It generates a reachability plot, revealing clusters with varying density and hierarchies. HDBSCAN [2] (Hierarchical Density-Based Spatial Clustering of Applications with Noise) leverages a hierarchical approach combined with a density-based one. It creates a tree of cluster relationships which allows the algorithm to find clusters with varying densities.

A completely different approach able to cope with clusters of varying density would be spectral clustering [9] which turn clustering into a graph-based problem and leverages the eigenvalues and eigenvectors of the similarity matrix to identify clusters. It transforms the data into a lower-dimensional space using the similarity matrix and runs a cluster algorithm on these structures.

While these also find clusters with varying densities, they fail to produce good results in situations with close high-density clusters together with distant low-density ones. All of these extensions to DBSCAN have in common, that they use a set of global parameters which cannot be locally adapted. Although these parameter choices may have local influences they are usually not very intuitive to determine.

Another approach that is able to achieve comparable results is C-DBSCAN [14], which uses pairwise *must-link* and *cannot-link* constraints to force pairs of points into either the same cluster or into different clusters. *Cannot-link* constraints can be used to separate a point from another cluster that it would normally belong to. It could be seen as locally setting μ or ε to such values that each of the points involved in the constraint is not contained in the others ε-neighborhood. *Must-link* constraints on the other hand merge clusters that would not be combined by DBSCAN because no two points' ε-neighborhoods overlap sufficiently. This can merge clusters that are close to each other (which could be seen as modifying μ or ε again) but also those which are too dissimilar to each other to be found by purely distance- based approaches. For both types of constraints these have to be given for each involved pair of data points. Creating a large set of constraint may therefore be prohibitively expensive or counter-intuitive.

Algorithm 1: Pseudocode of the λ-DBSCAN Algorithm

Input: DB: Database
Input: ε: Neighborhood size
Input: μ: MinPts
Input: $dist$: Distance function
Output: $label$: Points labels, initially undefined

1:    **for** $\vec{p} \in DB$
2:        **if** $label(\vec{p}) \neq undefined$ **then continue**
3:        Neighbors $N \leftarrow RangeQuery(DB, dist, \vec{p}, \boxed{\varepsilon(\vec{p})})$
4:        **if** $|N| < \mu$ **then**
5:            $label(\vec{p}) \leftarrow$ Noise
6:            **continue**
7:        **end if**
8:        $c \leftarrow$ next cluster label
9:        $label(\vec{p}) \leftarrow c$
10:      Seed set $S \leftarrow N \setminus \{\vec{p}\}$
11:      **for** $\vec{q} \in S$
12:          **if** $label(\vec{q}) =$ Noise **then** $label(\vec{q}) \leftarrow c$
13:          **if** $label(\vec{q}) \neq undefined$ **then continue**
14:          Neighbors $N \leftarrow RangeQuery(DB, dist, \vec{q}, \boxed{\varepsilon(\vec{p})})$
15:          $label(\vec{q}) \leftarrow c$
16:          **if** $|N| < \mu$ **then continue**
17:          $S \leftarrow S \cup N$
18:      **end for**
19:    **end for**

---

3 Implementation

When we want to extend DBSCAN in a way that allows us to incorporate prior knowledge (or different understandings of which points should be able to form a cluster and which not) the only parameters within the original algorithm are ε and μ and thus our first target for adjusting DBSCAN's clustering behaviour.

The λ-DBSCAN algorithm is based on the DBSCAN implementation. We still distinct between core, border, and noise points. The parameter ε which is used to determine which of these classes a point belongs to can be chosen for individual points rather than using one global value for each point, respectively. For the sake of this paper we will not alter the meaning of μ as our adaption of ε is sufficient. Technically a user of λ-DBSCAN could decide for each point individually which class a point belongs to (without calculating the membership to each class) but this will turn any practical application infeasible, of such a decision would have to be made for many points, points in a high-dimensional space, or both. Instead we select these parameters for whole regions of points, which enables the algorithm to be reasonably autonomous. The algorithm only relies on provided domain knowledge, if needed, except for one initial global choice of parameters.

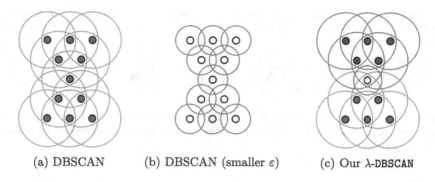

(a) DBSCAN   (b) DBSCAN (smaller ε)   (c) Our λ-DBSCAN

Fig. 1. Clustering of equidistant points with the classical DBSCAN and λ-DBSCAN.

The effect of this can be seen in Fig. 1 where we see 11 points arranged on a grid of equilateral triangles such that for every choice of ε and μ DBSCAN must either assign all points to the same cluster or to no cluster at all. Our Proposed algorithm λ-DBSCAN can however be parameterized in a way that the upper five points and the lower five points can be separated.

Algorithm 1 outlines an implementation with a global μ and individual ε based on [15, Chapter 2.2]. The changes have been marked in Algorithm 1. Note that we use a function $\varepsilon(p)$ which returns individual εs for each data point. Hence the name λ-DBSCAN, coming from the Python unnamed `lambda` functions. Regarding a practical implementation each individual ε could be pre-computed and stored in a list. This only adds a linear complexity to the algorithm (w.r.t. to the complexity class of $\varepsilon(p)$ which will mostly be in $\mathcal{O}(1)$) and therefore does not impact the performance by much (cf. Table 1). However, this computation is only performed once before the clustering starts and will diminish with larger datasets.

It is important to notice here that Equation (5) does not necessarily hold anymore. If set accordingly, a point $\vec{p}$ could be in the ε-neighborhood of another point $\vec{q}$, without $\vec{q}$ being the in the ε-neighborhood of $\vec{p}$. We see three possible solutions:

1. Rely on the order in which the algorithm traverses the dataset.
2. Keep the original interpretation of Equation (5) intact and only consider $\vec{p}$ and $\vec{q}$ as *directly density reachable* if they are mutually contained in each other's ε-neighborhoods
3. Alter the interpretation of Equation (5) such that $\vec{p}$ and $\vec{q}$ as *directly density reachable* if either is contained in the other's ε-neighborhood

We discard the first approach, since the clustering should be independent of the iteration of the observations. We tested implementations 2. and 3., both yielded the same result for all datasets in this paper.

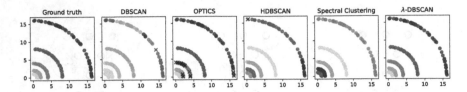

Fig. 2. Different clustering results for a dataset with clusters that form concentric rings. The number of observations per cluster are the same, the densities vary due to the different radii.

4   Evaluation

In the following, we compare results of λ-DBSCAN with selected algorithms on a variety of artificial and two real world datasets. Table 2 shows the parameters we used for each algorithm to cluster the synthetic datasets. Table 1 shows the runtimes and scores.

4.1   Synthetic Datasets

Rings: Fig. 2 shows different cluster results for an artificial concentric ring data. All rings contain the same number of elements, which results in different densities because of their varying radii. Such datasets pose a challenge for density based cluster algorithms. In order to separate the two inner rings one has to choose a rather small value for ε which in turn leads to the points on the outermost ring to not be assigned to the same cluster. With the variation in density in side of the same clusters even HDBSCAN and OPTICS sometimes either detect noise where none should be or separates one cluster into two different clusters. Only spectral clustering yields a perfect result with the nearest neighbor affinity.

λ-DBSCAN allows us to define a ε parameter based on the location of the points. Since it is known a priori that the clusters get sparser the farther away they are from the origin, one can set the ε parameter to a value depending on the distance of the point to the origin – in this case to one third of the distance of a point to the origin. Spectral clustering however needed more than 28 times longer to compute this result than DBSCAN or λ-DBSCAN (see Table 1).

Spirals: Figure 3 displays cluster results for an artificial spiral data. The spirals are meet at the origin, which makes them indistinguishable for any basic density based cluster algorithms. The points are farther away from each other the further they are away from the origin, resulting in clusters with varying densities. DBSCAN and OPTICS both fail to yield a result comparable to the ground truth, not matter how the parameter are chosen. Spectral clustering finds three clusters (because it has to given the parameter choice) but those have virtually nothing in common with the desired result. This stems from the underlying cluster model that is implicitly encoded in the algorithm itself.

λ-DBSCAN can use knowledge about local regions and thus separate the clusters around the origin. At the same time, it can use the knowledge about the

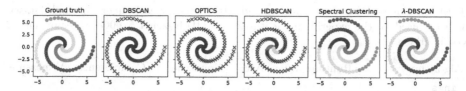

Fig. 3. A dataset with clusters that form concentric spirals. The number of observations per cluster are the same, the densities within each cluster decreases the further away a point is from the origin.

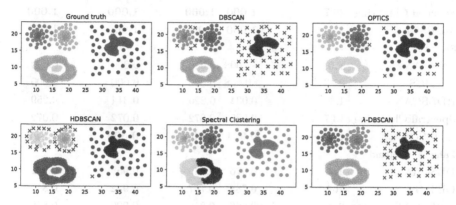

Fig. 4. The ground truth and five different clusterings of the compound dataset.

spirals to solve the issue of varying densities in the clusters itself. Figure 3 and Table 2 show that the way we calculate the individual ε is rather simple. A more sophisticated choice might even lead to a perfect clustering result around the origin.

Zahn's Compound: The Compound dataset [4] represents a challenge for spectral clustering, since it can not clearly distinguish the two clusters where one cluster encapsulates another cluster with different densities. OPTICS fails to distinguish them at all. Since DBSCAN uses the ε and μ globally it fails to cluster the shapes without noise. λ-DBSCAN produces results better than DBSCAN while taking the same amount of time (Fig. 4).

4.2 Real World Datasets

GPS Pickup Data: The GPS pickup dataset is a collection of roughly 2'800 GPS coordinates of pick-up and drop-off locations of customers of a ride pooling company. The aim is to find a suitable selection of locations that represent a pick-up or drop-off location for a group of users – or in other terms: find clusters in the GPS data. The dataset contains groups of GPS points with varying densities and member counts, as well as different distances between the clusters (Fig. 5).

While [7] finds clusters with varying densities and few points, they fail to split close clusters with high densities. The train station, for example, is one

Table 1. Comparison of the runtime and ARI scores for each algorithm and dataset. The PC used for the experiments features an AMD Ryzen 7 5700G processor and 32 GB DDR4-RAM.

Algorithm	Runtime in ms	ARI	Homogeneity	Completeness	V-measure
Rings					
DBSCAN	**1.3**	0.689	0.828	0.820	0.824
OPTICS	106.4	0.867	0.938	0.854	0.894
HDBSCAN	2.6	0.940	**1.000**	0.915	0.956
Spectral Clustering	36.7	**1.000**	**1.000**	**1.000**	**1.000**
λ-DBSCAN	**1.3**	**1.000**	**1.000**	**1.000**	**1.000**
Spirals					
DBSCAN	**1.0**	−0.008	0.000	0.000	0.000
OPTICS	63.3	−0.008	0.000	0.000	0.000
HDBSCAN	1.3	0.034	0.220	0.314	0.259
Spectral Clustering	13.7	0.065	0.072	0.072	0.072
λ-DBSCAN	1.2	**0.921**	**0.898**	**0.898**	**0.898**
Zahn's Compund					
DBSCAN	**1.4**	0.976	0.949	0.953	0.951
OPTICS	164.1	0.788	0.767	0.942	0.845
HDBSCAN	2.9	0.826	0.819	0.900	0.858
Spectral Clustering	83.2	0.585	0.776	0.809	0.792
λ-DBSCAN	1.5	**0.997**	**0.993**	**0.992**	**0.992**

big cluster, while it could possibly be an aggregation of four to five close but separate clusters. None of the referenced algorithms in [7] can cluster these close bigger clusters effectively, leaving the customers with only one pick-up and drop-off location for a rather large area. λ-DBSCAN can select these regions in the data with appropriate ε and μ values.

Application in Flow Visualization. We also applied λ-DBSCAN to a practical example in the field of flow visualization. We consider the *finite-time Lyapunov exponent (FTLE)* for a 2D flow. FTLE fields are scalar fields, its ridges are lines which one can see as a skeleton of the underlying flow. This skeleton divides the flow into multiple regions. Material inside the same region which is moved by the flow over time stays comparably close to each other over time. The extraction of FTLE ridges can greatly assist in many modern fields, e.g. path planning for autonomous underwater vehicles [13], tracking the movement of air pollution and its effects [10], and understanding medical phenomenons like aortic valve calcifications [8]. [16] recently proposed a new efficient and reliable attempt for sampling FTLE ridges with a population of particles. It results in a uniform sampling of these structures. Unfortunately, ridges in FTLE fields tend to be sharp and closely located to each other which leads to numerical difficulties.

Table 2. Parameters for the artificial datasets.

Algorithm	Parameters
Rings	
DBSCAN	$\varepsilon = 1, \mu = 2$
OPTICS	$\mu = 26$
HDBSCAN	$\mu = 6$
Spectral Clustering	$k = 5, \text{affinity} = \text{nearest neighbors}$
λ-DBSCAN	$\varepsilon = |p|/3, \mu = 3$
Spirals	
DBSCAN	$\varepsilon = 0.5, \mu = 2$
OPTICS	$\mu = 10$
HDBSCAN	$\mu = 1$
Spectral Clustering	$k = 3, \text{affinity} = \text{nearest neighbors}$
λ-DBSCAN	$\mu = 3, \varepsilon = \begin{cases} |p|^{0.2} & |p| > 1 \\ 0.1 & else \end{cases}$
Zahn's Compund	
DBSCAN	$\varepsilon = 1.48, \mu = 3$
OPTICS	$\mu = 20$
HDBSCAN	$\mu = 7$
Spectral Clustering	$k = 5, \text{affinity} = \text{rbf}$
λ-DBSCAN	$\mu = 5, \varepsilon = \begin{cases} 1 & |p - (16.5, 9.5)^T| < 7 \vee p_0 > 25 \\ 0.0001 & 13.25 < p_0 < 17.5 \wedge 15 < p_1 \\ 2.5 & else \end{cases}$

[16] point out the need for an appropriate clustering method in order to to detect outliers and to connect the correct ridges. We applied the cluster methods to an extraction example for a 2D flow around a cylinder inside a channel [5,12]. In this dataset the flow streams from left to right and is bounded by solid walls.

The FTLE ridge extraction resulting in 22,000 particles is shown in Fig. 6. The particles create many lines on the left side of the cylinder, indicating a strong turbulence with thin coherent regions. Further to the right, there are fewer lines, thus, less turbulence. However, numerical problems lead to outliers and partly discontinuous structures.

Figure 6 also displays five different clusterings of the ridges dataset. DBSCAN with $\varepsilon = 0.0008$ and $\mu = 3$ does yield a meaningful clustering of ridges that are further away from the origin (approx. $x > 0.5$) with only some ridges merged. The ridges near the origin however can not be separated at all. Other parameter choices yield a better result for the ridges near the origin, but fracture the clusters further away. OPTICS with $\mu = 2$ and $\xi = 0.00001$ results in a lot of noise mislabeling, while other parameter choices merge many ridges into one. HDBSCAN with $\mu = 13$ does a good job of finding ridges in less turbulent regions,

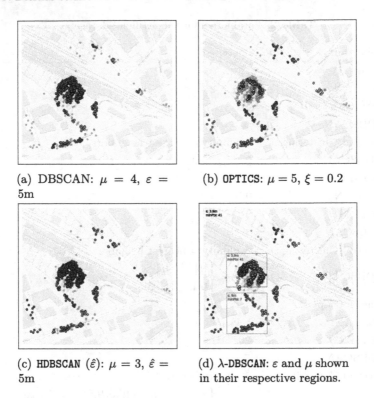

(a) DBSCAN: $\mu = 4$, $\varepsilon = $ 5m

(b) OPTICS: $\mu = 5$, $\xi = 0.2$

(c) HDBSCAN ($\hat{\varepsilon}$): $\mu = 3$, $\hat{\varepsilon} = $ 5m

(d) λ-DBSCAN: ε and μ shown in their respective regions.

Fig. 5. Different results for the GPS dataset. Parameters taken from [7] (except for λ-DBSCAN)

however it also is unable to differentiate the lines in turbulent regions. Spectral clustering with $k = 50$ and the nearest neighbors affinity splits and joins lines almost at random, also merging the lines in the turbulent regions.

Given the prior knowledge that turbulent regions form around the origin and contain more close ridges, while less turbulent regions form further away from the origin and contain fewer ridges which are further spread apart, one can set the parameters for the λ-DBSCAN algorithm accordingly. A clustering with

$$\mu = 3, \ \varepsilon(x,y) = \begin{cases} 0.0035 & x < 0.5 \\ 0.008 & else \end{cases}$$

is seen at the bottom of Fig. 6. The algorithm detects ridges around the origin, with only some being merged that are very close to each other.

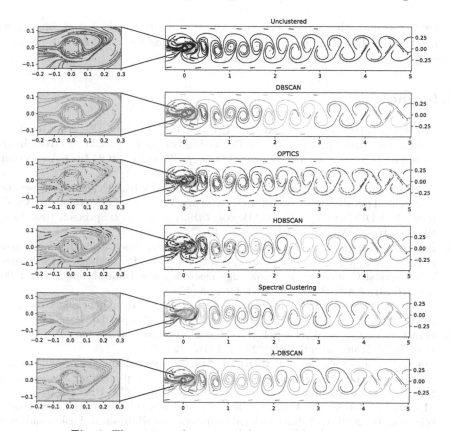

Fig. 6. The `ridges` dataset and five possible clusterings.

5 Conclusion and Future Work

In this paper, we proposed a new algorithm for using local specific knowledge λ-DBSCAN. Effectively our algorithms allows the user to run different instances of the DBSCAN algorithm on different portions of the dataset while seamlessly merging their outcomes into a single result.

The speed of λ-DBSCAN enables the algorithm to be used in interactive applications and be a drop-in for a pure DBSCAN implementation since it is usually not slower while yielding at least the same – if not better – results. A program that leverages the user as an oracle who marks regions that need parameter changes comes to mind allowing some form of semi-supervised clustering. The cluster result could update in real time (only re-cluster the parts that changed) and thus allow processing big datasets.

The initial regions in which different parameters for λ-DBSCAN might be needed could also be automatically be identified by running k-means with a sufficiently large k and use the density around the so-found proto-clusters to initialize the function $\varepsilon(\bar{p})$.

Source Code. The implementation of our algorithm is based on scikit-learn's [11] implmention of DBSCAN. Our changes are available at https://visual2.cs.ovgu.de/pubres/lambda-dbscan.

References

1. Ankerst, M., Breunig, M.M., Kriegel, H.P., Sander, J.: Optics: ordering points to identify the clustering structure. In: Proceedings of the 1999 ACM SIGMOD International Conference on Management of Data, p. 49-60. SIGMOD 1999, Association for Computing Machinery, New York, NY, USA (1999)
2. Campello, R.J.G.B., Moulavi, D., Sander, J.: Density-based clustering based on hierarchical density estimates. In: Pei, J., Tseng, V.S., Cao, L., Motoda, H., Xu, G. (eds.) PAKDD 2013. LNCS (LNAI), vol. 7819, pp. 160–172. Springer, Heidelberg (2013). https://doi.org/10.1007/978-3-642-37456-2_14
3. Ester, M., Kriegel, H.P., Sander, J., Xu, X.: A density-based algorithm for discovering clusters in large spatial databases with noise. In: Proceedings of the Second International Conference on Knowledge Discovery and Data Mining, KDD 1996, pp. 226-231. AAAI Press, Portland, Oregon (1996)
4. Fränti, P., Sieranoja, S.: K-Means properties on six clustering benchmark datasets. Appl. Intell. **48**(12), 4743–4759 (2018). http://cs.uef.fi/sipu/datasets/
5. Günther, T., Gross, M., Theisel, H.: Generic objective vortices for flow visualization. ACM Trans. Graph. **36**(4), 141:1–141:11 (2017)
6. MacQueen, J., et al.: Some methods for classification and analysis of multivariate observations. In: Proceedings of the Fifth Berkeley Symposium on Mathematical Statistics and Probability, vol. 1, pp. 281–297. Oakland, CA, USA (1967)
7. Malzer, C., Baum, M.: A hybrid approach to hierarchical density-based cluster selection. In: 2020 IEEE International Conference on Multisensor Fusion and Integration for Intelligent Systems (MFI), pp. 223–228 (2020)
8. Mutlu, O., Salman, H.E., Yalcin, H.C., Olcay, A.B.: Fluid flow characteristics of healthy and calcified aortic valves using three-dimensional Lagrangian coherent structures analysis. Fluids **6**(6), 203 (2021)
9. Ng, A., Jordan, M., Weiss, Y.: On spectral clustering: analysis and an algorithm. In: Dietterich, T., Becker, S., Ghahramani, Z. (eds.) Advances in Neural Information Processing Systems, vol. 14. MIT Press (2001)
10. Nolan, P.J., Foroutan, H., Ross, S.D.: Pollution transport patterns obtained through generalized Lagrangian coherent structures. Atmosphere **11**(2), 168 (2020)
11. Pedregosa, F., et al.: Scikit-learn: machine learning in python. J. Mach. Learn. Res. **12**, 2825–2830 (2011)
12. Popinet, S.: Free computational fluid dynamics. ClusterWorld **2**(6), 2–8 (2004)
13. Ramos, A., et al.: Lagrangian coherent structure assisted path planning for transoceanic autonomous underwater vehicle missions. Sci. Rep. **8**(1), 4575 (2018)
14. Ruiz, C., Spiliopoulou, M., Menasalvas, E.: C-DBSCAN: Density-Based Clustering with Constraints. Rough Sets, Fuzzy Sets, Data Mining and Granular Computing. LNCS(4482), 2007
15. Schubert, E., Sander, J., Ester, M., Kriegel, H.P., Xu, X.: DBSCAN revisited: why and how you should (Still) use DBSCAN. ACM Trans. Database Syst. **42**(3), 19 (2017)
16. Stelter, D., Wilde, T., Rössl, C., Theisel, H.: A Particle-Based Approach to Extract Dynamic 3D FTLE Ridge Geometry (under review)

Putting Sense into Incomplete Heterogeneous Data with Hypergraph Clustering Analysis

Vishnu Manasa Devagiri[1]([✉])[iD], Pierre Dagnely[2], Veselka Boeva[1][iD], and Elena Tsiporkova[2]

[1] DIDA, Blekinge Institute of Technology, Karlskrona, Sweden
{vmd,vbx}@bth.se
[2] EluciDATA Lab, Sirris, Brussels, Belgium
{pierre.dagnely,elena.tsiporkova}@sirris.be

Abstract. Many industrial scenarios are concerned with the exploration of high-dimensional heterogeneous data sets originating from diverse sources and often incomplete, i.e., containing a substantial amount of missing values. This paper proposes a novel unsupervised method that efficiently facilitates the exploration and analysis of such data sets. The methodology combines in an exploratory workflow multi-layer data analysis with shared nearest neighbor similarity and hypergraph clustering. It produces overlapping homogeneous clusters, i.e., assuming that the assets within each cluster exhibit comparable behavior. The latter can be used for computing relevant KPIs per cluster for the purpose of performance analysis and comparison. More concretely, such KPIs have the potential to aid domain experts in monitoring and understanding asset performance and, subsequently, enable the identification of outliers and the timely detection of performance degradation.

Keywords: Clustering · Heterogeneous data · Missing values · Hypergraph · Shared nearest neighbor similarity

1 Introduction

The majority of real-world data related to monitoring the performance of industrial equipment or of production and engineering processes is typically collected from multiple diverse sources and, thus, is very heterogeneous. It is a challenging task [13] to analyze and derive meaningful insights, e.g., evaluate and compare operational performance across sources, from such data. Moreover, industrial data usually contains a lot of missing entities due to different reasons such as incomplete metadata records, lack of standardization, equipment malfunctioning, registration errors, communication issues, etc. This missing data usually

V. M. Devagiri's and V. Boeva's research is partly funded by the Knowledge Foundation, Sweden, through the Human-Centered Intelligent Realities (HINTS) Profile Project (contract 20220068). P. Dagnely's and E. Tsiporkova's research received funding from the Flemish Government (AI Research Program).

affects small but varying sets of parameters/measurements, while the rest of the data records are available for analysis. However, mining information from complex multi-source data with missing values can be challenging since many state-of-the-art algorithms are not designed to handle missing values. In order to enable the use of such algorithms, common practices are to either impute missing values or to completely remove that particular instance or feature (the ones with a high degree of missing values). Both of these approaches can negatively affect the data quality though [8]. In addition, high-dimensional data is often composed of entries of a very diverse nature, and it might occur that some interesting, specific properties are associated only with a certain subset/type of features. There exists a risk of missing these when all the available features are explored together. Studies like [6,7,11], highlight the importance of viewing data from different perspectives or views (i.e., considering relevant feature subsets).

To address these challenges, we propose here a hybrid clustering methodology, realizing a multi-layer data analysis workflow, which is employing shared nearest neighbor similarity (SNNS) and hypergraph clustering. The approach is capable of extracting meaningful insights from high-dimensional data and, at the same time, is efficient at handling missing values without losing valuable information. More concretely, the method allows to organize the assets into separate (overlapping) groups such that the assets in each group share similar properties and, thus, are expected to exhibit comparable performance. In this way, it facilitates the complex task of monitoring and making sense of the performance of a large portfolio of heterogeneous in nature industrial assets. The potential of the proposed methodology is validated on a real-world data set with a substantial amount of missing values originating from the condition monitoring of a portfolio of industrial assets. These assets are very diverse in terms of technical specifications and functionalities, are used in different application contexts, and are produced by different manufacturers.

2   Related Work

Mining of data constructed across multiple domains or modalities (categorical, numerical, transactional, etc.) has been receiving high attention in the last decades due to the continuous increase of the variety in data sources [1]. This kind of data analysis is required within the context of grouping and understanding multi-view, heterogeneous, or multi-modal data in many real-world scenarios. The traditional clustering techniques usually fail to identify the cluster of objects with different characteristics due to difficulties in finding suitable similarity measures, or they are not capable of capturing the intrinsic structure of clusters.

To reduce these limitations, multi-view [6], multi-layered [7] or multi-type [11] clustering techniques are introduced. For example, multi-view clustering uses more than one set of attributes to improve the quality of generated clustering solutions [6]. In [7], a multi-layer clustering technique is introduced, originally designed for analysis of network data available in more than one layer [3]; where in contrast to multi-view clustering, conditional independence of layers is not assumed. In real-life scenarios, heterogeneous information networks (HINs) could be formed by the existence of multiple types of objects that are connected to each other through different kinds of links. In [15], the authors have studied the multi-type co-clustering problem in general HINs. They have proposed a clustering framework that can model general HINs and simultaneously generate clusters for all types of objects. In [11], a novel method that performs both clustering and classification tasks on HINs is proposed. The proposed technique is able to group heterogeneous objects in a network together and assign labels to unlabeled objects.

A common industrial challenge that impacts these methods is the presence of missing data. It is usually addressed by removing entries with missing data or imputing the missing data, e.g., replacing missing values of a feature with the average of known values for that feature or predicting them based on other known features of that entry. For instance, in [14], Yang et al. have proposed a multi-view clustering methodology that tackles missing data through imputation but also addresses inconsistencies between views.

However, data imputation always carries the risk of imputing noisy data, especially for industrial assets that are often highly heterogeneous. The creation of multi-view clustering approaches that deal with missing data in their input is still in its infancy. In [5], the authors use an indicator matrix whose elements indicate which data entries are observed and assess cluster validity only on observed entries. However, this approach cannot easily be generalized to all clustering approaches. Moreover, the proposed methodologies deal with incomplete views or missing values with some constraints, but they struggle when all views have missing values and even when the samples just miss a few features in a view [4].

In this work, we propose a novel multi-view clustering approach, which exploits the power of multi-layer clustering data analysis to transform the highly-dimensional heterogeneous data set into a hypergraph. Subsequently, the final clustering solution is obtained by applying a creative hypergraph clustering methodology. Our method goes beyond the existing state-of-the-art in its ability to deal efficiently with missing values without losing any information and to produce overlapping multi-view partitions without imposing any constraints on the underlying multi-source data.

3   Hypergraph-Based Clustering Analysis Method

In this work, we study a real-world industrial use case considering the exploration and analysis of multi-source data originating from a large portfolio of heterogeneous assets (compressors). The available data set is composed of both metadata and sensor measurements (in the form of time series) and contains substantial quantities of missing data. Analyzing and interpreting data from different types of assets is challenging as they cannot be directly compared since they may substantially differ in technical specifications and other essential characteristics. Asset comparison can be facilitated by grouping the assets into more

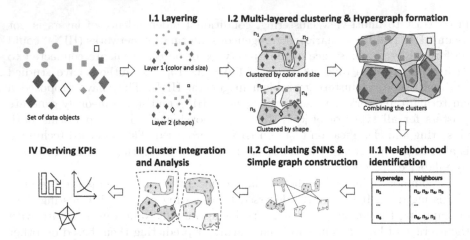

Fig. 1. Illustrative summary of the proposed approach using two layers. SNNS stands for shared nearest neighbor similarity.

homogeneous subsets (clusters), i.e., composed of assets with similar characteristics and settings. The high-dimensional metadata, describing assets' technical specifications and various other properties, is used for this purpose. The available metadata is very suitable for having a realistic evaluation of the application potential of our clustering approach.

In addition, the validity of the generated clustering solution is further evaluated on the time series data originating from the continuous monitoring of the assets during their operation in the field. The assets' performance is analyzed by estimating and comparing the evolution of diverse, performance-related, Key Performance Indicators (KPIs) (see Sect. 4.4 for more details).

The proposed clustering workflow for analyzing multi-source heterogeneous data is divided into four main steps as illustrated in Fig. 1 and outlined in detail in the subsections below.

3.1   Step I: Hypergraph Construction

Let us assume that multi-source information about a set D of data objects with missing entities, which needs to be grouped into a number of similar categories, is available. The data objects are described in terms of a set F of relevant features. The workflow of the different steps used to construct a hypergraph with the data objects acting as vertices is outlined below:

I.1: Layering. Initially, the features in F are categorized into L different thematic layers, each representing a particular aspect describing the data objects. Domain knowledge is used in this step to identify the different layers. Each layer $i \in L$ is represented by a feature set F_i, $F_i \subseteq F$. Generalizing the layer construction step to different use cases is not straightforward and requires investing time in engaging with domain experts and acquiring a good understanding of the phenomenon under study. Section 4.1 provides more details on how this step is performed for the use case studied here.

I.2: Multi-layered Clustering and Hypergraph Formation. Once the layers are identified, a hypergraph is constructed as it is explained below. A hypergraph is a generalized graph where edges, also referred to as hyperedges or nets can connect more than two nodes [12].

- In each layer i, for $i = 1, 2, \ldots, l$, data objects having missing values in feature set F_i are removed thus a subset of data objects $D_i \subseteq D$ is produced. It must be noted that these removed data objects are still considered in other layers where their features are completely captured. This allows not to exclude data objects with missing values from the analysis.
- Data objects D_i, for $i = 1, 2, \ldots, l$, in each layer are clustered to obtain a disjoint clustering solution represented by C_i.
- The produced clustering solutions from all the different layers are united to build an unweighted hypergraph $H = (V, N)$. The set of vertices V of this graph are the data objects, i.e., $V \equiv D$, and the set of hyperedges or nets N contains all the clusters identified by clustering the different layers, i.e., $N = C_1 \cup C_2 \cup \ldots \cup C_l$.

3.2   Step II: Transformation to Simple Graph

Once the hypergraph is built, it is transformed into an undirected weighted graph also known as a simple graph. This allows us to benefit further from the produced lower data granularity and facilitates grouping the data objects into the final clustering solution. In addition, this helps to handle missing entities, since the raw data are not used in the rest of the computations.

The new simple graph can be represented as $G = (N, E, s)$, where N, set of graph vertices, also the set of edges of the hypergraph H; E is the set of the edges of the simple graph, and s is a real value function assigning a weight to each graph edge presented by the SNNS between vertices connected by this edge.

II.1: Neighborhood Identification. For each hyperedge n_i, (for $i = 1, 2, \ldots, m$ and m the total number of hyperedges) its set of neighbors $\Gamma(n_i)$ is identified. A hyperedge is considered a neighbor of another hyperedge when there is a non-empty intersection between the two, i.e. if they have at least one common data object or pin (vertices in each hyperedge) of the hypergraph in both. Note that the hyperedge itself is also added to the list of neighbors (used while calculating the similarity). This can be formalized as:

$$\Gamma(n_i) = \{n_j \mid n_j \in N \wedge n_i \cap n_j \neq \emptyset\}. \tag{1}$$

Once the neighborhoods have been identified, the data objects will not be used since the rest of the computation is entirely conducted using this information.

II.2: Calculating Shared Nearest Neighbor Similarity. We use the shared nearest neighbor similarity (SNNS) to measure the resemblance between two hyperedges of a hypergraph by considering their neighborhoods. Once the neighbors are identified, SNNS (s) between each pair of hyperedges n_i and n_j, for $i, j \in \{1, 2, \ldots, m\}$ is calculated as follows (inspired from [10] and adapted):

$$s(n_i, n_j) = \begin{cases} \frac{|\Gamma(n_i) \cap \Gamma(n_j)|}{|\Gamma(n_i) \cup \Gamma(n_j)|}, & \text{if } n_i, n_j \in \Gamma(n_i) \cap \Gamma(n_j). \\ 0, \text{otherwise} \end{cases} \tag{2}$$

3.3   Step III: Cluster Integration and Analysis

III.1: Partitioning into Overlapping Clusters. Once the graph $G = (N, E, s)$ is constructed, any clustering technique can be used to divide its vertices into different clusters, e.g., k-medoids or some graph-based clustering algorithm. In the obtained clustering solution, the vertices N of the simple graph (which are also the hyperedges of the hypergraph) are replaced with the vertices V of the hypergraph to generate the final clustering output. Note that we obtain an overlapping final clustering solution of the data objects $D \equiv V$, i.e., each data object can be assigned to more than one cluster. This is a desired outcome as, in many application scenarios, it is difficult to categorize an object into just a single category.

III.2: Deriving Peak Density Hyperedges. Furthermore, we can identify the peak density vertices (hyperedges) in G similarly to the idea introduced in [10]. Namely, the SNNS between different vertices of $G = (N, E, s)$, can be used to calculate local density ρ of each vertex n_i, for $i = 1, 2, \ldots, m$, as follows: $\rho(n_i) = \sum_{n_j \in \Gamma(n_i)} s(n_i, n_j)$. Subsequently, a threshold t can be used such that vertices having a local density greater than or equal to t, i.e., $\rho(n_i) \geq t$, are considered as the peak density points (hyperedges). These hyperedges can be interpreted as the most representative (typical) groups of objects.

3.4   Step IV: Deriving KPIs to Analyze Performance

The obtained clustering solution defines several different profiles, each characterized by the specific feature values defining each cluster. Operational data collected from continuous condition monitoring can then be used to enrich these profiles, e.g., to characterize them. In most real-world scenarios hardly any labeled data on the actual performance of the assets is collected. Instead, general indicators like the mean time between failures, the percentage of unplanned maintenance, or the overall maintenance cost are taken into account. These KPIs could be used to link the profiles to asset health. However, often, as in our use case, such data is not available and moreover, these indicators are not always directly linked to operational efficiency and overall performance in general but just to failures. We suggest instead using KPIs that allow us to compare operational behavior and performance across assets. In Sect. 4.4, different KPIs are described, being identified together with the domain experts, related to concrete performance indicators concerning the use case.

4   Evaluation in Industrial Use-Case

This section presents a detailed overview of how the proposed methodology has been evaluated on a real-world industrial use case concerned with the condition monitoring of a fleet of compressors. Due to the heterogeneity of the available data, it is very challenging to derive useful insights about the operational performance of the different compressors. The research methodology proposed in this paper allows to overcome these challenges by facilitating incremental data exploration, resulting in partitioning the heterogeneous compressor population into relatively homogeneous overlapping groups sharing similar characteristics. Thanks to this partitioning, sensor data can be efficiently used to study and compare the operational performance within and across homogeneous compressor groups.

The data set used in our validation study has been offered to us by our industrial partners in the context of a research project exploring how to augment conventional analytics with log data. The company manages data from compressors installed worldwide and used in a wide variety of conditions, e.g., in factories and subjected to harsh conditions, or in hospitals and required to comply with very tight tolerance levels. The data set provided contains information about 265 compressors, characterized by 393 different parameters, e.g., brand, age, multitude of technical specifications. In addition, each of the compressors is monitored in the field by a wide range of sensors. This study focuses on four commonly used sensors: ambient temperature, compressor outlet temperature and pressure, and internal pressure. The sensor data is reported at a granularity of one second, but the amount of data varies considerably from one compressor to another, from 6 months to 7 years.

4.1   Step I: Hypergraph Construction

Initially, the data set has been cleaned and analyzed to identify relevant features, as using redundant features can negatively impact the final result. Over the 393 different parameters, only 24 relevant features have been retained using various techniques such as domain knowledge, dropping columns with a high percentage of NaNs ($> 60\%$), uniqueness of the feature values among different compressors, and correlation between the features. Subsequently, the selected features have been grouped into different conceptual layers using expert knowledge. For instance, all features related to pressure tolerance are grouped together. In addition, features of similar types (categorical, binary, and numerical) are kept together. In this use case, a total of ten layers have been created, consisting of 1 to 4 features per layer. The layers are finalized after being validated by domain experts, detailed information is given in Table 1. Based on the type of data available in each layer, different clustering techniques have been used (see Table 1).

Table 1. An overview of the multi-layered clustering phase of hypergraph construction.

Layer	Description	Instances with NaNs	Instances Used	Type of Features	Clustering based on	Clusters
1	Supplier	42	223	Categorical	Categories	25
3	Cooling type	47	218	Binary		4
10	Activated temp. sensors	19	246			4
9	Act. pressure sensors	19	246			11
2	Pressure tolerance	115	150	Numerical	K-means	4
5	Outlet temp. tolerance	99	166			5
6	Output settings	87	178			2
8	Ambient temp. tolerance	118	147			2
4	Installation type	11	254		Domain knowledge	2
7	Age	79	186		Binning	4

In layers where the k-means clustering technique is used, the optimal number of clusters is identified by applying four different cluster validation measures, namely Silhouette [2], Calinski Harabasz [2], Davies Bouldin [2], and Connectivity [9]. Once the clustering in each layer is finalized, an unweighted hypergraph is obtained by combining the clustering solutions of different layers. The obtained hypergraph has 63 edges (the sum of the number of clusters in each layer).

4.2   Step II: Transformation to Simple Graph

The obtained hypergraph in the previous step is transformed into a weighted simple graph by considering the hyperedges as the vertices of the new graph. The weight of each edge of the simple graph is obtained by using the SNNS between the two vertices (hyperedges) the edge connects, which can also be represented using a 63×63 SNNS adjacency matrix. The obtained simple graph is presented in Fig. 2 (left). It can be observed that the central region of the graph is very dense. Interesting to note that the four vertices far away from the center $(7, 13, 21, 23)$ are the singletons obtained in the final clustering solution.

4.3   Step III: Cluster Integration and Analysis

The 63×63 SNNS matrix calculated above, converted into a dissimilarity matrix, is the input for k-medoids clustering algorithm. The optimal number of clusters, twelve, has been identified by Silhouette (0.08) and Connectivity (92.3) validation indices. In addition to this, we have also used the dissimilarity matrix, with edges sorted based on the cluster they belong to (see Fig. 2, right), to visually validate the obtained clustering output. Well-formed similarity patterns can be observed along the diagonal confirming to some extent the validity of the obtained partition. Once the clustering is obtained, the vertices of the simple graph are replaced with those of the hypergraph, thus resulting in an overlapping clustering solution, as different hyperedges have common vertices.

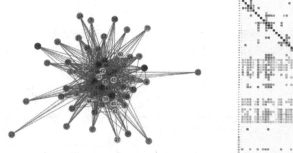 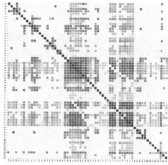

Fig. 2. Left: Visual representation of the simple graph, different layers are distinguished by the color of vertices. Right: Dissimilarity matrix based on SNNS between different vertices ordered according to their cluster belonging. White cells represent values closer to one.

Figure 3 depicts the number of compressors per cluster and their uniqueness in regard to how many clusters they have been assigned to. The clusters can be grouped into three categories: 1) two large, heavily overlapping, clusters (clusters 6 and 7) with 237 and 245 compressors, respectively; 2) six medium-sized clusters with between 6 and 63 compressors (clusters 0, 1, 3, 4, 9 and 11); 3) four singleton clusters not visualized in Fig. 3. These four singletons are also part of the two big clusters, 6 and 7. It is interesting to note that these singletons capture four unique compressor brands, only registered for these 4 compressors, thus confirming the potential of the approach to identify even small unique cluster groups. These singletons are not considered in the further discussion.

The peak-density vertices (defined in Sect. 3.3, III.2) have been also identified in the resulting simple graph. It is interesting to notice that all the twelve peak density points lie within the two big clusters (6 and 7) of the k-medoids clustering solution. This confirms that all the dense regions of the graph are situated together, thus making it difficult to identify robust clusters if a density-based clustering algorithm is used. This is also visualized in Fig. 2 (left), where one can see that the center of the graph is very densely populated.

It is interesting to investigate how the derived clusters differ in terms of the importance of the different features used in the construction of the initial hypergraph. The kernel density estimation of the different features in each cluster has been calculated for this purpose and visualized for three features in Fig. 4.

It can be observed that there is quite some variability across the clusters per feature, e.g., cluster 9 appears to have newer compressors, while the compressors in cluster 0 have been in use longer; cluster 11 is characterized by a higher *motor casing temperature high* in comparison to cluster 1; clusters 1 and 9 are characterized with higher *ambient temperatures high*, than clusters 3 and 11.

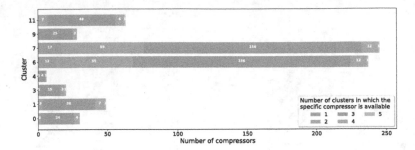

Fig. 3. Number of compressors (and their uniqueness) per cluster. The colors represent the compressor uniqueness within the clusters, e.g., one compressor represented in lime-green is present in 5 clusters $(1, 3, 6, 7, 11)$. (Color figure online)

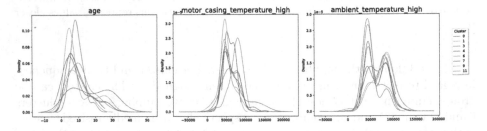

Fig. 4. Kernel density distributions of few selected features per cluster.

4.4   Step IV: Deriving KPIs to Analyze Performance

We also have a large sensor dataset at our disposal, which captures the compressors' performance in the field. Combining both data types (metadata and sensor time series) allows to derive performance-related KPIs, e.g., assess the percentage of time that the compressors are operating within the expected range as defined originally in the metadata. Four sensors are considered: *outlet pressure*, *internal pressure*, *outlet temperature*, and *ambient temperature*, and the results are shown in Fig. 5. The first three capture the compressor usage/internal behavior, while *ambient temperature* reflects the operational context.

Considering clusters 6 and 7 are the largest, they can be regarded as representing the typical (baseline) situation. It can also be observed that very few compressors operate at the desired/recommended ranges for *ambient temperature*. The KPIs based on other sensors show fewer periods outside the expected ranges, except for cluster 9 which appears to meet specifications for only around 60% of operating time for *outlet pressure*, and with the highest standard deviation for that feature. It is interesting to observe this is also the cluster that has the lowest compliance, 21% of the time, with respect to *ambient temperature* limits. These two deviations might be linked to one another, which is already an interesting insight to be subjected to further investigation by our industrial partners.

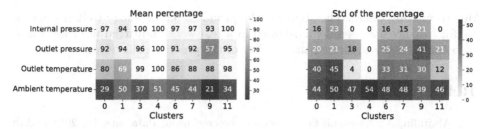

Fig. 5. The percentage of time, mean, and standard deviation (std), the compressors have been operating within the expected range for the different sensors.

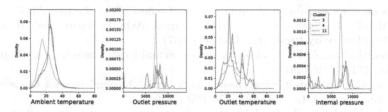

Fig. 6. Kernel density distribution of the sensor data of selected clusters.

In general, three different groups of clusters can be distinguished: 1) clusters 6 and 7 representing regular/baseline behavior; 2) clusters 3, 4, and 11 are almost always within range, indicating that these compressors are probably used in an environment where it is of high importance to work within the expected ranges; 3) clusters 0, 1 and 9 exhibit many more periods outside the expected ranges, probably containing compressors operating in less constrained contexts. Further characterization of these groups can be performed by examining the kernel density estimation of the different sensor values. For readability purposes, only group 2 (clusters 3, 4, and 11) is showcased in Fig. 6. It can be observed that cluster 4 is characterized by lower *ambient temperature* but higher *outlet temperature*, while these are the opposite for clusters 3 and 11. Cluster 4 is also characterized by very stable *outlet* and *internal pressures* (around 7000) supporting the hypothesis that some of these compressors are probably being used in some strictly regulated environments.

5   Conclusion

This work has presented a hybrid clustering methodology for analyzing and making sense of complex, multi-source heterogeneous data sets emerging nowadays from real-world industrial applications. Such data sets are typically, if not always, characterized by a high rate of missing values, which makes their analysis by traditional machine learning methods challenging. We have conceived a novel data exploration workflow incorporating concepts such as multi-layered clustering, hypergraph, shared nearest neighbor similarity, and k-medoids, allowing us to arrive at a multi-dimensional data partition without compromising data size due to missing values. The ability of the proposed methodology to derive meaningful insights has been demonstrated in a real-world industrial use case, which also convincingly confirmed the validity of the produced clustering solution.

References

1. Abdullin, A., Nasraoui, O.: Clustering heterogeneous data sets. In: 2012 Eighth Latin American Web Congress, pp. 1–8 (2012)
2. Ashari, I.F., et al.: Analysis of elbow, silhouette, Davies-Bouldin, Calinski-Harabasz, and rand-index evaluation on k-means algorithm for classifying flood-affected areas in Jakarta. J. Appl. Inform. Comput. **7**(1), 95–103 (2023)
3. Caldarelli, G.: Scale-Free Networks: Complex Webs in Nature and Technology. Oxford University Press, Inc., Oxford (2007)
4. Chao, G., Sun, S., Bi, J., et al.: A survey on multi-view clustering. arXiv preprint arXiv:1712.06246 (2017)
5. Chao, G., et al.: Multi-view cluster analysis with incomplete data to understand treatment effects. Inf. Sci. **494**, 278–293 (2019)
6. Fu, L., Lin, P., Vasilakos, A.V., Wang, S.: An overview of recent multi-view clustering. Neurocomputing **402**, 148–161 (2020)
7. Gamberger, D., Mihelčić, M., Lavrač, N.: Multilayer clustering: a discovery experiment on country level trading data. In: Džeroski, S., Panov, P., Kocev, D., Todorovski, L. (eds.) DS 2014. LNCS (LNAI), vol. 8777, pp. 87–98. Springer, Cham (2014). https://doi.org/10.1007/978-3-319-11812-3_8
8. de Goeij, M.C., et al.: Multiple imputation: dealing with missing data. Nephrol. Dial. Transplant. **28**(10), 2415–2420 (2013)
9. Handl, J., Knowles, J., Kell, D.B.: Computational cluster validation in post-genomic data analysis. Bioinformatics **21**(15), 3201–3212 (2005)
10. Liu, R., Wang, H., Yu, X.: Shared-nearest-neighbor-based clustering by fast search and find of density peaks. Inf. Sci. **450**(C), 200–226 (2018)
11. Pio, G., Serafino, F., Malerba, D., Ceci, M.: Multi-type clustering and classification from heterogeneous networks. Inf. Sci. **425**, 107–126 (2018)
12. Schlag, S., Heuer, T., Gottesbüren, L., Akhremtsev, Y., Schulz, C., Sanders, P.: High-quality hypergraph partitioning. ACM J. Exp. Algorithmics **27**, 1–39 (2023)
13. Wenz, V., Kesper, A., Taentzer, G.: Clustering heterogeneous data values for data quality analysis. J. Data Inf. Qual. **15**(3), 1–33 (2023)
14. Yang, M., et al.: Robust multi-view clustering with incomplete information. IEEE Trans. Pattern Anal. Mach. Intell. **45**(1), 1055–1069 (2022)
15. Zhang, X., et al.: Multi-type co-clustering of general heterogeneous information networks via nonnegative matrix tri-factorization. In: IEEE ICDM (2016)

Optimization

Efficient Lookahead Decision Trees

Harold Kiossou$^{(\boxtimes)}$, Pierre Schaus , Siegfried Nijssen , and Gaël Aglin

UCLouvain, Louvain-la-Neuve, Belgium
{harold.kiossou,pierre.schaus,siegfried.nijssen,gael.aglin}@uclouvain.be

Abstract. Conventionally, decision trees are learned using a greedy approach, beginning at the root and moving toward the leaves. At each internal node, the feature that yields the best data split is chosen based on a metric like information gain. This process can be regarded as evaluating the quality of the best depth-one subtree. To address the shortsightedness of this method, one can generalize it to greater depths. Lookahead trees have demonstrated strong performance in situations with high feature interaction or low signal-to-noise ratios. They constitute a good trade-off between optimal decision trees and purely greedy decision trees. Currently, there are no readily available tools for constructing these lookahead trees, and their computational cost can be significantly higher than that of purely greedy ones. In this study, we introduce an efficient implementation of lookahead decision trees, specifically LGDT, by adapting a recently introduced algorithmic concept from the MurTree approach to find optimal decision trees of depth two. Additionally, we utilize an efficient reversible sparse bitset data structure to store the filtered examples while expanding the tree nodes in a depth-first-search manner. Experiments on state-of-the-art datasets demonstrate that our implementation offers remarkable computation-time performance.

Keywords: Decision Trees · Lookahead · Optimization

1 Introduction

Decision tree-based learning algorithms are among the most widely used methods in machine learning, both for their predictive performance and because humans can relatively easily interpret these models. Finding an optimal tree (that minimizes the learning error) on a data set is an NP-hard problem. This is why learning a decision tree is usually done greedily, for example, using algorithms such as CART [5] and C4.5 [18]. A greedy algorithm selects an attribute in a node locally, starting from the root, down to the leaf nodes. The decision to select an attribute is made based on a single split and is never reconsidered later. In recent years, thanks to optimization solvers and new algorithmic ideas, approaches to infer optimal decision trees have been introduced [1–3,8,14,16,20,21].

These approaches have sparked excitement in the scientific community due to empirical evidence showing improved classification rates on unseen data [3]. Despite recent algorithmic enhancements, they struggle with inferring trees on

I. Miliou et al. (Eds.): IDA 2024, LNCS 14642, pp. 133–144, 2024.
https://doi.org/10.1007/978-3-031-58553-1_11

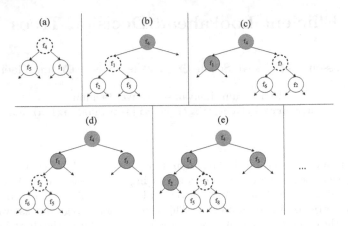

Fig. 1. Illustration of iterations to discover the feature in each node using level-two optimal decision trees. The level-two decision trees are rooted at dashed white nodes, while the fixed features are represented in dark-gray nodes. The tree is grown level-wise in this example. However, building it in a depth-first search order would result in the same final decision tree since each decision splits the data into two separate subproblems.

medium-sized datasets within reasonable computational times (e.g., German-credit dataset [2]). To address the limitations of both greedy and optimal decision trees, lookahead decision trees present a potential solution. These trees aim to overcome greedy decision tree shortcomings by considering future decisions during construction. A lookahead algorithm determines the next feature's decision in a less myopic, yet still greedy, manner based on sliding subtrees, as illustrated in Fig. 1 with a depth of two. Standard approaches like C4.5 use heuristics to decide each node's best one-level subtree or stump [13]. This concept can be extended to more than one level in lookahead decision trees [11]. Here, the attribute selected at a node corresponds to the root of the optimal or suboptimal two-level decision tree before moving to the next level. This allows the algorithm to anticipate the next step and make a more informed choice regarding the subsequent attribute. Setting the lookahead level to the maximum depth enables the identification of the optimal decision tree.

Despite the potential benefits of using lookahead approaches their adoption has been limited due to various factors. One of the primary concerns is the additional computation cost associated with lookahead, which requires substantial processing power and memory resources. Furthermore, while some studies have reported better performance using lookahead compared to greedy methods, there is no consistent evidence that lookahead consistently outperforms greedy counterparts across a range of applications. [11] argued that lookahead is more advantageous when there is high attribute interaction which was later confirmed and reinforced by [10], whose work demonstrated that the superior performance of lookahead decision trees is more pronounced when nonlinear relationships between feature pairs exist and when the signal-to-noise ratio is particularly low.

The enhanced capability of lookahead decision trees in handling feature interdependencies can result in more accurate predictive models. Despite their potential, there's a lack of widely available tools for researchers and practitioners to experiment with these methods. To address this, we introduce an efficient lookahead algorithm with remarkable computation runtimes. Our two-level lookahead approach achieves competitive error rates comparable to optimal trees, scaling to larger depths similar to standard greedy decision tree algorithms like CART or C4.5. We leverage an algorithmic concept from the MurTree algorithm [8] to identify optimal level-two decisions, adapting it for both information gain and misclassification error rate. Experiments on standard benchmarks show that this hybrid, less greedy approach strikes an excellent balance between tree learning speed and error rate. The paper structure includes a discussion of related work in Sect. 2, technical background in Sect. 3, an exploration of our methods in Sect. 4, and presentation of benchmark results in Sect. 5.

2   Related Work

Tree-based algorithms like CART and C4.5 often rely on heuristics for creating greedy decision trees, where attributes are chosen based on metrics like information gain or the Gini index. Despite their scalability, these trees may lack accuracy due to their myopic nature. To address this, lookahead searches, aim at optimizing upcoming iterations rather than just the next one. Norton [17] and Ragavan and Rendell [19] demonstrated successful results with lookahead, with the latter excelling in high attribute interaction scenarios at the cost of increased computational complexity. Esmeir and Markovitch [11] introduced ID3-k and LSID3 as lookahead strategies. ID3-k calculates k-level information gain (gain-k) at each node, selecting the attribute maximizing it for further splits. LSID3, using dynamic lookahead and a shallower tree-favoring criterion, surpasses greedy algorithms, especially with more available time. Relying on this, Donick et al. [10] introduced a random forest approach that is a stepwise lookahead variation. It considers three split nodes simultaneously in tiers of depth two, enhancing the identification of feature interdependencies. The lookahead algorithm outperforms the greedy algorithm in cases involving non-linear relationships between feature pairs and a low signal-to-noise ratio.

Thanks to hardware advancements and optimization solvers, various approaches for learning optimal decision trees have emerged. These fall into categories like mixed integer programming [1,3,4,21], constraint programming [20], SAT solvers [15], and dynamic programming [2,8,16]. Among these, dynamic programming methods, like DL8 [16], are considered the fastest and most accurate in a depth-constrained setting. DL8 uses a caching technique to save obtained subtrees, optimizing performance. DL8.5 [2] enhances DL8 with an upper-bound strategy and a lower-bound technique to efficiently explore sub search spaces. Further advancements by MurTree [8] include limiting tree nodes, an efficient depth-two tree computation, and a novel similarity-based lower bounding approach.

3   Technical Background

We consider binary datasets in which each feature is a value in the set $\{0,1\}$. Let $\mathcal{F}$ be the set of n features and $\mathcal{C} = \{+,-\}$ the set of two classes[1]. A binary dataset is defined by: $\mathcal{D} = \{(\boldsymbol{f}_i, c_i), \forall i \in \{1, 2, \ldots |D|\}\}$, where $\boldsymbol{f}_i = (f_{(i,1)}, f_{(i,2)}, \ldots, f_{(i,n)}) \in \{0,1\}^n$ is a feature vector of length n and $c_i \in \mathcal{C}$ is the class label for the i-th instance. Thus, each instance in $\mathcal{D}$ is represented by a feature vector and a corresponding class label, and the feature vectors are composed of elements from the set of features $\mathcal{F}$. Given a feature vector $\boldsymbol{f}$, $f_k = 1$, indicating the presence of f_k, is denoted as $f_k \in \boldsymbol{f}$ and $\bar{f}_k \notin \boldsymbol{f}$ otherwise for its absence (we ignore i for simplicity). The binary dataset $\mathcal{D}$ can be partitioned into the positive class of instances $\mathcal{D}^+$ and the negative class $\mathcal{D}^-$ such that $\mathcal{D} = \mathcal{D}^+ \cup \mathcal{D}^-$. The MurTree algorithm [8] for finding optimal decision trees of arbitrary depth uses a specialized algorithm for level-two decision trees whenever it reaches the one level before the last layer (the final trees have a fixed known limit on depth). This specialized level-two optimal algorithm is presented next for the sake of completeness.

3.1   Level-Two Specialized Algorithm

MurTree uses a dynamic programming approach with the upper-bound notion of DL8.5 to determine the optimal decision trees. It works in two phases. In the first phase, it computes the frequency for each pair of features (f_i, f_j). Let $FQ^+(f_i)$ and $FQ^+(f_i, f_j)$ be the frequency counts in positive instances for a single feature f_i and a pair of features (f_i, f_j), respectively. $FQ^-(f_i)$ and $FQ^-(f_i, f_j)$ are defined analogously for negative instances. Note that, based on $FQ(f_i)$ and $FQ(f_i, f_j)$, it is possible to compute $FQ(\bar{f}_i)$, $FQ(f_i, \bar{f}_j)$, $FQ(\bar{f}_i, f_j)$, and $FQ(\bar{f}_i, \bar{f}_j)$ without explicit counting. The frequency counts FQ^+ can be computed in $\mathcal{O}(|\mathcal{D}+| \cdot m_+^2)$ time with m_+ the maximum number of features in a single positive instance [8]. In the second phase, the tree is computed using the missclassification score $MS(f_i, f_j) = \min \{FQ^+(f_i, f_j), FQ^-(f_i, f_j)\}$ for a couple of features (f_i, f_j). Given a depth two decision tree, let MS_{left} and MS_{right} denote the misclassification scores of the left and right subtrees:

$$MS_{left}(f_{root}, f_{left}) = MS(\bar{f}_{root}, \bar{f}_{left}) + MS(\bar{f}_{root}, f_{left}) \tag{1}$$

$$MS_{right}(f_{root}, f_{right}) = MS(f_{root}, \bar{f}_{right}) + MS(f_{root}, f_{right}), \tag{2}$$

where f_{root} is the feature of the root node, f_{left} and f_{right} are the features of the left and right-child nodes.

---

[1] All the formula of this section can easily be adapted for multi-class contexts.

The procedure then iterates over all pairs of (f_i, f_j) of $\mathcal{F}$ to find the triplet $(f_{root}, f_{left}, f_{right})$ independently optimized for the left and right branches using the following equation:

$$
\min_{f_{root}, f_{left}, f_{right} \in \mathcal{F}} MS(f_{root}, f_{left}, f_{right})
$$
$$
= \min_{f_{root}}[\min_{f_{left} \in \mathcal{F}} MS_{left}(f_{root}, f_{left}) \tag{3}
$$
$$
+ \min_{f_{right} \in \mathcal{F}} MS_{right}(f_{root}, f_{right})].
$$

The algorithm iterates through each pair (f_{root}, f_{child}), computes the misclassification score of the left subtree using Eq. (1), updates the best left child for the feature f_{root}, and performs the same operation for the right child. Each subtree can be computed in $\mathcal{O}(|\mathcal{F}^2|)$ time. The global complexity of the algorithm is then $\mathcal{O}(|\mathcal{D}| \cdot m^2 + |\mathcal{F}^2|)$ with m the upper limit on the number of features in any single positive and negative instance.

3.2 Level-Two Lookahead Information Gain

In our algorithm, we can also use a level-two lookahead on information gain. This is based on the work of [11], where they developed ID3-k, which uses a level-k lookahead for information gain at each node to evaluate each feature. To fully take advantage of the level-two specialized algorithm, we only do a lookahead of two as it is easy to compute the different frequencies using $FQ^+(f_i)$ and $FQ^+(f_i, f_j)$ and $FQ^-(f_i)$ and $FQ^-(f_i, f_j)$. The information gain is then computed using the following Eq. (4):

$$
\max_{f_{root}, f_{left}, f_{right} \in \mathcal{F}} IG(f_{root}, f_{left}, f_{right})
$$
$$
= \max_{f_{root}}[\max_{f_{left} \in \mathcal{F}} IG_{left}(f_{root}, f_{left}) \tag{4}
$$
$$
+ \max_{f_{right} \in \mathcal{F}} IG_{right}(f_{root}, f_{right})],
$$

with:

$$
IG(f_{root}, f_{left}) = H(f_{root}) - \sum_{i=1}^{2} \frac{|leaf(f_{root}, f_{left})_i|}{|root|} H(leaf(f_{root}, f_{left})_i), \tag{5}
$$

$$
IG(f_{root}, f_{right}) = H(f_{root}) - \sum_{i=1}^{2} \frac{|leaf(f_{root}, f_{right})_i|}{|root|} H(leaf(f_{root}, f_{right})_i), \tag{6}
$$

and

$$
H(S) = - \sum_{i=1}^{c} p_i \log_2(p_i), \tag{7}
$$

where $IG(f_{root}, f_{child})$ is the depth two information gain of a branch from f_{root} to f_{child} and $child \in \{left, right\}$. $leaf(f_{root}, f_{child})_i$ corresponds to the leaves

---

Algorithm 1: LGDT(D, *minsup*, *maxdepth*)

---

1 **if** *maxdepth* ≤ 2 **then return** opt_dt(D, *maxdepth*)
2 *solution* $\leftarrow$ Tree()
3 *sub_tree* $\leftarrow$ opt_dt(D, *2*)
4 *solution*.root $\leftarrow$ make_tree(*sub_tree*.root)
5 Recursion(*solution*.root, *maxdepth* $- 1$, D)
6 **return** *solution*
7 **Procedure** Recursion(*node, depth, D*)
8   **if** *depth* > 0 **and** *node*.error > 0 **then**
9   *left* $\leftarrow$ *node*.left
10   *right* $\leftarrow$ *node*.right
11   **if** $FQ(left) \geq minsup$ **and** $FQ(right) \geq minsup$ **then**
12   *ws* $\leftarrow$ min(2, *depth*)
13   D.save()
14   D.project(*left*)
15   *sub_tree* $\leftarrow$ opt_dt(D, *ws*)
16   *node_tree* $\leftarrow$ make_tree(*sub_tree*.root)
17   *child* $\leftarrow$ *tree*.set_node(*left, node_tree*)
18   Recursion(*child, depth* $- 1$, D)
19   D.restore()
20   D.save()
21   D.project(*right*)
22   *sub_tree* $\leftarrow$ opt_dt(D, *ws*)
23   *node_tree* $\leftarrow$ make_tree(*sub_tree*.root)
24   *child* $\leftarrow$ *tree*.set_node(*right, node_tree*)
25   Recursion(*child, depth* $- 1$, D)
26   D.restore()
27   end
28   end

---

of branches (f_{root}, f_{child}), $H(S)$ the entropy of a node or leaf S with p_i the probability that an element belongs to the class i. We independently maximize the information gain of each branch of the tree to obtain a tree with the highest information gain.

4   Less Greedy Decision Trees

This section presents Less Greedy Decision Trees (LGDT), a more informed decision tree algorithm than classical greedy algorithms. The pseudocode of LGDT is described in Algorithm 1.

It fixes the feature of each node of the depth-limited tree in a depth-first way. In each node, it relies on the use of a generated depth-two decision tree based on the current data using the opt_dt method(lines 1, 3, 15 and 22). It uses D, the current data subset in a node, then computes FQ^+ and FQ^- and uses both to build a level-two decision tree based on one of the following approaches:

- The MurTree level-two specialized algorithm which returns an optimal decision tree minimizing the misclassification rate;
- a depth-two information gain tree taking advantage of the level-two specialized algorithm.

When the maximum depth is 1, the sliding window returns the depth-one tree with the feature optimizing the evaluated metric. At first, when building up to a depth-two tree, it is enough to return the tree generated by the sliding window (line 1). On the other hand, for deeper trees, a subtree is generated using D and the sliding window (line 3). The root node of the solution tree is set to the root of the subtree (line 4). The make_tree function generates for the feature and the data its two leaves and computes the misclassification error. Recursion is then called in a depth-first search fashion starting from the root node to build the decision tree. An internal node is only refined if the depth constraint is respected and the node is not pure (line 8). Otherwise, the recursion is stopped, and the node is considered as a leaf node. Moreover, before expanding a node the search ensured the number of data falling in its left and right branches (9-10) respects the minimum support constraint at line 11.

The algorithm will then be called recursively on the left and right parts. Lines 14 and 21 will update the data representation D to be able to list all examples falling respectively in *left* and *right*. Using the updated D the sliding window will build a subtree for the left and right (lines 15 and 22). The algorithm will append the sub-node trees (lines 16 and 23) to the parent node (lines 17 and 24). The recursive method is then called on the child with the current data representation D and an increment to the depth (lines 18 and 25).

In our implementation, we use a special data representation to efficiently store and process the input data. By using this representation, we aim to improve the performance and scalability of the algorithm, while reducing the memory consumption allowing us to efficiently generate decision trees for large datasets.

Data representation. Table 1 summarizes various dataset representations. Each example, identified by a unique *tid*, is linked to a feature set (Feats). The *tid-list* groups tids into example subsets. The boolean representation (Table 1a) uses 0s and 1s to show feature presence (1) or absence (0) per example. The horizontal representation (Table 1b) records present features. The vertical representation [12] (Table 1c) aligns rows with features, simplifying *tid-list* length computation. Intersection operations on *tid-lists* [12] locate transactions covered by two features, e.g., A, C arises from $t_2, t_3 \cap t_1, t_2, t_3 = t_2, t_3$.

We use intersection operations to significantly reduce data processing, saving time and memory. In this work, using the vertical data representation enhances algorithm efficiency, and the reversible sparse bitset structure reduces unnecessary computation and enables fast bitwise operations.

Bitsets and Bitwise Operations. In the vertical representation, the feature *tid-list* can be stored using arrays or bitsets. Arrays group all integer values (*tid*) associated with a feature, while bitsets use bits to represent each possible *tid* value.

Table 1. Dataset representations.

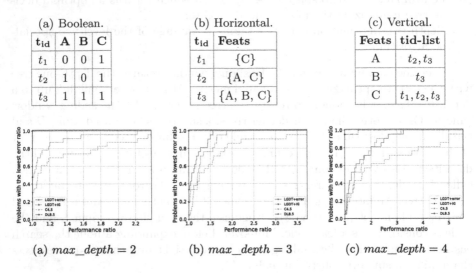

(a) Boolean.

t_{id}	A	B	C
t_1	0	0	1
t_2	1	0	1
t_3	1	1	1

(b) Horizontal.

t_{id}	Feats
t_1	{C}
t_2	{A, C}
t_3	{A, B, C}

(c) Vertical.

Feats	tid-list
A	t_2, t_3
B	t_3
C	t_1, t_2, t_3

(a) $max\_depth = 2$ (b) $max\_depth = 3$ (c) $max\_depth = 4$

Fig. 2. Performance Profile plots comparing the error rate of the two versions of LGDT against C4.5 and DL8.5.

A bitset is the size of the dataset ($|D|$), with each bit at index i indicating the presence (1) or absence (0) of the example at $tid = i$ in the feature *tid-list*. For a dataset with eight examples ($tid \in \Omega = \{1, 2, \ldots, 8\}$), an array retains valid values post-operations, while a bitset maintains the same size, setting unnecessary values to 0.

When dealing with large sets, bitsets are more memory-efficient than arrays due to each integer in an array requiring at least 32 or 64 bits, while a bitset needs just 1 bit per integer. Bitsets are especially useful in frequent itemset mining algorithms [6], where bitwise operations like counting and intersection are essential and the number of elements important than the data. These operations are often optimized for 64-bit processing, making bitsets ideal for storing parameters that require such operations. As a result, many data structures use an array of 64-bit bitsets, commonly referred to as *words*, to take advantage of these optimizations.

Reversible Sparse Bitset. The decision tree undergoes expansion via a recursive depth-first search. After data operations such as the `project` in Algorithm 1, the number of elements in the bitset decreases resulting in a sparser bitset. Elements are restored on backtracking from the recursive calls.

The Reversible Sparse Bitset (RSBS) [7] exploits bitset sparsity down the tree and is able to efficiently restore elements during backtracking. Operations are performed per 64-bit word for improved performance. RSBS employs sparse bitsets, separating empty from non-empty words, eliminating unnecessary counting of empty words. This sparsity aids intersection operations, ensuring unnecessary

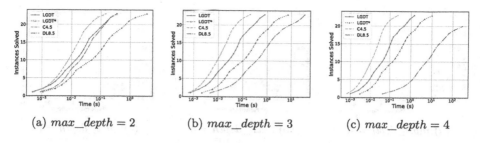

(a) *max_depth* = 2   (b) *max_depth* = 3   (c) *max_depth* = 4

Fig. 3. Cumulative number of dataset (y-axis) for which the decision tree algorithm has terminated within the time limit (x-axis).

intersections with a 0 bit are avoided. Moreover RSBS employs a reversibility technique from constraint programming solvers, enabling the data structure to recover previous states during searches, thus avoiding the overhead of copying parameters between parent and child nodes.

When backtracking, the RSBS can revert to a previous state using its trail of changes, which is implemented with a stack. In practice, only the size of the number of non-empty words needs to be restored to retrieve the correct partitioning between empty and non-empty words. Throughout the search, the entire algorithm operates with a single instance of the data structure, facilitating incremental changes at each step and ensuring consistency when exploring multiple paths. The stack chronicles successive changes, and during backtracking, the top layers are removed to revert to prior states.

LGDT uses a single instance of the RSBS(D) to maintain the data in the current node. Before going further down, the state is saved (lines 13 and 20) and the state bit vector is projected (using bitwise AND operations) with the next node feature (a fixed precomputed bit vector) (lines 14 and 21). Each projection reduces the dataset by filtering the examples that satisfy the condition of the selected feature. Before exploring a node and proceeding to the second split, the previous state must be restored, as indicated in lines 19 and 26. This save-and-restore mechanism leverages the internal stack state of the RSBS. Every save action corresponds to a push on the stack, and every restore action equates to a pop, allowing for the restoration of the partitioning between empty and non-empty words in constant time.

5   Results

This section presents the results of the experiments we have conducted. The source code and data used for the experiments in the paper are available at https://github.com/haroldks/pytrees. Our experiments compare three decision tree learning algorithms: C4.5, DL8.5[2], and LGDT. LGDT had two implementations based on data representation. The first, labeled as LGDT, uses a reversible

---

[2] https://github.com/aia-uclouvain/pydl8.5.

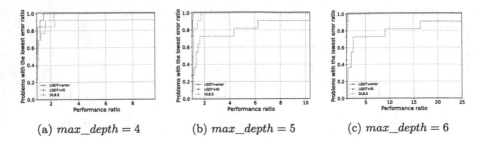

(a) $max\_depth = 4$ (b) $max\_depth = 5$ (c) $max\_depth = 6$

Fig. 4. Performance Profile plots comparing the error rates of LGDT and DL8.5 on large datasets.

sparse bitset structure, while the second, `LGDT*`, follows the original algorithm by Esmeir and Markovitch (2004) using a double loop and boolean data view. For comparison, we include a scikit-learn implementation of C4.5[3]. DL8.5 was executed with a 10-minute time limit. All experiments were conducted on an Intel i5-1245U machine with 16 GB RAM running Arch Linux. The study involved 23 discretized datasets from CP4IM[4], with a minimum support of 1.

First, we proceed with an experiment to determine the proximity of LGDT tree errors to optimality. Using a performance profile [9], we contrast error rates of two LGDT versions (`LGDT+error` using a level-two specialized optimal tree algorithm, and `LGDT+IG` with level-two information gain optimization) with those of C4.5 and DL8.5 on the aforementioned datasets, based on the training set. This aims to ascertain LGDT's viability as an alternative to C4.5 and to gauge its divergence from DL8.5. Figure 2 illustrates performance profiles for instances with maximum depths of 2, 3, and 4. The performance profile is a cumulative distribution of an algorithm's $s \in S$ enhanced performance versus other algorithms in set S across a problem set P: $p_s(\tau) = \frac{1}{|P|} \times | \{p \in P : r_{p,s} \leq \tau\} |$, with the performance ratio $r_{p,s} = \frac{t_{p,s}}{\min\{t_{p,s}|s\in S\}}$ where $t_{p,s}$ signifies each algorithm's error rate. The performance profile is then visualized by plotting, for each algorithm, the proportion of solved instances with the lowest error rate on the y-axis against the difference from the lowest error rate on the x-axis. Results reveal that, regardless of depth, LGDT consistently achieves superior error rates compared to C4.5 across all datasets. At depth 2, both `LGDT+error` and DL8.5 immediately secure the lowest error rates as they yield optimal trees in this context. `LGDT+IG` holds the lowest error rate for about 50% of instances, while C4.5 achieves it on around 30%. Moreover, permitting an error rate roughly double the best, `LGDT+IG` solves all instances, whereas C4.5 remains unable. With increasing depth, DL8.5 consistently outperforms others with the lowest error rate, except for depth 4, where it times out on some. The error gaps widen with depth, reflecting escalating performance ratios. Notably, the LGDT-C4.5 gap widens faster than DL8.5-LGDT, showcasing LGDT's greater reliability than C4.5. This may

---

[3] https://scikit-learn.org/.

[4] https://dtai.cs.kuleuven.be/CP4IM/datasets/.

be due to C4.5's inclination to make erroneous or sub-optimal decisions with deeper trees, increasing the capabilities of LGDT to mitigate underfitting.

Figure 3 illustrates the cumulative termination count of each algorithm. Notably, C4.5 exhibited the fastest performance, consistently producing decision trees within a second, regardless of depth. At a depth limit of 2, LGDT and DL8.5 displayed similar performance, as both returned optimal trees at this depth. DL8.5 and LGDT performed similarly due to their shared utilization of the level-two specialized algorithm and reversible sparse bitset data structure. LGDT* lagged behind the RSBS implementation, attributed to individual example checks for supported examples during each projection. With increasing depth, C4.5 maintained its speed edge, while DL8.5 progressively slowed and ultimately couldn't solve all instances within a 10-minute window starting from depth 4. Furthermore, when comparing LGDT implementations across depths, LGDT outperformed LGDT* by an average factor of 10.

To evaluate LGDT's effectiveness compared to optimal decision trees on large datasets, we conducted a final experiment using 15 classification datasets from the UCI Repository. These datasets had at least 30 features, binarized to create datasets with a minimum of 300 features. Experiments covered tree depths of 4, 5, and 6, with DL8.5 constrained to a 30-second runtime to highlight challenging scenarios. The performance profile on the runtime in Fig. 4 revealed that at depth 4, DL8.5 outperformed in about 40% of instances, but this advantage diminished and disappeared at depth 6. DL8.5 consistently lagged behind LGDT, and as depth increased, the error gap widened, showcasing LGDT's superior reliability for deeper trees. DL8.5's declining performance with depth is attributed to its tendency to become trapped in deeper search tree sections, prioritizing optimal features over comprehensive search space coverage. This focus results in substantial unexplored segments, leading to suboptimal outcomes compared to LGDT approaches.

6 Conclusion

In this paper, we proposed an efficient lookahead algorithm for constructing decision trees that can capture feature interdependencies within binary datasets. To offer the best computation time, and become a viable and practical alternative over pure greedy methods, the algorithm relies on two algorithmic ideas: the Murtree level-two specialized algorithm and the reversible sparse-bitset data structure also used in DL8.5. Through experiments on various datasets, we compared the performance of our algorithm with two state-of-the-art decision tree methods, C4.5 and DL8.5. Our results suggest that lookahead decision trees can be a valuable addition to the toolkit of data scientists. In future research, we aim to explore effective techniques for managing continuous features instead of binarizing them in advance.

References

1. Aghaei, S., Gómez, A., Vayanos, P.: Strong optimal classification trees. ArXiv Preprint ArXiv:2103.15965 (2021)
2. Aglin, G., Nijssen, S., Schaus, P.: Learning optimal decision trees using caching branch-and-bound search. Proc. AAAI. **34**, 3146–3153 (2020)
3. Bertsimas, D., Dunn, J.: Optimal classification trees. Mach. Learn. **106**, 1039–1082 (2017)
4. Boutilier, J., Michini, C., Zhou, Z.: Shattering inequalities for learning optimal decision trees. In: International Conference on Integration of Constraint Programming, Artificial Intelligence, and Operations Research, pp. 74–90 (2022)
5. Breiman, L., Friedman, J., Olshen, R., Stone, C.: Classification and regression trees. Wadsworth Int. Group. **37**, 237–251 (1984)
6. Burdick, D., Calimlim, M., Gehrke, J.: MAFIA: a maximal frequent itemset algorithm for transactional databases. In: Proceedings 17th International Conference On Data Engineering, pp. 443–452 (2001)
7. Demeulenaere, J., et al.: Compact-table: efficiently filtering table constraints with reversible sparse bit-sets. In: Rueher, M. (ed.) CP 2016. LNCS, vol. 9892, pp. 207–223. Springer, Cham (2016). https://doi.org/10.1007/978-3-319-44953-1_14
8. Demirović, E., et al.: MurTree: optimal decision trees via dynamic programming and search. J. Mach. Learn. Res. **23**, 1–47 (2022)
9. Dolan, E., Moré, J.: Benchmarking optimization software with performance profiles. Math. Program. **91**, 201–213 (2002)
10. Donick, D., Lera, S.: Uncovering feature interdependencies in high-noise environments with stepwise lookahead decision forests. Sci. Rep. **11**, 9238 (2021)
11. Esmeir, S., Markovitch, S.: Lookahead-based algorithms for anytime induction of decision trees. In: Proceedings Of The Twenty-first International Conference On Machine Learning, p. 33 (2004)
12. Holsheimer, M., Kersten, M., Mannila, H., Toivonen, H.: A perspective on databases and data mining. In: KDD, vol. 95, pp. 150–155 (1995)
13. Iba, W., Langley, P.: Induction of one-level decision trees. Mach. Learn. Proc. **1992**, 233–240 (1992)
14. Lin, J., Zhong, C., Hu, D., Rudin, C., Seltzer, M.: Generalized and scalable optimal sparse decision trees. In: ICML, pp. 6150–6160 (2020)
15. Narodytska, N., Ignatiev, A., Pereira, F., Marques-Silva, J., Ras, I.: Learning optimal decision trees with SAT. In: IJCAI, pp. 1362–1368 (2018)
16. Nijssen, S., Fromont, E.: Mining optimal decision trees from itemset lattices. In: KDD, pp. 530–539 (2007)
17. Norton, S.: Generating better decision trees. In: IJCAI, vol. 89, pp. 800–805 (1989)
18. Quinlan, J.: C4.5: Programs for Machine Learning. Elsevier (2014)
19. Ragavan, H., Rendell, L.: Lookahead feature construction for learning hard concepts. In: ICML (1993)
20. Verhaeghe, H., Nijssen, S., Pesant, G., Quimper, C., Schaus, P.: Learning optimal decision trees using constraint programming. Constraints **25**, 226–250 (2020)
21. Verwer, S., Zhang, Y.: Learning optimal classification trees using a binary linear program formulation. Proc. AAAI. **33**, 1625–1632 (2019)

Learning Curve Extrapolation Methods Across Extrapolation Settings

Lionel Kielhöfer[1], Felix Mohr[2], and Jan N. van Rijn[1]($\boxtimes$)

[1] Leiden Institute of Advanced Computer Science (LIACS), Leiden University,
Leiden, Netherlands
jvrijn@liacs.nl

[2] Universidad de La Sabana, Chia, Colombia

Abstract. Learning curves are important for decision-making in supervised machine learning. They show how the performance of a machine learning model develops over a given resource. In this work, we consider learning curves that describe the performance of a machine learning model as a function of the number of data points used for training. It is often useful to extrapolate learning curves, which can be done by fitting a parametric model based on the observed values, or by using an extrapolation model trained on learning curves from similar datasets. We perform an extensive analysis comparing these two methods with different observations and prediction objectives. Depending on the setting, different extrapolation methods perform best. When a small number of initial segments of the learning curve have been observed we find that it is better to rely on learning curves from similar datasets. Once more observations have been made, a parametric model, or just the last observation, should be used. Moreover, using a parametric model is mostly useful when the exact value of the final performance itself is of interest.

Keywords: Learning Curves · AutoML · Supervised Learning

1 Introduction

Learning curves are used for various types of decision-making in supervised learning. They show how the performance of a machine learning algorithm develops over a given resource, for example, the number of epochs, run time, or number of data points used for training. In this work, we look specifically at learning curves across data points. They are generally used in the following three decision-making situations [13, 20]:

1. Data-acquisition [10, 21]; in this situation the focus is on predicting if acquiring additional data would significantly increase the performance of a machine learning algorithm.
2. Early stopping [7, 17]; for determining when training a given machine learning model on more budget would not cause significant improvement.

© The Author(s), under exclusive license to Springer Nature Switzerland AG 2024
I. Miliou et al. (Eds.): IDA 2024, LNCS 14642, pp. 145–157, 2024.
https://doi.org/10.1007/978-3-031-58553-1_12

3. Early-discarding [8,14,19]; this situation focuses on picking the best learning algorithm from a set of options. Instead of training all of the learning algorithms on the entire dataset, which is computationally costly, they are trained on increasing budgets. The learning algorithms that are predicted to perform the worst can then be discarded early on.

In all decision situations, we typically have an incomplete learning curve and want to extrapolate how the performance of the algorithm develops when more budget is provided. Various model types are capable of extrapolating such curve segments, for example, parametric models [5], meta-learning models [2,11] or specialized classifiers [9]. Parametric models are usually fit to the points of the curve segment and can then be used to make extrapolations. A popular parametric model used for learning curves is the inverse power law (IPL) [4,7]. Alternatively, meta-learning models utilize learning curves from earlier encountered datasets [11,18]. For example, MDS considers a given pair of algorithms, utilizes learning curves of the same algorithms from related datasets, and combines these to extrapolate the learning curves on the current dataset [11]. Mohr et al. [15] performed an in-depth comparison between parametric models for the extrapolation of learning curves. They empirically evaluate several parametric models over various datasets, and conclude that the IPL model is outperformed by the Morgan-Mercer Flodin model (MMF). This extended the work of Gu et al. [5], who had previously performed this comparison on a smaller database with fewer parametric models and came to the opposite conclusion.

This paper complements the previous studies of Mohr et al. [15] by comparing parametric extrapolation with meta-learning-based extrapolation. We investigate when it is beneficial to use a parametric model versus a meta-learning model across various extrapolation settings, which are defined by the available learning curve segment, the prediction target (the training set size to be extrapolated to) and the prediction objective. We consider the following questions:

1. How does the available *curve segment* (i.e., the size of the learning curve that is already determined) influence the performance of parametric and meta-learning models for the extrapolation of learning curves?
2. How does the *prediction target* (i.e., the point towards which the learning curve needs to be extrapolated, this is typically determined by the size of the dataset) influence the performance of the aforementioned models?
3. In some situations we need the exact performance prediction, resulting in a regression task, and in other situations it suffices to select the best algorithm from a set of several options, resulting in a classification task; how does this *decision situation* influence the performance of the extrapolation models?

To compare these two extrapolation model types we take the best-performing parametric model according to current insights from the literature, namely MMF [5]. We compare it to the Meta-Learning on Datasamples (MDS) model developed by Leite and Brazdil [11]. This is the most fundamental model we are aware of that utilizes learning curves from other datasets. As an additional baseline, we include a simple model that horizontally extrapolates from the last observation in the curve segment. Of course, our results are conditioned on the choice of model per category (i.e., MDS, MMF, and the baseline), which leaves room for a more general study that includes more model types per category.

We find that using learning curves on other datasets with the meta-learning model is beneficial when only a few parts of the learning curve have been observed. However, when more observations have been made, a parametric model, or just the last observation, outperforms the meta-learning model. We further find that using a parametric model is in particular beneficial when the objective is to predict the exact performance of a classifier. When the objective is to pick the best algorithm from a set of two algorithms, the surprising finding is that simply choosing the algorithm with the best score on the last curve segment is better than picking the best one according to an extrapolation obtained from a parametric model, no matter how few observations have been made.

2   Related Work

Learning curve extrapolation is performed in many different contexts of machine learning. Various methods have been developed in this regard. Our main focus is on learning curve extrapolation methods that can be used in general, and which are usually tested on simple machine learning algorithms. In this section we show research done into learning curve extrapolation in the different contexts.

Parametric Models for Learning Curves: Learning curves have often been modelled by low-parameter models. Power-law models have successfully been used in model selection to efficiently allocate resources to promising models [14]. They have also been used for early-stopping to stop the training of models once it is highly probable that their performance will not significantly increase [7]. Theoretical works into the shapes of learning curves back up this line of work, as they suggest that learning curves usually have power-law behaviour, however, they also suggest that learning curves can have exponential behaviour [1,6]. Mohr et al. [15] compare all the parametric models used for learning curves that they could find in the literature. They perform this comparison by looking at the model selection capabilities of each learning curve model. They find that there are parametric models that can outperform power-law and exponential models.

Alternatively, Klein et al. [9] developed a classifier specialized for learning curve extrapolation, i.e., a Bayesian neural network that has built-in prior information about the aforementioned parametric models.

Meta-Learning Models for Learning Curves: Meta-learning models try to leverage knowledge across datasets [2]. Given a partial learning curve of a given algorithm on a given dataset and several complete learning curves of this algorithm on different datasets, the goal is to extrapolate the partial learning curve. The Meta-Learning on Datasamples (MDS) method addresses this problem [11]. It introduces a distance measure between datasets, to select datasets for which

Dataset: kr-vs-kp, Learner: SVCs, Last anchor in curve segment: 256, Target anchor: 1448

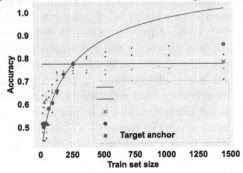

Fig. 1. Learning curve extrapolation performed by MMF, MDS (k=4) and Last for the given curve segment and target anchor. The blue line is the extrapolation performed by MMF. The green line is the extrapolation performed by Last. The smaller circles represent the k-nearest curves MDS uses to make its prediction, which is given by the red cross. (Color figure online)

the completed learning curve is known. They use this to pick out the k closest learning curves. The mean of these k curves at the target is then taken as the extrapolation. Leite and Brazdil [12] build upon their work by also including meta-features of the datasets. These are features that describe the dataset and can be used to find similarities between datasets. Van Rijn et al. [18] build on this work by introducing an algorithm that can rank a portfolio of classifiers. Chandrashekaran and Lane [3] introduce a method similar to MDS [11] for hyperparameter optimization. The main difference is that the distance measure is to completed learning curves on other hyperparameter settings instead of datasets.

3   Extrapolation Methods

We follow Mohr and van Rijn [13] in using the term *anchor point* or *anchor* to refer to a specific dataset size. We denote the performance value (accuracy in our case), for a learning algorithm a at an anchor point s of dataset d as $\hat{C}_d(a, s)$, or simply $\hat{C}(a, s)$ in case the dataset is implicitly clear. This value is the estimated out-of-sample accuracy of a model obtained with learning algorithm a and using s training examples, which is acquired using a 5-fold Monte-Carlo Cross Validation (MCCV) [15]. The extrapolation methods that we use are: Morgan-Mercer Flodin (MMF), Meta-Learning on Datasamples (MDS) and the performance at the last available anchor. We refer to the baseline that predicts the performance on the last seen anchor as 'Last'. Next, we will give an overview of these methods. Figure 1 shows each method used on a learning curve.

Morgan-Mercer Flodin (MMF): MMF is a parametric model given by $f_\theta = (ab + cx^d)/(b + x^d)$ where $\theta := (a, b, c, d)$. Given the anchor points $s_1, ..., s_m$ for a curve segment, the loss associated with parameters θ for the learning curve of some learning algorithm a is $loss(\theta) = \sum_{i=1}^{m} w_i(\hat{C}(a, s_i) - f_\theta(s_i))^2$ where w_i is the weight given to anchor s_i. Here, we follow Mohr et al. [15] and set $w_i = 2^i$, which implies that the i-th anchor is more important than *all* the anchors before s_i together. Exploring other weighing schemes could be interesting for future work. Mohr et al. [15][1] use the Levenberg-Marquadt algorithm, we use the AdamW optimizer in PyTorch [16] to optimize for the parameters of the model instead. We experimentally found that this performs better.

Meta-Learning on Datasamples (MDS): To predict the better performing learning algorithm out of two, MDS [11] uses learning curves from these algorithms on other datasets.[2] This approach is parameter-free, so predictions can only be made for anchors that have an observed performance on the other datasets. As such, it is required that the curves obtained from other datasets contain performance information for the target anchor. Among all the available curves of other datasets, the most relevant ones for the prediction task are determined based on the similarity of the already observed curve segments for both learners. Suppose that we want to predict which of algorithms a_1 and a_2 performs better at a given target anchor, we have measured their performance on dataset d for anchors $s_1, ..., s_m$. Then the (lack of) relevance of another dataset d' for the prediction task can be assessed through the similarity of the learning curves, which in turn can be measured as the squared anchor-wise deviations summed for both learners $R_{a_1, a_2}(d, d') = \sum_{i=1}^{m}(\hat{C}_d(a_1, s_i) - \hat{C}_{d'}(a_1, s_i))^2 + \sum_{i=1}^{m}(\hat{C}_d(a_2, s_i) - \hat{C}_{d'}(a_2, s_i))^2$. Given that the k nearest datasets are $d_1, ..., d_k$, MDS takes the mean performance of a_1 and a_2 at the target anchor s on these datasets and makes the extrapolations $\frac{1}{k}\sum_{i=1}^{k}\hat{C}_{d_i}(a, s)$ for both algorithms. The better performing algorithm is then predicted to be the one with the better extrapolated performance. Leite and Brazdil [11] improve their method by scaling curves on other datasets before using the distance measure. Given a learning curve on a dataset $d' \neq d$, each point of that curve is multiplied by the following constant $f = \frac{\sum_{i=1}^{N}(\hat{C}_d(a, s_i) \cdot \hat{C}_{d'}(a, s_i)) \cdot w_i)}{\sum_{i=1}^{N}(\hat{C}_{d'}(a, s_i) \cdot w_i)}$. We also include this scaling in our implementation. The weighting mechanism used by the authors is given by $w_i = i^2$. However, we again use $w_i = 2^i$ as initial experiments suggested this improves performance.

It should be noted that the proposed method is intended for the task of (binary) algorithm selection. We also explore the situation in which we have a single algorithm, and want to predict the performance at the target anchor. For this, we simply remove the dependence of the second algorithm by excluding it from the sum $R_a(d, d') = \sum_{i=1}^{m}(\hat{C}_d(a, s_i) - \hat{C}_{d'}(a, s_i))^2$. We use a value of $k = 4$ in our experiments as it provides the best results in preliminary experiments. The curve scaling could be applied before or after the computation of nearest neighbours. We apply it before the computation of the neighbours.

The Last Anchor Baseline: A trivial baseline is to extrapolate the known part of the learning curve simply with a horizontal line at the performance of

---

[1] The authors use the name mmf4 for the version of MMF that is used in our work.

[2] The authors use the name A_MDS for the version of MDS used in our work.

the last available anchor. Accordingly, we call this baseline 'Last' as was done by Mohr et al. [15]. For any target anchor, it predicts the value of the largest anchor point in the given curve segment. Similar to the other two methods, this prediction can utilize learning curves for both performance prediction as well as algorithm selection tasks.

4   Experimental Setup

The extrapolation can be an intermediate step in a comparison between various learning algorithms. In this case, the only result of interest is what learning algorithm will perform better at the target anchor, essentially making this a classification task. In other cases, we are interested in the exact performance value, essentially making this a regression task. We consider the following two learning curve tasks.

1. Performance prediction (regression task). Given a partial learning curve of an algorithm, predict the performance of this algorithm at the target anchor.
2. Binary algorithm selection (classification task). Given the partial learning curve of two algorithms, predict which of the two algorithms performs best at the target anchor.

The performance of the extrapolation methods depends on the context of the extrapolation setting. We define the context by the following variables: (1) **target anchor** (the anchor point to extrapolate to); (2) **curve segment** (the anchor points for which the curve is already given, and extrapolation is performed from this segment to the target anchor); and (3) **prediction objective** (what the extrapolation is used for). Whether the extrapolation method is used for performance prediction or binary algorithm selection. (4) **Metadata.** The type of additional data that can be used for the extrapolation task. MDS utilizes this meta-data, MMF does not. We published all experimental data online.[3]

Our analysis is based on the Learning Curve Database (LCDB) [15]. It contains learning curves of 20 learning algorithms on 248 unique datasets. To get a balanced overview of how the extrapolation methods work in different extrapolation settings, we consider a broad set of anchors. For each of 248 datasets and each of the 20 algorithms, LCDB contains scores for anchor points at sizes $\lceil 2^{\frac{7+k}{2}} \rceil$ with $k \in \{1, 2, ..\}$ until the maximum dataset size is reached. We pick the anchor points that are present in at least half the datasets (i.e., leaving out anchor points at higher values that are only obtained when using the larger datasets). This leaves us with the following anchor points $\{16, 23, 32, 45, 64, 91, 128, 181, 256, 362, 512, 724, 1024, 1448, 2048, 2896, 4096\}$. We do not consider the first anchor (16) as a target anchor, as we need at least one anchor point in the curve segment to make a prediction. Similarly, we do not include the last anchor (4096) in our curve segments. Regarding the curve segments, we start with just the first anchor and keep adding the next anchor. This results in 16 possible curve segments.

---

[3] See: https://github.com/ADA-research/LearningCurveExtrapolationSettings.

We use three distinct metrics in our comparison. Let s be some target anchor and a some learning algorithm (or a_1, a_2 two learning algorithms that are being compared). If the estimated and true prediction performances of a at anchor s are given by $\hat{y}(a, s)$ and $\hat{C}(a, s)$ respectively, then the loss of the extrapolation model is defined as:

- Absolute error $(a, s) = |\hat{y}(a, s) - \hat{C}(a, s)|$
- Binary error $(a_1, a_2, s) = \begin{cases} 0 & \text{if best learner was predicted} \\ 1 & \text{else} \end{cases}$
- Risk $(a_1, a_2, s) = \begin{cases} 0 & \text{if best learner was predicted} \\ |\hat{C}(a_1, s) - \hat{C}(a_2, s)| & \text{else} \end{cases}$

The absolute error is associated with the performance prediction task, while the binary error and risk are associated with the binary algorithm selection task. We use two metrics for this task, as we are interested in (1) how often the extrapolation methods pick the wrong learning algorithm and (2) the loss in performance if that learning algorithm were used instead of the better one.

5 Results and Discussion

Relative Performances of Extrapolation Methods for a Fixed Curve Segment and Target Anchor: We first discuss the performance of the extrapolation methods on the binary algorithm selection task. We use all combinations of 2 classifiers from the 20 classifiers of LCDB, resulting in 190 binary algorithm selection tasks. Figure 2 shows a matrix with the relative performance on each binary algorithm selection task of the three extrapolation methods, for a target anchor size of 4096 and a fixed curve segment of 724. Each cell represents a performance comparison between two extrapolation methods on such a task. The results are symmetric, and the diagonal is empty. For each comparison between extrapolation methods 'A vs B', the binary error of B is subtracted from the binary error of A (for each pair of learning algorithms). As we do this over all datasets considered in LCDB, the mean is taken over the performance obtained over 248 datasets. Positive values (red) indicate that extrapolation method B obtained a lower average error across datasets, while negative values (blue) indicate that extrapolation method A performed better. Each of these plots gives a global overview of which extrapolation method performed better in this setting, although this is just by visual inspection. The distributions of these means are then displayed as boxplots (right bottom).

The boxplots reveal that, for this particular curve segment and target anchor, the last anchor baseline (Last) clearly outperforms Morgan-Mercer Flodin (MMF) and Meta-Learning on Datasamples (MDS). This can be seen as the means and median are well above the 0 line. We also see that in the comparison between MMF and MDS in this specific extrapolation setting there is a slight advantage for MMF as the mean and median are slightly above the 0 line. However, we note that these results are depending on the extrapolation setting that was selected. The conclusions might differ in other settings (e.g., a different target anchor and a curve segment).

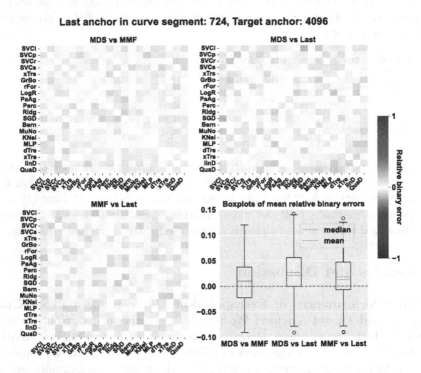

Fig. 2. Relative performances of the extrapolation methods for a fixed curve segment with a largest anchor of size 724, and fixed target anchor of size 4096. Shown are the results for the binary algorithm selection task with binary error as the metric. The colour bar refers to the figures in the left column and the top right figure. Names of learning algorithms are abbreviated (our GitHub project code contains the full name mapping).

Performances of Extrapolation Methods for Increasing Curve Segments and a Fixed Target Anchor: Figure 3 now offers a slightly more generalized view on the situation than Fig. 2 by varying over different curve segments (over the x-axis). The target anchor remains fixed at a size of 4096. The left and middle columns are for the binary algorithm selection task and the right column is for the performance prediction task. The left column uses the risk as metric, the middle the binary error, and the right the absolute error. As such, note that the metric in the middle column corresponds to Fig. 2. Results are aggregated by first computing the mean over the datasets per pair of classifiers (as in Fig. 2) or per individual learners (for the performance prediction task). We ignore the very first value in the performance prediction task for MMF and any comparison with MMF because it is unable to perform an extrapolation based on only a single curve segment.

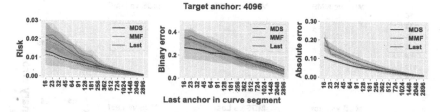

Fig. 3. Performances of extrapolation methods for moving curve segments and a fixed target anchor of size 4096. The lines represent the average, the dotted lines represent the median, and the shaded areas show the interquartile range of the performances according to the respective measure.

We see that MMF outperforms Last on average for the performance prediction task but is outperformed by Last on the binary algorithm selection task. For the performance prediction task, the interquartile ranges suggest Last outperforms MMF at times at some smaller curve segments. The reason for this is that there are quite a few learning curves in LCDB that descend at the beginning and then start rising at a later anchor point. When only the descending part of the learning curve is available MMF will extrapolate in the wrong direction. In these cases, another parametric model should be used. Even so, at smaller curve segments an improvement of around $0 - -4\%$ accuracy can be expected for the prediction. For the binary algorithm selection task, we see that MMF never improves over Last even at smaller curve segments. By looking at the interquartile ranges we see that for any curve segment, Last outperforms MMF for around 75% of cases. Last has quite a high error at lower curve segments, larger than 30%. However, it seems that MMF cannot improve on this.

Furthermore, we see that MDS seems to outperform both MMF and Last on smaller curve segments. For the performance prediction task, we can also see that MDS is a lot more precise and accurate at these smaller curve segments. It starts at an absolute error of around 0.1 and has closer quartiles than the other methods. As the learning curves represent accuracy, this means that MDS can predict the accuracy of a learning algorithm within ±10% with only the first anchor point. For the binary algorithm selection task, we see that the binary error of MDS is still quite high at the lower curve segments, but it stays about 5–10% lower than the other methods. Even though the binary error is large here we see that the risk is under 1.5%, which is quite low. This shows that even though MDS predicts wrong over 20% of times at lower curve segments when it predicts wrong the loss in accuracy is only under 1.5%.

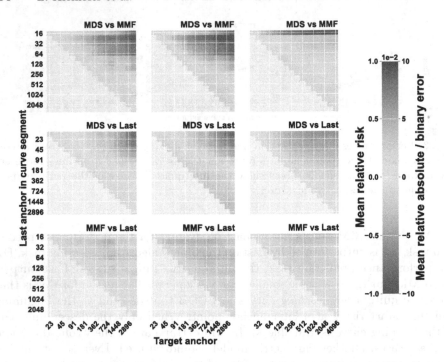

Fig. 4. Relative performances of extrapolation methods for increasing curve segments (indicated by the row) and target anchors (indicated by the column). A comparison of extrapolation methods 'A vs B' means that the risk of B is subtracted from the risk of A. The left column displays the results with the risk as metric, the middle column displays the binary error, and the right column displays the absolute error. The left side of the colour bar is for the risk and the right side is for both binary and absolute error.

Relative Performances of Extrapolation Methods for Increasing Curve Segments and Target Anchors: Figure 4 now further generalizes the previous view by additionally ranging over different target anchor sizes. This results in a matrix, where the row indicates the final curve segment, and the column indicates the target anchor. Each cell indicates the difference in performance between two extrapolation methods for that setting, aggregated over all datasets and pairs of classifiers. The colour bar shows two different scales, the left side is used for the risk values and the right for the absolute and binary error values. Values falling outside of the bounds given by the colour bar have been set to the value of the nearest bound. As in the previous section, for the performance prediction task, we ignore the curve segment that only contains the anchor 16 for any comparison that includes MMF. From this figure, we see that the trend identified in the previous section occurs for most target anchors. MDS outperforms both MMF and Last on smaller curve segments. However, the smaller the target anchor the smaller the anchor point at which Last or MMF becomes better than MDS. This is because, in the setting where we regard smaller target anchors, fewer observations are needed to complete the learning curve, thus smaller curve segments give more information than they would for larger target anchors. This shows that a meta-learning model that uses other learning curves is useful when there is little information about the learning curve. As a rough estimate we see that when the last anchor in the curve segment is under a 5th or 10th the size of the target anchor, MDS performs better than the other two extrapolation methods. It becomes clear that it is not beneficial to use a parametric model for the binary algorithm selection task. This is because MMF is outperformed by simply using only the last anchor in the curve segment. We see that as more data on the learning curve becomes available the last anchor becomes a better estimation of the target anchor and thus outperforms MDS. For the performance prediction task, it is better to use a parametric model over just the last anchor. At larger curve segments MMF will slightly outperform MDS. Thus, as more data on the learning curve becomes available it is better to use a parametric model if a high accuracy of prediction is required.

6   Conclusion

In this paper, we have compared two distinct learning curve extrapolation methods, i.e., the parametric model MMF and the meta-learning model MDS. Additionally, we have included a baseline which takes the performance at the last anchor of the learning curve as its prediction. We have performed this comparison across different extrapolation settings. In these settings, we vary the availability of data for the extrapolation, the size of the target anchor, and the prediction objective. Most importantly, we show that depending on the prediction task, different extrapolation methods can be considered the best. In particular, MDS is the better performing extrapolation method when little information is available on the learning curve. Due to the large amount of settings it is hard to exactly quantify, but loosely speaking MDS performs better than the other methods when the largest anchor in the curve segment is up to a 5th or 10th the size of the target anchor. Once more data becomes available, both MMF and the baseline generally outperform MDS. This goes to show that, with more data available it is better to rely on a parametric model, or just the last anchor, and with less data available it is better to rely on a meta-learning model. We find that, when predicting which of two learning algorithms is better, the parametric model is often outperformed by just using the information available on the last anchor. However, this is not the case when the objective is to find the exact value of the learning curve at the target. In that case, the parametric model outperforms using just the last anchor in almost all cases.

Interesting further research could be: (1) A conditional analysis that depends on the learners. (2) Including other parametric models in the analysis. (3) To study the impact of weights in the methods. Here we used an exponential decay of older anchors. (4) Performing the evaluation not in terms of curve segment/target anchor combinations, but in terms of difficulty. (5) Including a multi-class objective in the analysis, where the goal is to predict a ranking of the classifiers.

References

1. Bousquet, O., Hanneke, S., Moran, S., van Handel, R., Yehudayoff, A.: A theory of universal learning. In: STOC 2021: 53rd Annual ACM SIGACT Symposium on Theory of Computing, pp. 532–541. ACM (2021)
2. Brazdil, P., van Rijn, J.N., Soares, C., Vanschoren, J.: Metalearning: Applications to Automated Machine Learning and Data Mining, 2nd edn. Springer (2022). https://doi.org/10.1007/978-3-030-67024-5
3. Chandrashekaran, A., Lane, I.R.: Speeding up hyper-parameter optimization by extrapolation of learning curves using previous builds. In: Ceci, M., Hollmén, J., Todorovski, L., Vens, C., Džeroski, S. (eds.) ECML PKDD 2017. LNCS (LNAI), vol. 10534, pp. 477–492. Springer, Cham (2017). https://doi.org/10.1007/978-3-319-71249-9_29
4. Cortes, C., Jackel, L.D., Solla, S.A., Vapnik, V., Denker, J.S.: Learning curves: asymptotic values and rate of convergence. Adv. Neural Info. Proc. Syst. **6**, 327–334 (1993)
5. Gu, B., Hu, F., Liu, H.: Modelling classification performance for large data sets. In: Wang, X.S., Yu, G., Lu, H. (eds.) WAIM 2001. LNCS, vol. 2118, pp. 317–328. Springer, Heidelberg (2001). https://doi.org/10.1007/3-540-47714-4_29
6. Hutter, M.: Learning curve theory. CoRR abs/2102.04074 (2021)
7. John, G.H., Langley, P.: Static versus dynamic sampling for data mining. In: KDD'96: Proceedings of the Second International Conference on Knowledge Discovery and Data Mining, pp. 367–370. AAAI Press (1996)
8. Klein, A., Falkner, S., Bartels, S., Hennig, P., Hutter, F.: Fast bayesian optimization of machine learning hyperparameters on large datasets. In: International Conference on Artificial Intelligence and Statistics, AISTATS 2017. Proceedings of Machine Learning Research, vol. 54, pp. 528–536. PMLR (2017)
9. Klein, A., Falkner, S., Springenberg, J.T., Hutter, F.: Learning curve prediction with bayesian neural networks. In: International Conference on Learning Representations (ICLR) (2017)
10. Last, M.: Improving data mining utility with projective sampling. In: ACM SIGKDD International Conference on Knowledge Discovery and Data Mining, pp. 487–496. ACM (2009)
11. Leite, R., Brazdil, P.: Predicting relative performance of classifiers from samples. In: International Conference on Machine Learning (ICML 2005). ACM International Conference Proceeding Series, vol. 119, pp. 497–503. ACM (2005)
12. Leite, R., Brazdil, P.: Active testing strategy to predict the best classification algorithm via sampling and metalearning. In: ECAI 2010 - 19th European Conference on Artificial Intelligence. Frontiers in Artificial Intelligence and Applications, vol. 215, pp. 309–314. IOS Press (2010)
13. Mohr, F., van Rijn, J.N.: Learning curves for decision making in supervised machine learning – a survey. CoRR abs/2201.12150 (2022)
14. Mohr, F., van Rijn, J.N.: Fast and informative model selection using learning curve cross-validation. IEEE Trans. Pattern Anal. Mach. Intell. **45**(8), 9669–9680 (2023)
15. Mohr, F., Viering, T.J., Loog, M., van Rijn, J.N.: LCDB 1.0: An extensive learning curves database for classification tasks. In: Amini, M.-R., Canu, S., Fischer, A., Guns, T., Kralj Novak, P., Tsoumakas, G. (eds.) Machine Learning and Knowledge Discovery in Databases: European Conference, ECML PKDD 2022, Grenoble, France, September 19–23, 2022, Proceedings, Part V, pp. 3–19. Springer Nature Switzerland, Cham (2023). https://doi.org/10.1007/978-3-031-26419-1_1

16. Paszke, A., et al.: PyTorch: an imperative style, high-performance deep learning library. Adv. Neural Inf. Proc. Syst. **32**, 8024–8035 (2019)

17. Provost, F.J., Jensen, D.D., Oates, T.: Efficient progressive sampling. In: ACM SIGKDD International Conference on Knowledge Discovery and Data Mining, pp. 23–32. ACM (1999)

18. van Rijn, J.N., Abdulrahman, S.M., Brazdil, P., Vanschoren, J.: Fast algorithm selection using learning curves. In: Fromont, E., De Bie, T., van Leeuwen, M. (eds.) IDA 2015. LNCS, vol. 9385, pp. 298–309. Springer, Cham (2015). https:// doi.org/10.1007/978-3-319-24465-5_26

19. Swersky, K., Snoek, J., Adams, R.P.: Freeze-thaw bayesian optimization. CoRR abs/1406.3896 (2014)

20. Viering, T., Loog, M.: The shape of learning curves: a review. IEEE Trans. Pattern Anal. Mach. Intell. **45**(6), 7799–7819 (2022)

21. Weiss, G.M., Tian, Y.: Maximizing classifier utility when there are data acquisition and modeling costs. Data Min. Knowl. Disc. **17**(2), 253–282 (2008)

Efficient NAS with FaDE on Hierarchical Spaces

Simon Neumeyer[ID], Julian Stier[✉], and Michael Granitzer[ID]

University of Passau, Passau, Germany
`neumeyer.simon@gmx.de`, `julian.stier@uni-passau.de`

Abstract. Neural architecture search (NAS) is a challenging problem. Hierarchical search spaces allow for cheap evaluations of neural network sub modules to serve as surrogate for architecture evaluations. Yet, sometimes the hierarchy is too restrictive or the surrogate fails to generalize. We present FaDE which uses differentiable architecture search to obtain relative performance predictions on finite regions of a hierarchical NAS space. The relative nature of these ranks calls for a memory-less, batch-wise outer search algorithm for which we use an evolutionary algorithm with pseudo-gradient descent. FaDE is especially suited on deep hierarchical, respectively multi-cell search spaces, which it can explore by linear instead of exponential cost and therefore eliminates the need for a proxy search space.

Our experiments show that firstly, FaDE-ranks on finite regions of the search space correlate with corresponding architecture performances and secondly, the ranks can empower a pseudo-gradient evolutionary search on the complete neural architecture search space.

Keywords: darts · hierarchical neural architecture search · automl · differentiable structure optimization

1 Introduction

Automatically finding structures of deep neural architectures is an active research field. The exponentially growing space of directed acyclic graphs (DAGs), their complex geometric structure and the expensive architecture performance evaluation make **neural architecture searches (NAS)** a challenging problem. Methods such as evolutionary and genetic algorithms, bayesian searches and differentiable architecture searches compete for the most promising automatic approaches to conduct neural architecture searches [4].

Differentiable architecture search (DARTS) is a successful and popular method to relax the search space into a differentiable hyper-architecture. This relaxation allows to use differentiable search methods to learn both model weights and architectural parameters to evaluate sub-paths of a hyper-architecture [8]. While the combined and weight-shared hyper-architecture allows for a very fast training of few GPU days, the search space is limited to subspaces of the defined hyper-architecture. Evolutionary searches, in contrast, are way

I. Miliou et al. (Eds.): IDA 2024, LNCS 14642, pp. 158–170, 2024.
https://doi.org/10.1007/978-3-031-58553-1_13

more dynamic in the way they restrict the search space. Without any weight-sharing tweaks, this usually comes with a significant higher computational time.

We present a *FAst Darts Estimator on hierarchical search spaces* (*FaDE*) that aims at optimizing chained like hierarchical architectures while not resorting to a proxy domain. By iteratively fixing a finite set of sub-module architectures per neural sub-module and using DARTS for training, architecture ranks α are estimated based on the relative performance within an iteration. As the estimations are of relative nature, we require a state-less, batch-wise optimization algorithm to determine from those estimations a new finite set of cell architectures per cell. To this end, we apply an evolutionary approach which incorporates a pseudo-gradient descent for candidate generation. *FaDE* runs one independent optimization algorithm per cell which allows it to optimize additional depth with linear instead of exponential cost.

Our contributions contain the first usage of differentiably obtained ranks for neural architecture search in an open-ended search space. The usage is justified with a correlation analysis. We provide code and data of our experiments for reproducibility in a github repository. *FaDE* might be generalized to further types of hierarchical search spaces and could also be employed with other state-less, batch-wise search strategies in open-ended search spaces.

2   Fast DARTS Estimator

We construct chained hierarchical search spaces for neural architectures and use their (relative) estimated performance as architectural ranks. The obtained *FaDE*-ranks guide a search on the complete (but open) search space. The structure of finite regions is not arbitrary, but bounded to the set of architectures contained in a hyper-architecture. We use DARTS to train such a hyper-architecture, *FaDE* to predict the corresponding region of the search space from a trained hyper-architecture, and a mapping of the search space into Euclidean space together with a pseudo gradient descent to guide the exploration of the search space.

On Chained Hierarchical NAS Spaces. Motivated by repeated motifs in hand-crafted architectures, [12] introduce hierarchy to NAS spaces by considering an architecture to be constructed from several structurally identical sub-modules, so called *cells*. By *identical* the same architecture, yet each cell with its proper weights, is meant. A cell typically consists of several convolutional layers, each with a variable operation type and with variable connections between layers. Optimizing the *cell architecture* is often equivalent to finding the type of convolutional operation for a fixed number of layers and determining which layers are being connected. The *macro-architecture* determines how multiple, often identical, cells are being stacked to one complete architecture. Quite often, the stacking is done in e.g. a simple chain structure [12]. Many search strategies on hierarchical search spaces fix the macro-architecture and solely optimize the architecture of a single cell which is often done on a shallower *proxy domain*. For example, in image classification, a proxy domain might consist in switching from *CIFAR-100* to *CIFAR-10*. Even though such search strategies can achieve very good

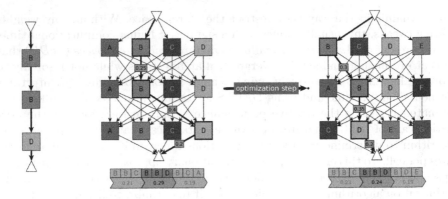

Fig. 1. (left) Discrete architecture $BBD \in \mathcal{S}^3$ featuring cell architectures $B, D \in \mathcal{S}$. (middle) BBD contained in a hyper-architecture $H \in \mathcal{H}_{3,4}(\mathcal{S})$ that allows for several cell architectures per row. Obtaining relative $FaDE$-ranks on trained hyper-architecture: factorizing architecture parameters along the corresponding path of the hyper-architecture. (right) Each step in the outer NAS optimization discovers new cell architectures per row.

results, they are not always successful. On the one hand, the performance of a single cell might not generalize to the performance of an overall neural network architecture consisting of multiple structural copies of that cell.

2.1 DARTS

DARTS relaxes discrete architecture decisions within a differentiable hyper-architecture. Liu et al. [8] consider a neural network module consisting of a set of edges E, that densely connects several ordered vertices, and a finite set of convolutional operations O. The goal is to find the top $k \in \mathbb{N}$ incoming edges and respective operation per vertex. To this end, [8] dedicate one architecture parameter $\alpha_{e,o}$ per edge $e \in E$ and operation $o \in O$, and calculate the output for any edge $e \in E$ as

$$Softmax(\alpha_e)^\top o(.), o \in O \tag{1}$$

Finding the best architecture parameters α aims at solving the bi-level optimization problem

$$\arg\min_\alpha L_{val}(\alpha, \arg\min_\omega L_{train}(\alpha, \omega)) \tag{2}$$

with L_{val} and L_{train} denoting the neural network loss on a dedicated set of training samples each. [8] propose a *second-order* and a simplified *first-order* objective for solving Eq. 2 via gradient descent. The latter can be implemented by alternatingly fixing α and training ω on its respective split of training samples and vice verca. After training, they derive an optimal discrete model by keeping only those operations with large architecture parameters. A driving factor of [8] becoming the baseline work of differentiable architecture search consists in their bi-level optimization algorithm and in their choice of search space. Liu et al. work on a cell from [12] and further adopt their proxy concept, consisting of a smaller dataset and a smaller chain of cells during training.

Gumbel-Softmax. [1] criticize the Softmax in Eq. 1 learning a well performing combination of architectures to which no single architecture will generalize once selected after training. Instead, they propose a sampling mechanism in order to only activate the connections of a single architecture during any forward pass, while still applying back-propagation to the complete hyper-architecture. To this end, instead of Softmax they apply

$$Gumbel\text{-}Softmax_\tau(x) := Softmax_\tau(x + G^n) \tag{3}$$

to $x \in \mathbb{R}^n$ where $\tau > 0$ is a temperature parameter for Softmax and G^n is an i.i.d. vector of the *standard Gumbel* distributed random variable G. Gumbel-Softmax adds stochastic noise to a Softmax transformation while supporting differentiability. For their forward pass they use an additional one-hot encoding on Eq. 3. [6] show that it holds for any $\tau > 0, x \in \mathbb{R}^n$: $\mathbb{E}[Onehot \circ Gumbel\text{-}Softmax_\tau(x)] = Softmax(x)$

2.2 Training a Chained Hierarchical Architecture Using DARTS

For an abstract space of cell architectures $\mathcal{S}$ and a depth $d \in \mathbb{N}$, we build the chained search space $\mathcal{S}^d$, see Fig. 1. Given a window size $w \in \mathbb{N}$, we consider the corresponding space of hyper-architectures as $\mathcal{H} := \mathcal{H}_{d,w}(\mathcal{S}) := \mathcal{S}^{d \times w}$, see Fig. 1 (middle). We use a matrix notation for hyper-architectures for convenience. Note, that any hyper-architecture $H \in \mathcal{S}^{d \times w}$ can be identified with a subset of the search space by considering $\times_{i \leq d} \{H_{ij} \mid j \leq w\} \subset \mathcal{S}^d$, the cross product modelling the combinatorics of chaining cells along the depth of the search space. Hence, a row of $H \in \mathcal{S}^{d \times w}$ represents the pool of cells that H features at the corresponding depth.

Using differentiable architecture search (DARTS) [8], we endow a hyper-architecture $H \in \mathcal{H}$ with architecture parameters $\alpha \in [0, 1]^{d \times w}$, such that after training H with a bi-level optimization algorithm, architecture parameter α_{ij} reflects the performance of cell H_{ij} in the context of exclusive competition within each row of H. For convenience, $\alpha_i, i \leq d$, always denotes the architecture parameters after the Softmax transformation.

In our work, the cells themselves serve as building blocks for DARTS as opposed to [8] where DARTS is being applied to optimize the architecture of a single cell. Following [3] and [11] we choose Gumbel-Softmax [6] to incorporate the architecture parameters into the computation path. Usually, DARTS-based methods consider the architecture parameters as variables over a convex loss surface and optimize them in the same fashion as neural networks weights. As we want to use the trained architecture parameters as performance predictors, we found it reasonable to firstly regularize the architecture parameters and secondly use a constant learning rate to not interfere with the effect of the former. Our regularization should serve the purpose of preventing the hyper-architecture

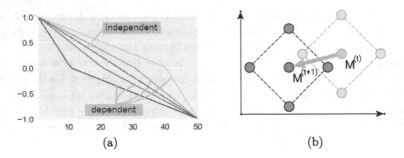

(a)   (b)

Fig. 2. 2a **Regularization** We apply two modes, cell-dependent and cell-independent regularization by means of a regularization factor $r_i \in \mathbb{R}$ (here in [-1,1]) along training epochs (up to 50). Cell-independent regularization applies a regularization factor r which linearly decreases with increasing epochs (linearly decreasing line in the middle). Cell-dependent regularization, however, applies a differently regularized loss per cell i: r_i decreases faster the smaller i. 2b Depiction of an outer optimization step with gradient descent with finite differences in $\mathcal{S}$ respectively $\mathcal{F}$. Compare Eq. 5 for the update of an anchor point $M^{(t)}$ to $M^{(t+1)}$. Any memory-less search strategy can make use of *FaDE*-ranks by making a step towards better ranked architectures.

from converging too early in favor of certain architectures while supporting such a convergence in later epochs. To this end, we add the maximum norm on the Gumbel-Softmax regularized architecture parameters to the loss function, scaled by a factor $r \in \mathbb{R}$ linearly decreasing over epochs. Hence, we update our architecture parameters $\alpha := Gumbel\text{-}Softmax(\alpha'_i)_{i \leq d}, \alpha' \in \mathbb{R}^{d \times w}$, according to the following objective gradient:

$$\nabla_{\alpha'}(L(\alpha, \omega) + r\|\alpha\|_\infty) \tag{4}$$

where $L(\alpha, \omega)$ denotes the neural network loss dependent on architecture parameters α and neural network weights ω.

Cell-dependent Regularization. We elaborate on the regularization factor r in Fig. 4 to regularize shallower cells earlier than deeper cells, compare Fig. 2a This approach is motivated by a cell only being able to learn if its input is somewhat stable.

Weight Sharing. In addition to implicit weight sharing within the hyperarchitecture, implied by a linear number of cells making up for an exponential number of contained architectures, weights might also be shared per row, requiring an implementation of the hyper-architecture where each row itself consists of a hyper-cell that coalesces several cells. In our experimental setup, for example, where cell architectures are modelled by graphs, we considered the largest graph G_{max} contained in the search space, fully connected, and implemented a hyper-architecture of which each row solely consists of the implementation of G_{max}, the hyper-cell. Now any hyper-architecture can be derived from this hyper-architecture by considering all graphs as sub-graphs of G_{max} and renouncing on devoting strictly separate computation paths to each cell, but instead only virtually separating them as different paths within the hyper-cell.

2.3 Deriving FaDE-Ranks on Hyper-Architecture

We use α to rank the subset of architectures contained in H based on

$$\psi_\alpha : H \to \mathbb{R}, (H_{1k_1}, \dots, H_{dk_d}) \mapsto \prod_{i \leq d} \alpha_{ik_i}$$

as shown in Fig. 1 (middle). Note that ψ_α is just one of many ways to apply the information encoded in α to a ranking of corresponding architectures. The benefit of this approach arises from the practical - not theoretical - assumption of independence along the depth of an architecture, that allows for predicting an exponential search space in linear time. Training several architectures $H_{val} \subset H \subset \mathcal{S}^d$ from scratch yields a validation function ρ on H_{val}. A rank correlation coefficient between ρ and ψ_α, the latter restricted to the validation set, documents how well ψ_α predicts relative performances of single architectures contained in H.

2.4 Joint Batch-Wise Pseudo Gradient Descent

A proper correlation between ρ and ψ_α indicates the usefulness of the information contained in the α parameters and motivates us to use ψ_α in guiding a memory-less, batch-wise search in $\mathcal{S}^d$. The downside of the hyper-architecture approach persists in the relative nature, implying that FaDE-ranks ψ_α can in general not be compared among new hyper-architecture evaluations. Contrary to Pham et al. [9] we argue though that changes in weights of a hyper-architecture eventually invalidate former architecture evaluations within that hyper-architecture and hence any information obtained, whether prediction or real evaluation, inhibits a relative nature anyways, its information content *fading* during search. Search methods such as gradient descent do not need memory and hence can use the FaDE-ranks to navigate a NAS. To apply gradient descent, we assume the search space to be Euclidean and use finite differences on a batch of FaDE-ranks to approximate a gradient. Details on the caveat of $\mathcal{S}$ to be Euclidean are being discussed further below.

The overall search works by iteratively sampling hyper-architectures $H^{(t)}, t \in \mathbb{N}$ where the i-th row of $H^{(t+1)}$, $i \leq d$ is solely dependent on the i-th row of the corresponding architecture parameter $\alpha^{(t)}$, obtained after training $H^{(t)}$. Compare Fig. 1. Formally, we consider d independent w-dimensional stochastic processes in $\mathcal{S}^w$ and we therefore refer to the search over $\mathcal{S}^d$ being a joint search over $\mathcal{S}$. The goal is to find hyper-architectures containing well-performing single architectures. For any row $i \leq d$, an *anchor* point $M_i^{(t)} \in \mathcal{S}$ is being maintained, with $M_i^{(1)}$ being randomly initialized. The cells of row i in $H^{(t+1)}$ are being derived from the standard unit vectors originating from the anchor:

$$H_i^{(t)} = \{M_i^{(t)}\} \cup \{M_i^{(t)} \pm \gamma * e_k \mid k \leq \frac{w}{2}\}$$

where the dimension of $\mathcal{S}$ is assumed to be $\frac{w}{2}$ and where e_k denote the standard unit vectors, $k \leq \frac{w}{2}$.

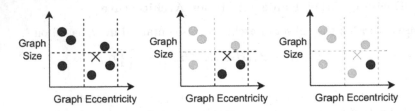

Fig. 3. Graph generation: a sample in embedding space determines a corresponding bucket from which a graph is drawn.

The hyper-parameter $\gamma > 0$ controls the width of the local environment around the anchor. After having obtained the architecture parameters for $H_i^{(t)}$ by training $H^{(t)}$, the anchor $M_i^{(t+1)}$ is being derived from descending $M_i^{(t)}$ according to the finite differences along the standard unit vectors:

$$M_i^{(t+1)} = M_i^{(t)} - \lambda \sum_{k=1}^{\frac{w}{2}} e_k(\beta_i(M_i^{(t)} + \gamma * e_k) - \beta_i(M_i^{(t)} - \gamma * e_k)) \qquad (5)$$

where $\lambda > 0$ controls the step size of gradient descent and $\beta_i(\cdot)$ mapping submodules to their corresponding architecture parameter after train iteration i.

Note that weight sharing could be considered among successive hyper-architectures. We tested pre-initializing the normal neural network weights of $H^{(k+1)}$ with the trained weights of $H^{(k)}$.

Search Space. We focus on the graph attributes of neural network cell architectures. Therefore we provide a bijective *embodiment* : $\mathcal{G} \rightarrow \mathcal{S}$ from a space of directed acyclic graphs to the space of cell architectures. We let *embodiment* map a directed acyclic graph to a cell architecture by first prepending an input vertex to vertices with no incoming edges and appending an output vertex to all vertices with no outgoing edges. The input vertex just serves as interface to distribute the input vector $\boldsymbol{x}$ to all source vertices. On all edges, except those originating from the input vertex, we place structurally identical convolution layers. Vertices with more than one input edge combine their inputs by summation and ReLU non-linearity. While channel count and feature map

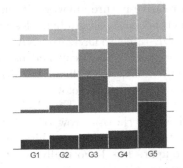

Fig. 4. Density of softmaxed architecture parameters: predicted ranks based on averaged architecture parameter per graph architecture per cell (dark=deep, light=shallow)

size within a cell are being fixed, between succeeding cells we approximately double the channel count while proportionally reducing the feature map size. To enable the pseudo gradient descent from Subsect. 2.4, we use a low-dimensional *feature space* $\mathcal{F}$ and a (stochastic) *generator* : $\mathcal{F} \to \mathcal{G}$ that generates graphs from Euclidean vectors. The generator function is desired to be surjective and smooth with regard to the performance of cell architectures in $\mathcal{S} \cong \mathcal{G}$. There are several possible choices for $\mathcal{G}, \mathcal{F}$ and *generator*, including $\mathcal{F}$ to be the parameter space of a graph generation algorithm. In our experiments we settle for simple choices, starting with a rather small $\mathcal{G}$ containing directed acyclic graphs with a low number of vertices. Hoping on smoothness we consider several scaled graph attributes that have been shown [10] to correlate with performance of the corresponding sub-modules in a certain embodiment and training setting, to construct the dimensions of $\mathcal{F}$. We construct an *embedding* : $\mathcal{G} \to \mathcal{F}$ by determining the required graph attributes for each graph of $\mathcal{G}$. As this embedding is not dense in $\mathcal{F}$, we propose a mapping from the feature space to the space of cell architectures as follows. $\mathcal{F}$ is being separated into disjunct $dim(\mathcal{F})$-dimensional intervals and define *interval* : $\mathcal{F} \to 2^{\mathcal{F}}$ to map a point in the feature space to its containing interval. We then define *generator* as $U \circ embedding^{-1} \circ interval$ where U is the discrete uniform sampling. Figure 3 visualizes the sampling process via G on an exemplary two-dimensional Euclidean feature space. In addition to potentially increasing sampling speed, dependent on the implementation, the sampling via intervals adds some appreciated noise to the generator. [1] extend DARTS [8] by sampling a single embedded architecture for each forward pass instead of taking a weighted sum which, besides reducing memory, might even result in more reliable training as the sparsity of a forward pass is the same for hyper-architecture and target architecture. Also [3] and [11] feature a sparse forward pass by using Gumbel-Softmax [6] instead of a weighted sum for aggregating parallel architecture choices. [2] progressively drop the weakest connections as training progresses to reduce memory footprint. They use the saved resources to progressively increase depth of the hyper-architecture w.r.t. stacked sub-modules and thus aim at closing the gap between proxy and target domain. However, they feature only copies of the same sub-module. [5] propose a progressive approach that successively searches a stack of different sub-modules. **Progressive** approaches are also used without DARTS, e.g. by starting from a space containing very small network modules and end up in a space containing modules of desirable size [7]. This restricts search space exploration to a constant complexity in each iteration by step-wise building up from well performing smaller modules. **Weight sharing** comes implicit in DARTS but is also used in other work, i.e. [9] equip a recurrent NAS pipeline with a hyper-architecture such that instead of training discrete architectures proposed by the pipeline from scratch, they are initialized with the corresponding weights of the hyper-architecture. That enhancement is applicable out of the box to most NAS approaches and shows a significant decrease in resource usage while achieving comparable results.

3 Experiments

We consider a multi-cell search space consisting of $n_c = 4$ chained cells, each cell featuring a DAG with less than $n_v = 6$ nodes as cell architecture. In a first experiment we obtain *FaDE*-ranks on a single hyper-architecture and show that they correlate well with the actual performances of a small subset of architectures contained in the hyper-architecture. Another experiment iteratively trains hyper-architectures according to Subsect. 2.4 in order to search the complete search space. For the latter we present that search results improve over iterations, though not significantly.

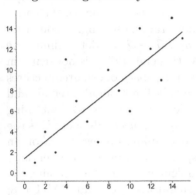

Dataset Preprocessing. All experiments are conducted on the CIFAR-10 image classification dataset. The employed dataset split is $1-1-4$, meaning one part was used for *testing*, one part for *architecture training* and four parts for *weight training*. Between succeeding cells, max pooling is applied for downsampling. All convolutional cells have been using the same kernel shape of 5×5. Gradient clipping during backpropagation was employed with an absolute value of 10.

Fig. 5. Correlation between predicted and evaluated ranks.

We used the following **hyperparameters** across all experiments and provide a github repository for detailed reproducibility: activation function is ReLU, loss is cross-entropy, DARTS aggregation is hard Gumbel-Softmax, the aggregation func. temperature is ten. We used 16 or 32 channels for deepest cells and Max() as pooling operation. For optimization we used a batch size of 128, Adam with $\beta_1 = 0.1, \beta_2 = 10^{-3}, \alpha = 10^{-3}, \epsilon = 10^{-8}$, a weight decay of 10^{-4}, Kaiming as weight initialization, $\mathcal{N}(0; 0.5)$ for the architecture initialization and a gradient clipping value of ten.

3.1 Validating FaDE-Ranks

We construct a hyper-architecture that with $n_g = 5$ manually selected DAGs as parallel computation paths per cell. Once embedded, each DAG comes with the same number of weights, approximately $0.25 \cdot 10^6$. The hyper-architecture H spans a finite search space with $n_g^{n_c} = 5^4 = 625$ architectures. After training the H, we obtain *FaDE*-ranks as described in Subsect. 2.3.

We consider the architecture parameter distributions per cell, averaged over several experiment repetitions, as independent marginals of a joint discrete distribution on the finite architecture search space. Figure 4 illustrates the predicted marginal distributions of the trained hyper-architecture.

Training a subset of 16 manually selected discrete architectures enables calculating a spearman rank correlation between *FaDE*-ranks and evaluation ranks, as shown in Fig. 5. The correlation coefficient of 0.8 is significant and shows that

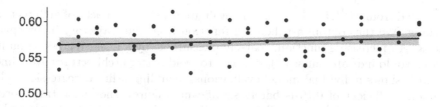

Fig. 7. Evaluation accuracy of architectures generated from points in the search trajectory: The evaluation performance is slightly increasing

our methodology predicts the relative performances within the small architecture subset quite well. That means, the obtained *FaDE*-ranks can be used to guide an open-ended search with local information.

From further experiments we observed that weight sharing decreases the obtained correlation compared to the results in Fig. 5. However, weight sharing yields quite significant correlation results already with a few number of training epochs. This may be owed to the fact, that weight sharing implies weights to be trained more often and thus counteracts the reduction in epochs. We will resort to a fewer number of training epochs while sharing weights in later experiments.

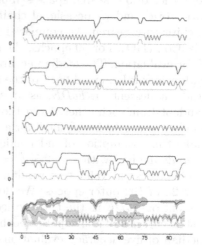

3.2 NAS on Iterative FaDE-Ranks

We aim at iteratively improving the cell architectures of the hyper-architecture from Subsect. 3.1 according to the methodology described in Subsect. 2.4. A pseudo gradient descent serves as optimization strategy on the feature space $\mathcal{F}$. We consider architectures with 4 cells, assign one feature space per cell and run independent, parallel pseudo gradient descent algorithms on them. The *joint feature space* refers to

Fig. 6. Search trajectories in $\mathbb{R}^3$ per cell per dimension: per cell the anchor point of the pseudo gradient descent is plotted for 100 epochs. The anchor point coordinates are color-coded by feature space dimension which share a common y-axis as their values are normed to the same interval. We observe a rough convergence of trajectories.

the product of the 4 feature spaces. Building on [10], $e : \mathcal{G} \to \mathcal{F}$ maps a DAG to three dimensions according to its normed eccentricity variance, degree variance and number of vertices, such that $\mathcal{F} = \mathbb{R}^3$. A single experiments runs with 100 epochs, each featuring five epochs of hyper-architecture optimization. For each cell $i = 1, \ldots, 4$, we obtain feature space trajectories $(M_i^{(t)})_{t=1,\ldots,100}$. In order to validate these trajectories, in regular intervals of t, we repeatedly construct an architecture from the DAGs generated from $M_i^{(t)}, i = 1, \ldots, 4$, and evaluate it using 30 epochs of architecture training. We thus obtain a validation function from the trajectory index space into $\mathbb{R}$. An increasing function, coarsely measurable by a linear regression on its index, would indicate our search strategy to yield better architectures over time. Figure 7 shows individual model evaluations, including a linear regression. The pearson coefficient of 0.16 is barely significant. Figure 6 shows the trajectories in all three dimensions per cell from shallow to deep and its bottom graphic shows the trajectories aggregated over cells. We interestingly notice that the trajectories already show convergence within 20 epochs. Note that convergence in *number of operations* towards the high end of the scale occurs in every cell. This convergence is in accordance with what one would naturally expect. We also observe that the deeper the cell the more divergent its trajectories. There are artifacts that could be attributed to the sparsity of our search space or poor heuristics of our search space sampling, for example the large step sizes of the trajectories and its occasional chainsaw pattern. To validate the outer Neural Architecture Search, we compared to random search (**RS**) and bayesian optimization (**BO**). For 50 epochs, the search methods propose a point in the joint $\mathcal{F}$ for evaluation. The median accuracy of five such architecture generations and evaluations is being fed back to the search algorithm. **BO**, similar to the pseudo gradient descent on *FaDE*, assumes its stochastic models per cell to be independent from each other. Even though, this is not the case, we argue that the number of epochs is too small for a more complex bayesian model that does not make this assumption of independence. For **BO**, we use the *Upper Confidence Bounds* method with $\kappa = 2.5$, $\xi = 0$. We compare the results with the pseudo gradient descent validation results, this time validating the trajectories for the first 25 of 100 outer epochs. We do not use 50 epochs for the pseudo gradient descent on *FaDE* as the trajectories already converge earlier. Figure 8 provides test accuracies of the top 10 epochs per search method. Points in Fig. 8 are a median of five accuracy evaluations, generated from the same point in $\mathcal{F}$. The distribution of accuracy evaluations of a single point represents an important criteria on the suitability of the search space. A wide-spread distribution indicates that the feature space does not capture well architecture features that correlate with the corresponding evaluation performance. The mean standard deviation across all data points in our plot is 0.014 which is quite high compared to the observed magnitudes in Fig. 8.

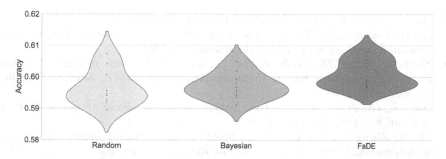

Fig. 8. Comparing the distribution of top architectures found by random search (**RS**), bayesian optimization (**BO**) and *FaDE* on $\mathcal{F}$ spanned by variances of eccentricity, degree and number of vertices. *FaDE* finds more well-performing architectures in less time. RS provides a baseline which clearly shows a higher standard deviation across found accuracy scores. BO is a common NAS method, especially when searching through a low-dimensional search space as we used it.

4   Conclusion and Future Work

We presented *FaDE*, a method to leverage differentiable architecture search to aggregate path decisions from a fixed hierarchical hyper-architecture into point estimates for an open-ended search. The aggregated estimates are called *FaDE*-ranks and show a positive rank correlation with individually trained architectures. Justified with this correlation, *FaDE*-ranks can be used to guide an outer search in a pseudo-gradient descent manner. The method is generalizable in a way that alternative strategies for the outer search can be employed as long as the relative nature of the ranks are respected. We see future work in both **1/** the analysis of the quality rank information for global search as well as **2/** experiments with more complex graph feature spaces, e.g. obtained from generative graph models.

References

1. Cai, H., Zhu, L., Han, S.: Proxylessnas: direct neural architecture search on target task and hardware (2018)
2. Chen1, X., Xie2, L., Wu1, J., Tian, Q.: Progressive darts bridging the depth gap between search (2019)
3. Dong, X., Yang, Y.: Searching for a robust neural architecture in four GPU hours. In: Proceedings of the IEEE/CVF Conference on Computer Vision and Pattern Recognition, pp. 1761–1770 (2019)
4. Elsken, T., Metzen, J.H., Hutter, F.: Neural architecture search: a survey. J. Mach. Learn. Res. **20**(1), 1997–2017 (2019)
5. Hao, J., Zhu, W.: Layered feature representation for differentiable architecture search (2021)
6. Jang, E., Gu, S., Poole, B.: Categorical reparameterization with Gumbel-Softmax. In: International Conference on Learning Representations (2017)

7. Liu, C., et al.: Progressive neural architecture search (2018)
8. Liu, H., Simonyan, K., Yang, Y.: Darts: differentiable architecture search. In: International Conference on Learning Representations (2018)
9. Pham, H., Guan, M., Zoph, B., Le, Q., Dean, J.: Efficient neural architecture search via parameters sharing. In: International Conference on Machine Learning, pp. 4095–4104. PMLR (2018)
10. Stier, J., Granitzer, M.: Structural analysis of sparse neural networks. Procedia Comput. Sci. **159**, 107–116 (2019)
11. Xie, S., Zheng, H., Liu, C., Lin, L.: SNAS: stochastic neural architecture search. In: International Conference on Learning Representations (2018)
12. Zoph, B., Vasudevan, V., Shlens, J., Le, Q.V.: Learning transferable architectures for scalable image recognition. In: Proceedings of the IEEE Conference on Computer Vision and Pattern Recognition, pp. 8697–8710 (2018)

Investigating the Relation Between Problem Hardness and QUBO Properties

Thore Gerlach[1]([✉]) [iD] and Sascha Mücke[2] [iD]

[1] Fraunhofer IAIS, Sankt-Augustin, Germany
thore.gerlach@iais.fraunhofer.de
[2] Lamarr Institute, TU Dortmund University, Dortmund, Germany
sascha.muecke@tu-dortmund.de

Abstract. Combinatorial optimization problems, integral to various scientific and industrial applications, often vary significantly in their complexity and computational difficulty. Transforming such problems into Quadratic Unconstrained Binary Optimization (QUBO) has regained considerable research attention in recent decades due to the central role of QUBO in Quantum Annealing. This work aims to shed some light on the relationship between the problems' properties. In particular, we examine how the spectral gap of the QUBO formulation correlates with the original problem, since it has an impact on how efficiently it can be solved on quantum computers. We analyze two well-known problems from Machine Learning, namely Clustering and Support Vector Machine (SVM) training, regarding the spectral gaps of their respective QUBO counterparts. An empirical evaluation provides interesting insights, showing that the spectral gap of Clustering QUBO instances positively correlates with data separability, while for SVM QUBO the opposite is true.

Keywords: QUBO · Spectral Gap · Quantum Computing

1 Introduction

Combinatorial optimization problems lie at the core of many NP-hard problems in Machine Learning (ML) [7]: Clustering a data set comes down to deciding, for every point, if it belongs to one cluster or another. Training a Support Vector Machine (SVM) involves identifying the subset of support vectors (SVs). These decisions are highly interdependent, making the tasks computationally complex.

In the advent of Quantum Computing (QC), combinatorial problems have gained renewed attention due to the possibility of solving them through the exploitation of quantum tunneling effects. Particularly, the Ising model and the equivalent QUBO problem have become the central target problem class for Quantum Annealing (QA) [6,9]. In QUBO, a parameter matrix Q is given that parametrizes a loss function over binary vectors, which we want to minimize. It can be shown that, in general, this problem is NP-hard [15]. The value of QUBO lies in its versatility: Many NP-hard problems can be reduced to it by means of

I. Miliou et al. (Eds.): IDA 2024, LNCS 14642, pp. 171–182, 2024.
https://doi.org/10.1007/978-3-031-58553-1_14

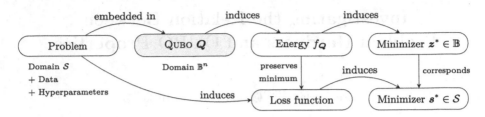

Fig. 1. Workflow of embedding a problem into QUBO. Every solution candidate s in the problem domain S has an easily computable corresponding $z \in \mathbb{B}^n$.

computing Q from input data and hyperparameters, thus QUBO has seen many applications in various domains [2, 8, 10, 11, 13, 14, 16]. Solving the QUBO formulation yields a minimizing binary vector, which maps back to an optimal solution of the original problem. Figure 1 shows a schematic view of this workflow.

However, despite the promise of quantum speedup, not every instance of QUBO is equally easy to solve. It was shown that certain instances require exponential annealing time [1], which may render solving them on quantum computers equally infeasible as by brute force. A central determining factor is the minimal *spectral gap* (SG) of the corresponding Annealing Hamiltonian (AH), which in turn dictates the annealing speed (see Sect. 2): A small SG leads to a higher probability of obtaining sub-optimal results (see e.g. [17]). The SG is a physical property of the AH, and its connection to classical complexity theory is poorly understood. One would expect that classically hard problems tend to be more difficult to solve, even using non-classical methods like QA.

We investigate this connection, both by means of an empirical study and theoretical considerations. To this end, we take instances of optimization problems from ML, embed them into QUBO, and compare their properties to uncover correspondences. Our central research question is: **What is the relation between the hardness of a particular optimization problem and its corresponding QUBO formulation?** In the scope of this paper, "hardness" refers to data properties rather than complexity. We find that the relationship surprisingly not always aligns with intuition: With clustering, a stronger separation of data corresponds with a larger SG, while for SVM learning the opposite is true.

This paper is structured as follows: In Sect. 2, we give an overview on the background of Adiabatic QC. QUBO formulations for the two classical learning problems, Binary Clustering and SVM, can be found in Sects. 3.1 and 3.2. In Sect. 4, we conduct experiments and a conclusion is drawn in Sect. 5.

2   Background

In a QUBO problem, we are given an upper triangle matrix $Q \in \mathbb{R}^{n \times n}$, which parameterizes the *energy function* f_Q defined as

$$f_Q(z) := z^\top Q z = \sum_{i=1}^n \sum_{j=i}^n Q_{ij} z_i z_j \,, \tag{1}$$

where $z \in \mathbb{B}^n$ is a binary vector with $\mathbb{B} = \{0,1\}$. The objective is to find a vector z^* that minimizes f_Q, i.e., $\forall z \in \mathbb{B}^n : f_Q(z^*) \leq f_Q(z)$. The Ising model is almost identical to QUBO, but uses the binary set $\mathbb{S} = \{-1,+1\}$. The model's energy function can be obtained from f_Q through a simple change of $z \mapsto (s+1)/2$.

QA is a promising method for approximating the minimizing solutions of Ising models, first proposed by Kadowaki and Nishimori [9]. Instead of bits for variables it uses *qubits*, which, when measured, take either state from $\mathbb{S}$ with a certain probability. Further, systems of n qubits can exhibit arbitrary probability distributions over the space $\mathbb{S}^n$.

The rough quantum-mechanical equivalent of loss functions in ML are *Hamiltonians*, which are complex-valued hermitian matrices $H \in \mathbb{C}^{2^n \times 2^n}$ that describe the total energy of a system. The expected energy of an n-qubit state $|\psi\rangle$ is given by $\langle\psi|H|\psi\rangle$, where $|\psi\rangle$ is a 2^n-element complex vector describing the qubits' state, and $\langle\psi|$ its conjugate transpose. The Ising model Hamiltonian is diagonal and contains the energy values for all possible binary states $\mathbb{S}^n$, which are simultaneously its eigenvalues. The minimizing state corresponding to the smallest eigenvalue is called *ground state*. The Adiabatic theorem states that a system with a time-dependent Hamiltonian $H(t)$ tends to stay in its ground state, even if the Hamiltonian slowly changes over time [4]. At the core of QA lies the idea to prepare a quantum system in the ground state of an "easy" Hamiltonian H_I and slowly change it to the actual problem Hamiltonian H_P over time:

$$H(s) := f(s)H_I + g(s)H_P \,, \tag{2}$$

with $f, g : [0,1] \to \mathbb{R}_{\geq 0}$, such that $f(0) \gg g(0)$ and $f(1) \ll g(1)$. The speed at which the Hamiltonian can safely evolve without the system leaving its ground state depends on the minimal SG, which is the minimal difference between the two lowest eigenvalues over time. Let $\lambda_1(s), \ldots, \lambda_{2^n}(s)$ denote the eigenvalues of $H(s)$ in increasing order ($\lambda_i(s) < \lambda_{i+1}(s) \,\forall 1 \leq i < 2^n$), then $\gamma(H(s)) := \min_{s \in [0,1]} \lambda_2(s) - \lambda_1(s)$. When a Hamiltonian H is not time-dependent, we simply write $\gamma(H)$ to denote the difference between its lowest eigenvalues. A small SG requires a slow change rate, which leads to a long, potentially exponential (c.f. [1]) *annealing time*. It is therefore desirable to somehow increase the SG by choosing H_I, H_P, f and g accordingly. As H_I, f and g are usually prescribed by the annealing hardware at hand, the only free variable is H_P.

When talking about the SG of a QUBO instance Q, it is important to make the distinction between the parameter matrix Q and its corresponding Hamiltonian H_Q: The entries along the diagonal of the latter correspond to the values $z^\top Q z$ for every possible $z \in \mathbb{B}^n$. Therefore, the SG is simply the difference between the lowest and second-to-lowest values of f_Q (which is very hard to compute for large n). The eigenvalues of Q hold no particular relevance.

3 QUBO Formulations and Their Spectral Gaps

The minimal SG $\gamma(H(s))$ cannot be easily predicted from either $\gamma(H_I)$ or $\gamma(H_P)$. However, we can still make some statements about it using known results about eigenvalues of Hermitian matrices.

Theorem 1 (Weyl's Inequality). *Let M, N, R be $m \times m$ Hermitian matrices with $N + R = M$. Let μ_i, ν_i, ρ_i denote their respective eigenvalues in ascending order, i.e., $\mu_i \leq \mu_{i+1}$ $\forall 1 \leq i < m$, and for ν_i, ρ_i analogously. Then the following inequality holds for all $1 \leq i \leq m$:*

$$\nu_i + \rho_1 \leq \mu_i \leq \nu_i + \rho_m . \tag{3}$$

Applying this inequality multiple times we find the following bound on the SG of sums of two Hamiltonians:

$$\underbrace{\mu_2 - \mu_1}_{=\gamma(M)} \leq \underbrace{\nu_2 - \nu_1}_{=\gamma(N)} + \underbrace{(\rho_m - \rho_1)}_{:=\Gamma(R)} . \tag{4}$$

Recall the definition of the standard time-dependent AH given in Eq. (2). Assume that $f, g : [0, 1] \rightarrow [0, 1]$, $f(0) = g(1) = 1$, $f(1) = g(0) = 0$, f is monotonous decreasing and g is monotonous increasing. E.g., f and g can be chosen as $f(s) = 1 - s$ and $g(s) = s$ with $s = t/T_a$, where T_a is the total annealing time and t the current time in the annealing process. With the assumption $f(x) = 1 - g(x)$, we obtain $\min_{s \in [0,1]} af(s) + bg(s) = \min_{s \in [0,1]} af(s) + b(1 - f(s)) = \min\{a, b\}$, and from Eq. (4) follows

$$\gamma(H(s)) \leq \min\{\gamma(H_P), \gamma(H_I)\} \leq \gamma(H_P) . \tag{5}$$

This result provides motivation to increase $\gamma(H_P)$ when trying to improve QA performance, as it is an upper bound on $\gamma(H(s))$: Increasing it does not guarantee a larger minimal SG, but is a necessary precondition.

3.1   Kernel 2-Means Clustering

Our first QUBO embedding of interest is clustering: Assume we are given a set of n data points $\mathcal{X} \subset \mathbb{R}^d$, $|\mathcal{X}| = n$. We want to partition $\mathcal{X}$ into disjoint clusters $\mathcal{X}_1, \mathcal{X}_2 \subset \mathcal{X}$, $\mathcal{X}_1 \dot\cup \mathcal{X}_2 = \mathcal{X}$. We gather the data in a matrix $X := [x^1, \ldots, x^n]$, $\forall i : x^i \in \mathcal{X}$, and assume that it is centered, i.e., $X\mathbf{1} = \mathbf{0}$, where $\mathbf{1}$ denotes the n-dimensional vector consisting only of ones. A QUBO formulation was derived in [2], which minimizes the *within cluster scatter*:

$$\min_{s \in \mathbb{S}^n} -s^\top X^\top X s \Leftrightarrow \min_{z \in \mathbb{B}^n} -z^\top X^\top X z + \mathbf{1}^\top X^\top X z , \tag{6}$$

where $s = 2z - 1$. A value $z_i = 1$ indicates that data point x^i is in cluster $\mathcal{X}_1$, and in $\mathcal{X}_2$ for $z_i = 0$. Observing that $X^\top X$ is a Gram matrix leads to a possible application of the kernel trick. For this, we consider a centered kernel matrix $K \in \mathbb{R}^{n \times n}$ with elements $k(x^i, x^j)$, where $k : \mathbb{R}^n \times \mathbb{R}^n \to \mathbb{R}$ is a kernel function. $k(x^i, x^j)$ indicates how similar data points x^i and x^j are in some feature space. We can reformulate Eq. (6) to

$$\min_{z \in \mathbb{B}^n} \mathbf{1}^\top K z - z^\top K z \Leftrightarrow \min_{z \in \mathbb{B}^n} \sum_{i,j=1}^n K_{ij}(1 - z_i)z_j \Leftrightarrow \min_{\mathcal{X}_1, \mathcal{X}_2} \sum_{x \in \mathcal{X}_1, y \in \mathcal{X}_2} k(x, y). \tag{7}$$

We can attribute problem properties to effects on the SG of the resulting Hamiltonian of the QUBO formulation, by observing that the similarities between the different clusters are summed up. We see that the QUBO energy according to Eq. (7) is negatively correlated to the similarities within the clusters and positively correlated to the similarities between the clusters.

Thus, we claim that the SG of the QUBO formulation in Eq. (7) is (i) positively correlated to the *separability* (the inter-cluster distance), and (ii) negatively correlated to the *compactness* (the intra-cluster distances), which we validate in Sect. 4.1.

3.2   Simple Support Vector Machine Embedding

A linear Support Vector Machine (SVM) is a classifier that takes a labeled data set $\mathcal{D} = \{(\boldsymbol{x}^i, y_i)\}$ with $\boldsymbol{x}^i \in \mathbb{R}^d$ and $y_i \in \{-1, +1\}$ for $i \in [n] := \{1, \ldots, n\}$, and separates them with a hyperplane [5]. As there may be infinitely many such hyperplanes, an additional objective is to maximize the *margin*, which is the area around the hyperplane containing no data points, in order to obtain best generalization. The hyperplane is represented as a normal vector $\boldsymbol{w} \in \mathbb{R}^d$ and an offset or *bias* $b \in \mathbb{R}$. To ensure correctness, the optimization is subject to $\langle \boldsymbol{w}, \boldsymbol{x}^i \rangle - b) \cdot y_i \geq 1 - \xi_i$, i.e., every data point must lies on the correct side of the plane. As for real-world data perfect linear separability is unlikely, *slack variables* $\xi_i > 0$ allow for slight violations, which we want to minimize. This yields a primal objective function of

$$\text{minimize} \ \frac{1}{2} \|\boldsymbol{w}\|_2^2 + C \sum_i \xi_i$$

$$\text{s.t.} \ \forall i. \ (\langle \boldsymbol{w}, \boldsymbol{x}^i \rangle - b) \cdot y_i \geq 1 - \xi_i \ .$$

We optimize over $\boldsymbol{w}$, b and $\boldsymbol{\xi}$, while $C > 0$ is a hyperparameter controlling the impact of misclassification. Typically this problem is solved using its well-established Lagrangian dual

$$\text{maximize} \ \sum_{i=1}^n \alpha_i - \frac{1}{2} \sum_{i=1}^n \sum_{j=1}^n \alpha_i \alpha_j y_i y_j \langle \boldsymbol{x}^i, \boldsymbol{x}^j \rangle \tag{8}$$

$$\text{s.t.} \ \sum_i \alpha_i y_i = 0, \quad 0 \leq \alpha_i \leq C \ \forall i \ . \tag{9}$$

Following [14] we make the simplifying assumption that α_i can only take the values 0 or C, which allows us to introduce binary variables $z_i \in \mathbb{B}$ and write $\alpha_i = C z_i \ \forall i$. The condition Eq. (9) can be included in the main objective by introducing the penalty term $-\lambda \left(\sum_i \alpha_i y_i \right)^2$, which is 0 when the condition is fulfilled, and negative otherwise. Similarly to the Clustering case (see Sect. 3.1), we can apply the kernel trick and derive the following QUBO formulation:

$$\min_{z \in \{0,1\}^n} \ -\mathbf{1}^\top z + C z^\top \left(\frac{1}{2} (\boldsymbol{Y} \odot \boldsymbol{K}) + \lambda \boldsymbol{Y} \right) z \ , \tag{10}$$

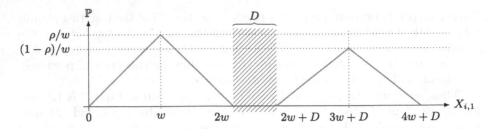

Fig. 2. Distribution of the first dimension of the 2-dimensional synthetic data used for our experiments (before applying the rotation): Two clusters are sampled such that there is a separating margin of at least size D between them. The parameter w controls the spread of data points, while r is the ratio between the number of data points in the first vs. the second cluster.

where $\boldsymbol{Y}$ has entries $Y_{ij} = y_i y_j \; \forall i, j \in [n]$ and $\odot$ denotes the entry-wise product. The parameter λ has to be chosen large enough to ensure Eq. (9) is fulfilled, however, if it is much larger than the objective in Eq. (8), the SG will be very small [3]. C controls how "soft" the margin is, i.e., how strongly misclassified data points are penalized. A large C does so heavily, which may result in overfitting.

We claim that the SG is negatively correlated (i) with C, (ii) with λ, and also (iii) with the *separability*, i.e., actual margin size of the data. We validate this in Sect. 4.2.

4   Experiments

At the core of this work we conduct an empirical evaluation of QUBO formulations and their properties. For each experiment, the steps we take are as follows: 1. Choose **problem type** and **hyperparameters**, 2. Sample **data set** with known properties, 3. Compute **QUBO parameters** and record the SG. Using the acquired data, we investigate the relationship between data parameters and SG, which has a high impact on problem hardness for QC, as we have shown before. We consider the two problems of binary clustering and SVM training, which we described in previous sections. As input data, we sample synthetic data sets for each repetition of the experiments. To this end, we consider two different types, CONES and CIRCLES, both of which have parameters allowing us to vary the resulting optimization problems' difficulty by adjusting the data class separation.

CONES. Let $n \in \mathbb{N}$ with $n \geq 2$ denote the number of data points, $\rho \in (0, 1)$ the cluster size ratio, $w > 0$ a spread parameter, and $D \in \mathbb{R}$ a separating margin size. We set $n_1 := \min\{1, \lfloor \rho n \rfloor\}$ and $n_2 := 1 - n_1$ as the cluster sizes. We create a matrix $\boldsymbol{X} \in \mathbb{R}^{n \times 2}$ where every entry X_{ij} is sampled i.i.d. from a triangular distribution within the interval $[0, 2w]$ and with mode w. We chose the triangular distribution over a normal distribution because it has no outliers, which allows

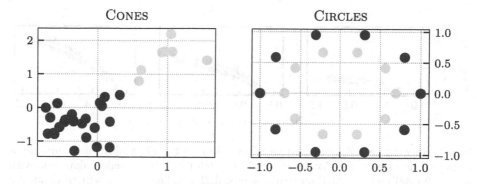

Fig. 3. Exemplary instances of the data sets used for our experiments.

us to define a hard lower bound on the separating margin between clusters. For all $i > n_1$ we then set $X_{i,1} \mapsto X_{i,1} + 2w + D$, which shifts all of these points such that $\|X_{i,\cdot} - X_{\ell,\cdot}\|_2 \geq D$ is tight for all $i \in [n_1]$ and $n_1 < \ell \leq n$. The distribution of $X_{\cdot,1}$ is visualized in Fig. 2; the distribution of $X_{\cdot,2}$ consists of just a single triangle from 0 to $2w$ with height $1/w$ at mode w. Next, we sample θ uniformly from $[0, 2\pi)$ and apply $X \mapsto X R(\theta)$, where $R(\theta)$ is a 2D rotation matrix. This rotation leaves the distances unchanged but introduces another degree of freedom. Lastly, we center the data by computing $\mu_j := \sum_{i=1}^n X_{ij}/n$ and applying $X_{ij} \mapsto X_{ij} - \mu_j$ for all $i, j \in [n] \times [2]$. The target vector $y \in \mathbb{S}^n$ is set to $y_i = -1$ for $i \in [n_1]$ and $y_i = +1$ for $n_1 < i \leq n$.

CIRCLES. As a second data set type, we consider two circles, which are not linearly separable in $\mathbb{R}^2$. The radius of the outer circle is fixed to 1 and for our experiments we vary the radius of the inner circle r. The circles consist of an equal number of points $n/2$ and Gaussian noise with standard deviation σ is added to every point (see Fig. 3, right). To bridge the gap to linear separability, we project the data set to a higher-dimensional feature space via a feature map $\phi : \mathbb{R}^2 \to \mathbb{R}^3$, $\phi(x) := (x_1, x_2, a\|x\|^2)^\top$. In this space, the data is linearly separable. A corresponding kernel function is given by $k(x, y) := \langle \phi(x), \phi(y) \rangle = \langle x, y \rangle + a^2 \|x\|^2 \|y\|^2$. For every experiment, we compare three different problem sizes, that is, we consider $n \in \{8, 20, 32\}$. To make SG comparable between QUBO instances of the same size, we scale each Q such that $\|Q\|_\infty = 1$.

4.1 Clustering

We first explore the Clustering QUBO in Eq. (7). Changing the maximum separating margin size between the two clusters changes the SG of the corresponding QUBO instance, as shown in Figs. 4 and 5: For Fig. 4, we sample 1000 different CONES data sets with varying cluster distances in $[0, 1]$ and fix $w = 0.2$, $\rho = 0.5$. We find that the SG is increasing with an increasing problem size n and that there is a clear quadratic positive correlation between the SG and the margin size.

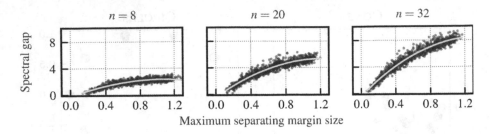

Fig. 4. Spectral gap of QUBO instances according to Eq. (7) against maximum separating margin size D for CONES; $w = 0.2$, $\rho = 0.5$ fixed, 1000 random data sets with $n \in \{8, 20, 32\}$ and $D \in [0, 1]$ uniformly sampled. The yellow curve is a fitted quadratic function. (Color figure online)

A similar setup can be found in Fig. 5, where 1000 different CIRCLES data sets are sampled with varying inner radius in $[0, 1]$ and $\sigma = 0.05$, $\rho = 0.5$. Again we find that the SG increases with an increasing margin size, but with a linear correlation. Since different kernels are used in Fig. 4 and Fig. 5, the exact correlation form is dependent on the exact data set and the used kernel.

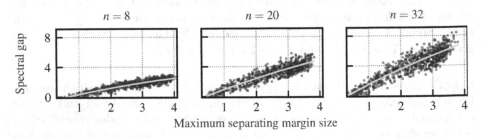

Fig. 5. Spectral gap of QUBO instances according to Eq. (7) against maximum separating margin size D for CIRCLES; $\sigma = 0.05$, $\rho = 0.5$ fixed, 1000 random data sets with $n \in \{8, 20, 32\}$ and $r \in [0, 1]$ uniformly sampled. The yellow curve is a fitted quadratic function. (Color figure online)

In Fig. 6, the effects of a varying cluster ratio in $\rho \in [0.1, 0.5]$ are depicted for the CONES setup. We again sample 1000 data sets with fixing $w = 0.2$ and $D = 0.5$. A positive correlation becomes evident between cluster ratio and SG. That is, the QUBO problem is easier to solve with QC when the clusters have the same size. The effect that the plots look like a step function for small n is due to the fact that there $n/2$ different configurations, e.g., for $n = 8$, we can have the four cases $n_1 = 1$, $n_1 = 2$, $n_1 = 3$ and $n_1 = 4$.

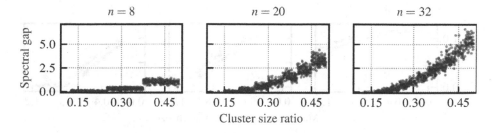

Fig. 6. Spectral gap of QUBO instances according to Eq. (7) against cluster ratio for CONES; $D = 0.5$, $w = 0.2$ fixed, 1000 random data sets for $n \in \{8, 20, 32\}$ and $\rho \in [0.1, 0.5]$ uniformly sampled.

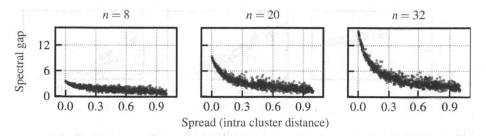

Fig. 7. Spectral gap of QUBO instances according to Eq. (7) against spread for CONES; $D = 0.5$, $\rho = 0.5$ fixed, 1000 random data sets for $n \in \{8, 20, 32\}$ and $w \in [0, 1]$ uniformly sampled.

In Fig. 7, we vary the spread $w \in [0, 1]$ for 1000 different data sets and fix $D = 0.5$, $\rho = 0.5$. We can see that the SG is negatively correlated to the spread of the data set.

Putting the results together we can deduce that the is SG positively correlated with the inter-cluster distance (separability) and negatively correlated with the intra-cluster distance, supporting our claims in Sect. 3.1.

4.2 Support Vector Machine

We move over to experiments with the SVM QUBO in Sect. 3.2. Again, we depict the effect of changing the maximum separating margin size between the two clusters on the SG of the corresponding QUBO in Figs. 8 and 9. We use the same parameters for the data sets as in Figs. 4 and 5. Interestingly, we now observe a negative correlation between the SG and the margin size, making the problem harder to solve with a quantum computer if the data is well separable. For $n = 32$ the spectral basically vanishes from a certain margin size for CIRCLES. Furthermore, we again observe that this correlation is quadratic for CONES and linear for CIRCLES, leading to a large dependence on the used kernel function and the data set at hand.

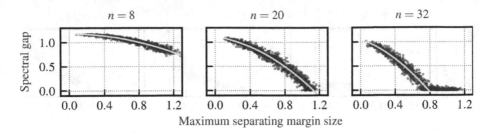

Fig. 8. Spectral gap of QUBO instances according to Eq. 10 against maximum separating margin size D for CONES. Same configuration as for Fig. 4.

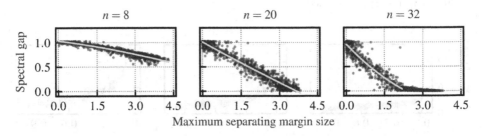

Fig. 9. Spectral gap of QUBO instances according to Eq. 10 against maximum separating margin size D for CIRCLES. Same configuration as for Fig. 5.

Note that we are considering a QUBO formulation for a soft-margin SVM: even though the SG might be very small, the second best solution might also be satisfactory for solving the original problem. In contrast, the second best solution of a Clustering QUBO is much worse: changing a single bit leads to a data point within the wrong cluster, which increases the energy more dramatically the further the two clusters are separated.

In Figs. 10 and 11, we show the effect of varying λ and C on the SG for CONES. We fix $\rho = 0.5$, $w = 0.2$, $D = 0.5$ and sample 10 000 data sets with $C \in [0, 0.1]$, $\lambda \in [0, 100]$ for Fig. 10 and $\lambda \in [0, 10]$ for Fig. 11, respectively. It is evident from Fig. 10 that the SG decreases as λ and C are increase. However, there are interesting intervals for λ when fixing C, such that the SG first increases and then decreases with increasing λ, forming a triangular shape when plotted, which gets more pointy with an increasing value of C – see Fig. 11 for a closer view. We observe a similar effect with CIRCLES.

Combining our observations we deduce that the SG is negatively correlated with the inter-cluster distance (separability) and the parameters λ, C except for a small region, supporting our claims from Sect. 3.2.

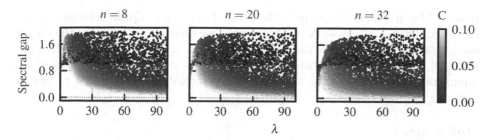

Fig. 10. Spectral gap of QUBO instances according to Eq. 10 against λ and C for CONES; $w = 0.2$, $\rho = 0.5$ fixed, 10 000 random data sets for $n \in \{8, 20, 32\}$ and $\lambda \in [0, 100]$, $C \in [0, 0.1]$ uniformly sampled.

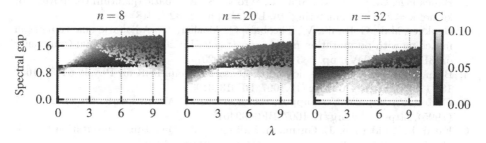

Fig. 11. Same as Fig. 10, zoomed in on $\lambda \in [0, 10]$.

5   Conclusion

In this paper, we investigated the connection between the problem hardness of classical ML problems and their solvability on quantum hardware. We considered QUBO formulations for the 2-Means Clustering and SVM learning. We highlighted that the SG of these formulations impact their solvability on quantum hardware, and showed how the SG behaves when adapting the problem parameters, which we underpinned with an empirical study.

We found that for 2-Means Clustering an easier problem also leads to a better solvability on quantum computers. Here, "easy" refers to the separability and compactness of the different classes. We found a positive correlation between these properties and the SG of the corresponding QUBO. Interestingly, this is not the case for the SVM problem, where we would expect a better solvability of a problem with a large separation of the classes. However, we found a negative correlation between separation and SG. Furthermore, other hyperparameters, such as the one controlling the softness of the margin avoiding overfitting, are negatively correlated to the SG. This is due to the balancing different objectives in one single QUBO problem. Combining these two insights, we conclude that the original problems' hardness is not directly connected to the solvability on quantum computers, as one might assume. Instead, it depends not only on the data set at hand, but also on specifics in the used QUBO formulation.

It would be insightful to compare the properties of more QUBO formulations of more problems in future work. Furthermore, investigating the effect of the QUBO parameters and not only the problem parameters is an interesting research direction [12]. This could strengthen our intuition about which problems are hard for quantum computers in particular, and the potential and limitations of QC in general.

References

1. Altshuler, B., Krovi, H., Roland, J.: Adiabatic quantum optimization fails for random instances of NP-complete problems. arXiv:0908.2782 [quant-ph], December 2009
2. Bauckhage, C., Ojeda, C., Sifa, R., Wrobel, S.: Adiabatic quantum computing for kernel k = 2 means clustering. In: LWDA, pp. 21–32 (2018)
3. Benkner, M.S., Golyanik, V., Theobalt, C., Moeller, M.: Adiabatic quantum graph matching with permutation matrix constraints. In: 2020 International Conference on 3D Vision (3DV), pp. 583–592. IEEE (2020)
4. Born, M., Fock, V.: Beweis des Adiabatensatzes. Zeitschrift für Physik 51(3), 165–180 (1928). https://doi.org/10.1007/BF01343193
5. Cortes, C., Vapnik, V.: Support-vector networks. Mach. Learn. 20(3), 273–297 (1995). https://doi.org/10.1007/BF00994018
6. Farhi, E., Goldstone, J., Gutmann, S., Sipser, M.: Quantum computation by adiabatic evolution. arXiv preprint quant-ph/0001106 (2000)
7. Guruswami, V., Raghavendra, P.: Hardness of learning halfspaces with noise. SIAM J. Comput. 39(2), 742–765 (2009)
8. Hammer, P.L., Shlifer, E.: Applications of pseudo-Boolean methods to economic problems. Theor. Decis. 1(3), 296–308 (1971)
9. Kadowaki, T., Nishimori, H.: Quantum annealing in the transverse Ising model. Phys. Rev. E 58(5), 5355 (1998)
10. Kochenberger, G., Glover, F., Alidaee, B., Lewis, K.: Using the unconstrained quadratic program to model and solve max 2-SAT problems. Int. J. Oper. Res. 1(1–2), 89–100 (2005)
11. Kochenberger, G., et al.: The unconstrained binary quadratic programming problem: a survey. J. Comb. Optim. 28(1), 58–81 (2014)
12. Mücke, S., Gerlach, T., Piatkowski, N.: Optimum-preserving QUBO parameter compression. arXiv preprint arXiv:2307.02195 (2023)
13. Mücke, S., Heese, R., Müller, S., Wolter, M., Piatkowski, N.: Feature selection on quantum computers. Quant. Mach. Intell. 5(1) (2023). https://doi.org/10.1007/s42484-023-00099-z
14. Mücke, S., Piatkowski, N., Morik, K.: Learning bit by bit: extracting the essence of machine learning. In: Proceedings of the Conference on "Lernen, Wissen, Daten, Analysen" (LWDA). CEUR Workshop Proceedings, vol. 2454, pp. 144–155 (2019)
15. Pardalos, P.M., Jha, S.: Complexity of uniqueness and local search in quadratic 0–1 programming. Oper. Res. Lett. 11(2), 119–123 (1992)
16. Piatkowski, N., Gerlach, T., Hugues, R., Sifa, R., Bauckhage, C., Barbaresco, F.: Towards bundle adjustment for satellite imaging via quantum machine learning. In: 2022 25th International Conference on Information Fusion (FUSION), pp. 1–8. IEEE (2022)
17. Somma, R.D., Boixo, S., Barnum, H., Knill, E.: Quantum simulations of classical annealing processes. Phys. Rev. Lett. 101(13), 130504 (2008)

Example-Based Explanations of Random Forest Predictions

Henrik Boström$^{(\boxtimes)}$ (ID)

KTH Royal Institute of Technology, Stockholm, Sweden
bostromh@kth.se

Abstract. A random forest prediction can be computed by the scalar product of the labels of the training examples and a set of weights that are determined by the leafs of the forest into which the test object falls; each prediction can hence be explained exactly by the set of training examples for which the weights are non-zero. The number of examples used in such explanations is shown to vary with the dimensionality of the training set and hyperparameters of the random forest algorithm. This means that the number of examples involved in each prediction can to some extent be controlled by varying these parameters. However, for settings that lead to a required predictive performance, the number of examples involved in each prediction may be unreasonably large, preventing the user from grasping the explanations. In order to provide more useful explanations, a modified prediction procedure is proposed, which includes only the top-weighted examples. An investigation on regression and classification tasks shows that the number of examples used in each explanation can be substantially reduced while maintaining, or even improving, predictive performance compared to the standard prediction procedure.

Keywords: Random forests · Explainable machine learning · Example-based explanations

1 Introduction

Random forests [2] is a very popular and competitive machine learning algorithm that is widely considered to produce black-box models; even if each individual tree in a forest is interpretable, it is very hard to grasp an explanation that consists of several hundred (and sometimes even more) paths, each leading from the root of a tree to a leaf node, often also providing conflicting predictions. Techniques for explaining predictions of black-box models have received a lot of attention in recent years, with LIME [12] and SHAP [7] being two prominent examples of model-agnostic approaches that explain predictions by feature scores. In addition to explaining random forest predictions using feature scores, e.g., using TreeSHAP [6], techniques have also been proposed to approximate the random forests by interpretable rule sets, e.g., [1,3,9,13].

In contrast to explaining predictions by feature scores and rule sets, example-based explanation techniques explain the predictions by sets of examples [10].

© The Author(s), under exclusive license to Springer Nature Switzerland AG 2024
I. Miliou et al. (Eds.): IDA 2024, LNCS 14642, pp. 185–196, 2024.
https://doi.org/10.1007/978-3-031-58553-1_15

The latter can be useful in particular when the features are difficult to interpret. Such techniques do however require that the training examples can be presented to the user in an accessible way, e.g., as images. Apart from research on counterfactual explanation techniques, see e.g., [5,15], which synthesize new examples that lead to changing a prediction, example-based explanation techniques for tree-based methods have received limited attention. One exception is the prototype selection approach proposed in [14], which applies clustering to find prototypical examples for each class to approximate a random forest by a nearest-neighbor procedure. In contrast to this approach and also to the previous rule-based approaches, we will in this work focus on exact (perfect fidelity) explanations; we hence do not rely on approximating the underlying model.

In [8], it was shown that a prediction of a random regression forest can be expressed as a scalar product of the labels of the training examples and a set of weights obtained from the leaf nodes into which the test object falls. In [8], the weights and labels were used to form cumulative distribution functions for quantile regression forests, while we will here instead consider them for explaining the predictions. As noted in [4], the weight attribution also applies to classification; class membership of each training example can be encoded by a binary vector, which can be readily used when computing the scalar product. Using such a formulation, we can hence identify exactly which, and to what extent, training examples contribute to a prediction of a random forest for both classification and regression tasks.

To the best of our knowledge, there has been no investigation of the effective number training examples used in the predictions of a random forest, i.e., the number of training examples with non-zero weights. This number may not only be dependent on the dimensionality of the training set, but also on the leaf and forest sizes. Even if we to some extent can control this number, potentially at the cost of reduced predictive performance, e.g., by keeping the number of trees in the forest small, we may still end up with a number of examples that is too large to be useful, e.g., interpreting hundreds of training examples may be as difficult as interpreting hundreds of paths. In this work, we propose to control this number by a modified prediction procedure; only the top-weighted training examples are used when forming the prediction for a test example. We hence end up with a procedure that is constrained in number (or weight) of the involved training examples, while providing an exact example-based explanation for each prediction, i.e., there is no approximation involved in how the actual prediction is computed. The main question that we will investigate is whether the effective number of examples can be reduced without sacrificing predictive performance.

In the next section, we describe the proposed approach in detail and in Sect. 3, we present results from an empirical investigation, where we first study how the way in which the random forest is formed may affect the effective number of training examples involved in the predictions, followed by an investigation, using both regression and classification datasets, of how controlling this number may impact the predictive performance. Finally, in Sect. 4, we discuss the main findings and outline some directions for future work.

2   Modifying the Prediction Procedure of Random Forests

We start out with some notation, before proceeding with the proposed modified prediction procedure of random forests.

2.1   Random Forests

Each training example consists of an object and a label; let $X = \{x_1, \ldots, x_n\}$ denote the set of training objects and $y = \{y_1, \ldots, y_n\}$ the set of labels. For a regression problem, each $y_i \in \mathbb{R}$. For a classification problem, where the class labels of the training objects are $\{y'_1, \ldots, y'_n\}$, with each $y'_i \in \{c_1, \ldots, c_k\}$, each label $y_i = \langle \mathbb{1}(y'_i = c_1), \ldots, \mathbb{1}(y'_i = c_k) \rangle$, i.e., a binary vector with zeros for all classes except the class label of the object.

Let $F = \{T_1, \ldots, T_s\}$ be a random forest; we refer to it as a classification forest if each T_t is a classification tree, and a regression forest if each T_t is a regression tree. Let $\hat{y}_t = T_t(x)$ denote the output (prediction) of a regression or classification tree T_t for a (test) object x; for a regression tree $\hat{y}_t \in \mathbb{R}$ and for a classification tree $\hat{y}_t = \langle p_1, \ldots, p_k \rangle \in [0,1]^k$, such that $\sum_{i=1}^{k} p_i = 1$, i.e., the output is a class probability distribution. The prediction of the random forest F for the test object x is:

$$F(x) = s^{-1} \sum_{t=1}^{s} T_t(x) \tag{1}$$

Note that for a classification forest, the prediction is a class probability distribution, similar to the individual trees in the forest. Following [8], the above can be equivalently expressed as the scalar product of the labels of the training objects (y) and a set of (non-negative) weights $w_x = \{w_{x,1}, \ldots, w_{x,n}\}$:

$$F(x) = y \cdot w_x \tag{2}$$

where each weight $w_{x,i}$ is defined by $w_{x,i} = s^{-1} \sum_{t=1}^{s} w_{x,i,t}$ and $w_{x,i,t}$ is defined by $w_{x,i,t} = \frac{b_{x,i,t}}{\sum_{j=1}^{n} b_{x,j,t}}$. Where $b_{x,i,t}$ denotes the number of occurrences of the (possibly duplicated) training object x_i in the leaf node of the tree T_t into which x falls. Note that in case a training object has not been part in the construction of a tree, i.e., it is out-of-bag for that tree, the corresponding weight will be zero independently of what leaf node the test object falls into. Note also that in case a training object does not occur in any of the leafs in any of the trees that the test object falls into, the total weight for the training example will be zero.

2.2   Modifying the Predictions

Without loss of generality, we may assume that the weights for a test object are sorted in decreasing order, i.e., we can form the random forest prediction by:

$$F(x) = \langle y_{\sigma_1}, \ldots, y_{\sigma_n} \rangle \cdot \langle w_{x,\sigma_1}, \ldots, w_{x,\sigma_n} \rangle \tag{3}$$

where $w_{x,\sigma_1}, \ldots, w_{x,\sigma_n}$ are the weights for the test object (x) sorted from the highest to the lowest with each σ_i denoting the original index. We will investigate two alternative ways of making a prediction with a reduced number of training examples; by choosing the k top-weighted objects only (Algorithm 1) and choosing a set of examples such that the cumulative weight exceeds a specified threshold (Algorithm 2); for brevity, we denote each weight $w_{x,i}$ by w_i in the algorithms. Note that in both algorithms the selected weights need to be normalized.

Algorithm 1: k top-weighted	**Algorithm 2:** Cumulative weight
Require: $y = \{y_1, \ldots, y_n\}$ $\quad w = \{w_1, \ldots, w_n\}$ $\quad 0 < k \leq n$	**Require:** $y = \{y_1, \ldots, y_n\}$ $\quad w = \{w_1, \ldots, w_n\}$ $\quad 0 < c \leq 1$
1: $\sigma_1, \ldots, \sigma_n \leftarrow$ $\quad SortedIndex(w)$	1: $\sigma_1, \ldots, \sigma_n \leftarrow SortedIndex(w)$
2: $z \leftarrow \sum_{i=1}^{k} w_{\sigma_i}$	2: $z_j \leftarrow \sum_{i=1}^{j} w_{\sigma_i}$, for $j = 1, \ldots, n$
3: $\hat{y} \leftarrow \langle y_{\sigma_1}, \ldots, y_{\sigma_k} \rangle \cdot$ $\quad \langle w_{\sigma_1}/z, \ldots, w_{\sigma_k}/z \rangle$	3: $k \leftarrow \min_{\{1,\ldots,n\}}$ s.t. $z_i \geq c$
4: **return** $\hat{y}$	4: $\hat{y} \leftarrow \langle y_{\sigma_1}, \ldots, y_{\sigma_k} \rangle \cdot$ $\quad \langle w_{\sigma_1}/z_k, \ldots, w_{\sigma_k}/z_k \rangle$
	5: **return** $\hat{y}$

3   Empirical Investigation

In this section, we first investigate the effect of hyperparameter settings and dimensionality of the dataset on the effective number of training examples needed to form the predictions, i.e., the number of training instances with non-zero weights. We then present results from controlling the effective number of training examples on two prediction tasks.

3.1   Observing the Effective Number of Training Examples

Experimental Setup. We have chosen the Lipophilicity dataset from MoleculeNet [16], which contains measurements of the octanol/water distribution coefficient for 4200 chemical compounds, represented by the simplified molecular-input line-entry system (SMILES). The Python package RDKit[1] is used to generate features from the SMILES strings, more specifically, *Morgan fingerprints* (binary vectors, all of length 1024, if not stated otherwise). In addition to considering the original regression problem, we also frame it as a binary classification problem, where the task is to predict whether the target is greater than or equal to the mean of the targets, and as a multiclass classification problem, by equal-width binning of the regression values into ten categories.

We employ 10-fold cross validation, using the same folds and random seeds for all generated forests. In the first of four investigations for the three tasks,

---

[1] https://www.rdkit.org.

we vary the number of training examples by subsampling from the available training set, where a larger subsample always includes a smaller. In the second investigation, we vary the number of features by considering Morgan fingerprints of different sizes. In the third investigation, we vary the number of trees in the forests, and finally, in the fourth investigation, we vary the minimum sample size in each leaf. In addition to the average number of training examples that are assigned a non-zero weight for each test example (N), we also report the predictive performance; root mean squared error (RMSE) and Pearson correlation coefficient (Corr.) for regression, and accuracy (Acc.) and area under ROC curve (AUC) for classification.

The regression and classification forests are generated using `scikit-learn` [11], with the default settings, except when stated otherwise. The methods to fit and apply the forests have been modified to allow for measuring and controlling the number of training examples used in the predictions. It has been verified that the generated predictions, when not limiting the number of involved training examples, are identical to those generated by the original implementation.

Results for Regression. In Table 1, the results from the four investigations on the regression task are shown. Table 1a shows that the predictive performance is improved, as expected, when increasing the number of training examples. More interestingly, the effective number of training examples can be observed to decrease when increasing the training set size. A similar effect can be observed when increasing the number of features, as shown in Table 1b; the predictive performance is improving while the effective number of training examples is decreasing. In contrast, increasing the number of trees in the forest leads to an increased number of training examples with non-zero weights, while the predictive performance is improved, albeit quite marginally, as seen in Table 1c. Finally, Table 1d shows that increasing the minimum leaf sample size has a detrimental effect on both predictive performance and the number of examples needed to explain the predictions, assuming that fewer examples are preferred.

Results for Binary Classification. In Table 2, the results for the binary classification task are shown. A first observation is that the number of training examples with non-zero weights are much larger for this task compared to the regression task; this can be attributed to the larger number of training examples falling into each leaf. In Table 2a, the predictive performance is again observed to be improved with the number of training examples, but in contrast to the regression task, the effective number of training examples is consistently increasing with larger training sets. The picture is a bit different when increasing the number of features, as seen in Table 2b; although the predictive performance is increasing with the number of features, the effective number of examples is instead changing non-monotonically, with a maximum reached at 1024 features. When it comes to increasing the number of trees in the forest, the results are similar to when considering the regression task; larger forests lead to larger number of training examples with non-zero weights; while the predictive performance

Table 1. Regression results for the Lipophilicity dataset

#Ex.	RMSE	Corr.	N
500	1.042	0.503	61.7
1000	0.977	0.587	59.8
1500	0.940	0.630	58.8
2000	0.904	0.665	57.6
2500	0.894	0.674	56.2
3000	0.876	0.689	55.5
3500	0.860	0.703	54.1

(a) No. of training examples

#Feat.	RMSE	Corr.	N
128	0.942	0.632	64.9
256	0.901	0.671	62.1
512	0.868	0.698	57.2
1024	0.848	0.713	53.6
2048	0.820	0.734	51.0
4096	0.806	0.744	48.7
8192	0.799	0.750	48.0

(b) No. of features

#Est.	RMSE	Corr.	N
100	0.852	0.710	53.6
250	0.846	0.716	105.1
500	0.846	0.716	167.4
750	0.845	0.716	215.1
1000	0.844	0.717	255.9
1250	0.844	0.717	291.7
1500	0.844	0.718	322.5

(c) No. of trees

#Samp.	RMSE	Corr.	N
1	0.849	0.712	53.5
5	0.872	0.698	267.2
10	0.910	0.664	466.3
15	0.938	0.638	632.0
20	0.960	0.616	767.9
25	0.977	0.598	879.3
30	0.992	0.580	978.3

(d) Minimum leaf sample size

(marginally) improves, as can be observed in Table 2c. Finally, Table 2d shows that increasing the minimum leaf sample size again has a detrimental effect on both predictive performance and the number of examples.

Results for Multiclass Classification. In Table 3, the results for the multiclass classification task are shown. The effective number of training examples used in the predictions can be observed to fall in between of regression forests and binary classification forests. Due to the more fine-grained class labels, the tree growth typically continues beyond that of the binary classification trees, resulting in leafs with fewer examples, which has a direct effect on the number of training examples with non-zero weights. Table 3a shows that the predictive performance improves when increasing the number of training examples, as observed also for the previous tasks, but in contrast to these, the effective number of training examples is not monotonically increasing or decreasing with larger training sets, but peaks near the middle of the considered range of training set sizes. Again, the predictive performance is increasing with the number of features, and similarly to the regression task, but different from the binary classification task, the effective number of involved training examples is decreasing with the

Table 2. Binary classification results for the Lipophilicity dataset

#Ex.	Acc.	AUC	N
500	0.683	0.747	306.8
1000	0.718	0.789	402.4
1500	0.744	0.819	455.2
2000	0.751	0.828	493.1
2500	0.768	0.843	521.0
3000	0.774	0.853	545.7
3500	0.772	0.857	560.5

(a) No. of training examples

#Feat.	Acc.	AUC	N
128	0.760	0.837	388.5
256	0.770	0.849	473.3
512	0.772	0.856	530.1
1024	0.785	0.863	565.5
2048	0.782	0.868	540.5
4096	0.791	0.874	489.7
8192	0.797	0.876	392.2

(b) No. of features

#Est.	Acc.	AUC	N
100	0.789	0.865	566.3
250	0.786	0.865	983.8
500	0.788	0.867	1382.4
750	0.791	0.867	1624.7
1000	0.792	0.868	1807.5
1250	0.790	0.869	1939.0
1500	0.791	0.868	2050.3

(c) No. of trees

#Samp.	Acc.	AUC	N
1	0.789	0.864	567.6
5	0.766	0.849	702.3
10	0.752	0.830	1295.9
15	0.740	0.816	1759.9
20	0.730	0.807	2135.4
25	0.721	0.800	2423.2
30	0.710	0.791	2668.6

(d) Minimum leaf sample size

dimensionality, as seen in Table 3b. As was observed for both the regression and binary classification tasks, larger forests consistently lead to increasing the effective number of used examples, while the predictive performance is marginally affected, as can be observed in Table 3c. Finally, Table 3d shows that similar to the previous two cases, an increased minimum leaf sample size results in lower predictive performance and larger number of examples.

Summary of the Findings. Two consistent patterns were observed across the three considered predictions tasks; increasing the number of trees in the forests leads to improved predictive performance and an increased number of training examples involved in the predictions, while increasing the minimum leaf sample size leads to deteriorated predictive performance and a substantial increase in the number of training examples with non-zero weight. The last finding suggests that the smallest possible minimum leaf sample size, i.e., 1, should be employed, which indeed is the default for random forests in scikit-learn. When it comes to the number of trees in the forests, there is a trade-off between the predictive performance and the effective number of examples; there may be reasons to use more than the default of 100 trees in scikit-learn, but the relatively small

Table 3. Multiclass classification results for the Lipophilicity dataset

#Ex.	Acc.	AUC	N
500	0.238	0.604	125.9
1000	0.257	0.637	136.0
1500	0.283	0.664	134.7
2000	0.292	0.679	134.7
2500	0.297	0.687	130.1
3000	0.303	0.701	128.1
3500	0.309	0.705	126.0

(a) No. of training examples

#Feat.	Acc.	AUC	N
128	0.311	0.694	132.1
256	0.310	0.704	138.5
512	0.320	0.708	134.3
1024	0.310	0.714	124.8
2048	0.320	0.719	107.8
4096	0.315	0.723	92.2
8192	0.324	0.722	77.7

(b) No. of features

#Est.	Acc.	AUC	N
100	0.318	0.715	123.7
250	0.320	0.723	256.2
500	0.316	0.724	421.5
750	0.313	0.725	549.9
1000	0.318	0.725	654.2
1250	0.319	0.727	743.3
1500	0.316	0.726	820.7

(c) No. of trees

#Samp.	Acc.	AUC	N
1	0.321	0.715	124.0
5	0.291	0.722	698.0
10	0.278	0.713	1308.7
15	0.268	0.704	1782.8
20	0.261	0.698	2176.4
25	0.261	0.692	2436.8
30	0.250	0.687	2694.8

(d) Minimum leaf sample size

improvements beyond 500 trees or so come at a quite substantial cost in the number of examples needed to explain the predictions.

The most surprising finding was that increasing the training set size may not only lead to improved predictive performance, as expected, but also to a reduced number of training examples used in the predictions, as was observed for the regression and multiclass classification tasks. This means that reducing the training set is not always a good strategy to minimize the effective number of examples. A similar finding was made with respect to the number of features; a higher dimensionality consistently lead to higher performance, and for the regression and multiclass classification tasks, the lowest number of examples were used when the highest number of features were considered. The two last findings suggest that using as many training examples and features as possible can, at least in some cases, be advisable as both the predictive performance and the number of training examples used in the explanations benefit from this.

3.2 Controlling the Number of Examples Used in the Predictions

Results for Regression. As in the previous section, we here consider the Lipophilicity dataset for the (original) regression task, using the largest number of features (8192). Again, we perform 10-fold cross-validation, here using a regression forest with 500 trees and with all other parameters set to the default.

Table 4. Regression results for the Lipophilicity dataset

k	N	W	RMSE	Corr.
1	1.0	0.222	0.988	0.656
3	3.0	0.398	0.850	0.723
5	5.0	0.481	0.822	0.737
10	10.0	0.586	0.803	0.747
15	15.0	0.644	0.796	0.751
20	20.0	0.685	0.793	0.752
30	29.9	0.743	0.789	0.755
50	48.6	0.818	0.788	0.756
100	87.5	0.915	0.791	0.755

(a) Varying number of examples

c	W	N	RMSE	Corr.
0.1	0.238	1.4	0.918	0.693
0.2	0.289	2.6	0.865	0.717
0.3	0.365	4.7	0.834	0.732
0.4	0.447	7.8	0.817	0.740
0.5	0.535	12.6	0.804	0.747
0.6	0.621	19.9	0.796	0.751
0.7	0.713	31.3	0.792	0.753
0.8	0.806	50.7	0.792	0.753
0.9	0.902	82.9	0.793	0.753
1.0	1.000	137.5	0.796	0.752

(b) Varying cumulative weight

In Table 4, the results from controlling the effective number of examples (Table 4a) and the cumulative weight of the examples (Table 4b) are presented. In column N, the effective number of training examples are shown; note that in the first sub-table, this number may be less than the specified number (k), in particular for large values of the latter, as the number of examples receiving a non-zero weight may be less than k. Column W presents the average observed cumulative weight of the examples; note that in the second sub-table, this number is typically larger than the specified cumulative weight (c), in particular for smaller values of c, as the latter provides a lower bound. The predictive performance of the standard regression forest is shown in the last row of Table 4b (where c = 1.0), where on average 137.5 training examples receive a non-zero weight. The results in Table 4a show that the same predictive performance as the original forest can be obtained with as few as 15–20 training examples, which corresponds to a reduction of 85–90% in the number of examples needed to explain the predictions. Interestingly, it can be observed in Table 4b that using a cumulative weight of 0.7 outperforms the original regression forest (as well as most other considered settings for the cumulative weight), while reducing the number of involved examples to less than a fourth.

In Fig. 1, we illustrate the use of the above model trained on 90% of the data and applied to a random test object, using the top five ($k = 5$) training examples for forming the predictions. Below the test object in Fig. 1a, the predicted ($\hat{y}$) and actual (y) values are shown. Below each of the training objects in Fig. 1b-f, the label (y) and the weight (w) are shown. Highlighted atoms in the training objects indicate parts that are missing in the test object. Even without knowledge about

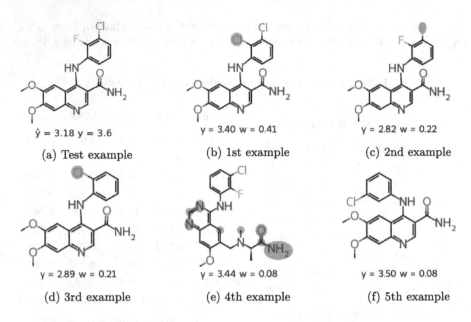

ŷ = 3.18 y = 3.6

(a) Test example

y = 3.40 w = 0.41

(b) 1st example

y = 2.82 w = 0.22

(c) 2nd example

y = 2.89 w = 0.21

(d) 3rd example

y = 3.44 w = 0.08

(e) 4th example

y = 3.50 w = 0.08

(f) 5th example

Fig. 1. Test example and $k = 5$ training examples

the particular features used by the black-box model, the user can inspect and reason about the actual objects that constitute the basis for the prediction.

Results for Classification. We here consider the MNIST dataset with 70 000 handwritten digits, represented by 784 features (28×28 pixel boxes) and for which the set of class labels is $\{0, \ldots, 9\}$. We employ ten-fold cross-validation and consider classification forests of 500 trees with all other parameters set to default.

In Table 5, the results from controlling the effective number of examples (k in Table 5a) and the cumulative weight of the examples (c in Table 5b) are presented, again with the columns N and W corresponding to the effective number and the cumulative weight of the examples, respectively. The predictive performance of the standard random forest is shown in the last row of Table 5b (where c = 1.0), where on average 5846.9 training examples receive a non-zero weight. The results in Table 5a show that the original forest can be outperformed with as few as $k = 10$ training examples; this corresponds to a reduction of 99.8% in the number of examples needed to explain the predictions. Similar results can be observed for several of the settings in Table 5b.

In Fig. 2, we illustrate the use of a classification forest trained on 90% of the data when forming predictions using the top five ($k = 5$) training examples. We have randomly selected one test object with label $y = 7$ incorrectly predicted as $\hat{y} = 2$, shown in Fig. 2a; the predicted label is chosen according to the predicted class probability distribution $\langle 0, 0, 0.75, 0.09, 0, 0, 0, 0.16, 0, 0 \rangle$ (over the

Table 5. Classification results for the MNIST dataset

k	N	W	Acc.	AUC
1	1.0	0.008	0.861	0.923
3	3.0	0.020	0.922	0.988
5	5.0	0.029	0.952	0.996
10	10.0	0.048	0.974	0.999
15	15.0	0.064	0.980	0.999
20	20.0	0.077	0.982	0.999
30	30.0	0.101	0.983	1.000
50	50.0	0.140	0.984	1.000
100	100.0	0.217	0.984	1.000

(a) Varying number of examples

c	W	N	Acc.	AUC
0.1	0.101	51.7	0.983	0.999
0.2	0.201	136.8	0.984	1.000
0.3	0.301	252.4	0.984	1.000
0.4	0.400	403.4	0.983	1.000
0.5	0.500	600.3	0.982	1.000
0.6	0.600	859.4	0.980	0.999
0.7	0.700	1209.9	0.979	0.999
0.8	0.800	1712.8	0.976	0.999
0.9	0.900	2537.9	0.975	0.999
1.0	1.000	5846.9	0.972	0.999

(b) Varying cumulative weight

labels $0, \ldots, 9$), which is defined by the weights (w) and labels (y) of the training objects in Fig. 2b-f. Again, the user may inspect the training examples that fully explain the prediction, i.e., no other examples are involved in forming it, and e.g., reason about whether the prediction is reliable or not.

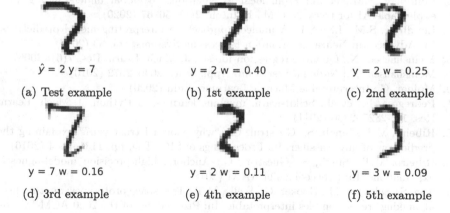

(a) Test example $\hat{y} = 2\ y = 7$ (b) 1st example $y = 2\ w = 0.40$ (c) 2nd example $y = 2\ w = 0.25$

(d) 3rd example $y = 7\ w = 0.16$ (e) 4th example $y = 2\ w = 0.11$ (f) 5th example $y = 3\ w = 0.09$

Fig. 2. Test example and $k = 5$ training examples

4 Concluding Remarks

An investigation of the number of training examples involved in random forest predictions has been presented, highlighting the impact of dataset properties and hyperparameter settings. An approach to controlling this number by including only the top-weighted examples has been proposed, and an empirical investigation shows that this approach may substantially reduce the effective number of training examples involved in the predictions, while maintaining, and even improving, the predictive performance compared to the standard prediction procedure.

Directions for future research include extending the empirical investigation, e.g., by considering more datasets and hyperparameter settings, and investigating other approaches to selecting examples based on the weights. Other directions concern studying the usability of the example-based explanations when solving practical tasks, and also exploring combinations of explanation techniques, e.g., complementing the example-based explanations with rules or feature scores.

References

1. Boström, H., Gurung, R.B., Lindgren, T., Johansson, U.: Explaining random forest predictions with association rules. Arch. Data Sci. Ser. A **5**(1), A05 (2018)
2. Breiman, L.: Random forests. Mach. Learn. **45**(1), 5–32 (2001)
3. Deng, H.: Interpreting tree ensembles with inTrees. Int. J. Data Sci. Anal. **7**(4), 277–287 (2019)
4. Geurts, P., Ernst, D., Wehenkel, L.: Extremely randomized trees. Mach. Learn. **63**, 3–42 (2006)
5. Guidotti, R.: Counterfactual explanations and how to find them: literature review and benchmarking. Data Mining Knowl. Discovery, 1–55 (2022)
6. Lundberg, S.M., et al.: From local explanations to global understanding with explainable AI for trees. Nat. Mach. Intell. **2**(1), 56–67 (2020)
7. Lundberg, S.M., Lee, S.I.: A unified approach to interpreting model predictions. In: Advances in Neural Information Processing Systems, vol. 30 (2017)
8. Meinshausen, N.: Quantile regression forests. J. Mach. Learn. Res. **7**(6) (2006)
9. Meinshausen, N.: Node harvest. Ann. Appl. Stat., 2049–2072 (2010)
10. Molnar, C.: Interpretable Machine Learning. Lulu (2020)
11. Pedregosa, F., et al.: Scikit-learn: machine learning in Python. J. Mach. Learn. Res. **12**, 2825–2830 (2011)
12. Ribeiro, M.T., Singh, S., Guestrin, C.: "why should I trust you?" explaining the predictions of any classifier. In: Proceedings of SIGKDD, pp. 1135–1144 (2016)
13. Ribeiro, M.T., Singh, S., Guestrin, C.: Anchors: high-precision model-agnostic explanations. In: Proceedings of AAAI (2018)
14. Tan, S., Soloviev, M., Hooker, G., Wells, M.T.: Tree space prototypes: another look at making tree ensembles interpretable. In: Proceedings of the 2020 ACM-IMS on Foundations of Data Science Conference, pp. 23–34 (2020)
15. Tolomei, G., Silvestri, F., Haines, A., Lalmas, M.: Interpretable predictions of tree-based ensembles via actionable feature tweaking. In: Proceedings of SIGKDD, pp. 465–474 (2017)
16. Wu, Z., et al.: MoleculeNet: a benchmark for molecular machine learning (2018)

FLocalX - Local to Global Fuzzy Explanations for Black Box Classifiers

Guillermo Fernandez[1]([✉]) [ID], Riccardo Guidotti[2] [ID], Fosca Giannotti[3] [ID],
Mattia Setzu[2] [ID], Juan A. Aledo[1] [ID], Jose A. Gámez[1] [ID], and Jose M. Puerta[1] [ID]

[1] Intelligent Systems and Data Mining Lab, Albacete, Spain
{Guillermo.Fernandez,JuanAngel.Aledo,Jose.Gamez,Jose.Puerta}@uclm.es
[2] University of Pisa, Pisa, Italy
{riccardo.guidotti,mattia.setzu}@unipi.it
[3] Scuola Normale Superiore, Pisa, Italy
fosca.giannotti@sns.it

Abstract. The need for explanation for new, complex machine learning models has caused the rise and growth of the field of *eXplainable Artificial Intelligence*. Different explanation types arise, such as *local explanations* which focus on the classification for a particular instance, or *global explanations* which aim to show a global overview of the inner workings of the model. In this paper, we propose FLocalX, a framework that builds a fuzzy global explanation expressed in terms of fuzzy rules by using local explanations as a starting point and a metaheuristic optimization process to obtain the result. An initial experimentation has been carried out with a genetic algorithm as the optimization process. Across several datasets, black-box algorithms and local explanation methods, FLocalX has been tested in terms of both fidelity of the resulting global explanation, and complexity The results show that FLocalX is successfully able to generate short and understandable global explanations that accurately imitate the classifier.

Keywords: XAI · Optimization · Metaheuristics · Fuzzy Rule-Based Systems · Local Explanations · Global Explanations

1 Introduction

In recent years, the increasing amount of data has allowed new, more complex models to be incorporated into a wide range of tasks [5,6,26]. However, the increasing complexity usually causes a decrease in model interpretability [2], which may not be advisable or suitable in certain critical fields, i.e., medicine, law, aviation, etc. Current European legislation also deals with this topic by means of the *right to explanation* included in the General Data Protection Regulation [18], which affects both humans and artificial intelligence techniques. *eXplainable Artificial Intelligence* (XAI) [3,9] aims to push the usage of interpretability and explainability in order to gain an understanding of complex black box models used in sensitive contexts and critical areas.

© The Author(s), under exclusive license to Springer Nature Switzerland AG 2024
I. Miliou et al. (Eds.): IDA 2024, LNCS 14642, pp. 197–209, 2024.
https://doi.org/10.1007/978-3-031-58553-1_16

Within the XAI taxonomy, one important distinction is whether a method generates *local* or *global* explanations. *Local explanations* are aimed at individual instances, and explain the decisions made by the model in a small neighborhood of the feature space around an instance, while *global explanations* aim to explain the entire behavior of the model. A common type of local explanation are factual and counterfactual explanations [7,8]. Factual explanations explain the reasoning behind a decision, while counterfactual explanations highlight the necessary changes to revert that decision. Focusing on decision rules as explanations, LORE (LOcal Rule-based Explainer) [8] is a well-known XAI algorithm that generates both factual and counterfactual local explanations by learning a proper neighborhood of the given instance, then inducing a *crisp* decision tree from which crisp rules are extracted. Further building on this idea, FLARE[1] instead leverages a *fuzzy* decision tree, extracting fuzzy, rather than crisp, rules. Due to their ease of extraction and high accuracy, local explanations have become a building block for global ones, blurring the line between the two. In [11] the authors turn local Shapley values into global explanations by means of functional decomposition. Other works merge local and global explanations through feature importance [15], concept relevance [19], saliency maps [20] and strategy summaries [13]. Most related to our application on rules as explanations, GLocalX [22], from which this paper takes inspiration, merges local crisp explanations to build a *global explanation theory*.

In this paper, we introduce *FLocalX*, a framework to create an agnostic global explanation theory for a black-box classifier in the form of a *fuzzy* rule-based system built using *local fuzzy explanations*. This global fuzzy explanation theory mimics the behavior of the underlying black-box classifier, and can be used to provide factual explanations for novel, previously non-explained instances whiel providing a general understanding of the model. This way, a user can better understand how the classifier works, and how it will behave upon new instances, rather than generate explanations *ex-novo*. Building a global theory with fuzzy, rather than crisp, rules leads to additional benefits, making the global explanation more understandable, flexible, and faithful to the black-box model. Fuzzy rules leverage linguistic labels, which improve their readability by associating high-level human-understandable concepts with their premises and have been widely used to design explainable systems [16,25] Fuzziness also allows us to infer several, rather than one, explanations per instance, effectively providing the user with alternative explanations. Performance-wise, fuzzy rule-based systems are particularly apt to leverage different types of local explanations [23].

The rest of the paper is structured as follows. Section 2 presents the problem and identifies the relevant elements. Section 3 illustrates the workflow of our proposal. Section 4 shows the experiments and behavior of FLocalX. Finally, Sect. 5 presents the conclusions and indicates some future research lines.

---

[1] https://dsi.uclm.es/descargas/technicalreports/DIAB-24-02-1/
FLARE_Tech_Rep.pdf.

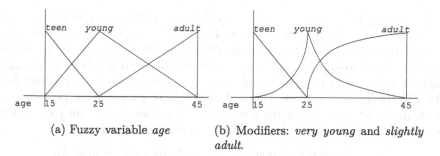

(a) Fuzzy variable *age*

(b) Modifiers: *very young* and *slightly adult*.

Fig. 1. Strong fuzzy partition for a fuzzy variable *age*

2   Setting the Stage

The Local to Global Explanation Problem [22] we aim to solve consists of finding a function g that, from a set of local explanations extracted from a black box classifier, yields a global explanation theory that describes its underlying logic.

First, let us revise some related concepts. In a classification problem, an instance $x = (x_1, \ldots, x_n) \in \mathcal{X}_1 \times \cdots \times \mathcal{X}_n$, where $\mathcal{X}_1, \ldots, \mathcal{X}_n$ are n sets of *input variables*, is mapped to a decision $y \in \mathcal{Y} = \{y_1, \ldots, y_n\}$ by a function (classifier) $f : \mathcal{X}_1 \times \cdots \times \mathcal{X}_n \to \mathcal{Y}$. We write $f(x) = y$ to denote the classification y given to x. Let us denote by n_{cont} (resp. n_{disc}) the number of continuous (resp. discrete) variables in $\mathcal{X}$, s.t. $0 \leq n_{cont}, n_{disc} \leq n$, $n_{cont} + n_{disc} = n$. Let us assume that, associated with each continuous input variable $\mathcal{X}_i$, there is a fuzzy (linguistic) variable $\mathcal{F}_i = \{v_{i,1} \ldots, v_{i,k_i}\}$ defined through a Ruspini partition [1] of k_i ordered fuzzy sets (see Fig. 1)[2]. We use v_{i,z_i} to denote both the fuzzy set and its corresponding associated linguistic label, indistinctly. A triangular fuzzy set is defined by a triple of real-valued points: (start, peak, end), i.e. *teen* = $(15, 15, 25)$ and *young* = $(15, 25, 45)$ in Fig. 1a. . If we know the minimum and maximum values of $dom(\mathcal{X}_i)$, the partition becomes specified by $k_i - 2$ values. Given a value $\delta \in dom(\mathcal{X}_i)$, let $\mu_i(\delta) = (\mu_{i,1}(\delta), \ldots, \mu_{i,k_i}(\delta))$ be the vector of membership degrees of δ to the k_i fuzzy sets of $\mathcal{F}_i$. In other words, $\mu_{i,z_i}(\delta)$ is the membership degree of δ to the set v_{i,z_i}. A linguistic hedge, or linguistic modifier, is a function that alters the membership function of a fuzzy set, which can modify the shape of the fuzzy set (see Fig. 1b). In this work, we use two of the most common linguistic hedges, "very" and "slightly": $\mu_{i,z_i}^{very}(x_i) = (\mu_{i,z_i}(x_i))^2$ and $\mu_{i,z_i}^{slightly}(x_i) = \sqrt{\mu_{i,z_i}(x_i)}$. Finally, for discrete variables, we can interpret each value as a linguistic label whose associated fuzzy set has membership degree 1 in case the instance takes that value and 0 otherwise.

Let $b()$ be a classifier whose decision-making process needs to be explained, i.e., a black-box model, learned from a training dataset $TR =$

---

[2] Triangular membership functions are used in this article to illustrate the proposed method for simplicity/convenience. The framework allows other types of membership functions (Gaussian, trapezoidal, etc.) to represent the underlying fuzzy sets. However, the partitions must cover the complete domain for Eq. 1 to be valid.

$\{(x_1^t, \ldots, x_n^t, y^t)\}_{t=1}^T$. Let $e = \{r_1, \ldots, r_e\}$ be a multi-rule explanation formed by one (or more) fuzzy decision rules. Each rule $r = P(r) \rightarrow y(r)$ consists of a set of premises in conjunctive form $P(r) = p_{s_1} \wedge \cdots \wedge p_{s_r}$ and an outcome $y(r) \in \mathcal{Y}$. Each premise $p_i = \langle \mathcal{F}_i, v_{i,z_i} \rangle$ is an attribute-value pair. For the continuous variables, $\mathcal{F}_i$ is a fuzzy variable and v_{i,z_i} is one of its corresponding fuzzy sets. For the discrete variables, $\mathcal{F}_i = \mathcal{X}_i$ and v_{i,z_i} is a value from its domain. As an example, let us consider the following explanation for a loan request for a user $x = \{(age = 30), (job = Accountant), (amount = 20k)\}$:

$$e = \{(r_1 = age\ is\ young \wedge job\ is\ Accountant \rightarrow accept),$$
$$(r_2 = age\ is\ adult \wedge amount\ is\ high \rightarrow accept)\}$$

One property of multi-rule explanations is that, given an explanation e that explains the instance x, then $y(r) = b(x)$ for all $r \in e$. Fuzzy rules differ from crisp rules in that, while a crisp rule has a binary (0 or 1) match with an instance x, a fuzzy rule r has a *matching degree* with the instance, $md(r, x)$, defined as:

$$md(r, x) = \min_{i \in \{s_1, \ldots, s_r\}} \{\mu_{i,z_i}(x_i)\} \in [0, 1]$$

An explanation theory $E = e_1 \cup \cdots \cup e_q$ consists of a union of explanations which may have different outcomes.

Thus, the Local to Global Explanation Problem can be defined as follows: Given a black box $b()$, a set of instances $X = \{x^1, \cdots, x^q\}$ and their local explanations $\{e_1, \cdots, e_q\}$, the Local to Global Explanation Problem consists in deriving a global explanation theory $E_G = e'_1 \cup \cdots \cup e'_{q'}$ that aggregates the local explanations in order to summarize the logic of b.

3   Fuzzy Local to Global Explanation Framework

In this paper we propose FLocalX, a Fuzzy Local to Global Explanation framework that generates a global explanation theory which mimics a black box classifier given an initial set of local explanations. FLocalX takes the following elements as input a set of instances X and an explanation theory $E_L = e_1 \cup \cdots \cup e_q$ formed by the union of the explanations of every instance in X, and generates the global explanation theory E_G by applying the following steps:

- First, it *transforms*, i.e., maps, the local fuzzy sets $\mathcal{F}^j$ defined for each $e_j \in E_L$ to a common definition of fuzzy sets $\mathcal{F}^C$. This ensures that all local explanations in E_L share the same set of fuzzy variables. We name this explanation theory with common fuzzy sets E_C.
- Second, it *encodes* E_C into a simple, unique representation that will be the initial configuration C^{E_C} of the optimization process.
- Third, it *generates* the global explanation theory E_G from C^{E_C} through an optimization process.

This process results in a global explanation theory E_G that closely resembles the behavior of $b()$, and can provide a factual explanation for novel instances. Factual explanations can be extracted from E_G by obtaining, for instance, the *minimum robust factual explanation* defined in [7].

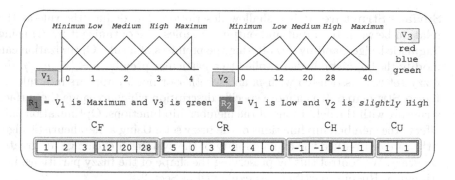

Fig. 2. Representation of the encoding of a FRBS

3.1 Local to Global Fuzzy Set Transformation

Depending on the method employed to extract the local explanations, they may not share the same fuzzy variable definitions, thus the same linguistic features may be defined by different fuzzy sets. For the sake of homogeneity, we uniform the fuzzy variable definitions $\mathcal{F}_i^1, \ldots, \mathcal{F}_i^{|E_L|}$ of a given variable $\mathcal{X}_i$, and establish a global fuzzy variable definition $\mathcal{F}^C$ by partitioning the domain of the numerical variables into equal-width sets, unless expert-provided sets are available.

Given two fuzzy sets $v_{i,z_i} \in \mathcal{F}_i$ and $v'_{i,z'_i} \in \mathcal{F}'_i$, we compute their similarity as

$$S(v_{i,z_i}, v'_{i,z'_i}) = A(v_{i,z_i} \cap v'_{i,z'_i})/A(v_{i,z_i} \cup v'_{i,z'_i})$$

where $A(v)$ is the area of the fuzzy set v. As usual in the literature, we use *min* as the intersection and *max* as the union. Then, given a variable $\mathcal{X}_i$, we define

$$M(v'_{i,z'_i}, \mathcal{F}_i) = \arg \max_{v_{i,z_i} \in \mathcal{F}_i} S(v_{i,z_i}, v'_{i,z'_i}), \qquad (1)$$

which takes a fuzzy set $v'_{i,z'_i} \in \mathcal{F}'_i$ and returns the set $v_{i,z_i} \in \mathcal{F}_i$ with the greatest similarity. We get E_C by applying Eq. 1 to every premise of each $e_i \in E_L$.

3.2 Global Fuzzy Set Theory Encoding

In FLocalX, we frame the objective of building a global explanation theory as the process of optimizing the Fuzzy Rule-Based System (FRBS) formed by the set of fuzzy decision rules in E_C. To this aim we need an encoding of E_C, this is, a representation of a potential solution to the problem which will be used by the metaheuristic algorithm in the optimization process.

The objective of the optimization process used by FLocalX is twofold. First, maintaining the degree in which the FRBS mimics the black-box classifier as accurate as possible. Second, making the FRBS as compact as possible (in terms of number of rules), to favor interpretability [9]. Inspired by [4], we design a procedure to tune FRBS maintaining interpretability by using a genetic algorithm. To this aim, there are two elements of the FRBS that must be optimized:

- **Surface Structure**. It is a shallow description that defines the rule as the relation between the input and output variables. We optimize it by *(i)* using linguistic hedges, and by *(ii)* altering the premises of a rule. Optimization can modify the linguistic hedge applied to a particular premise $p = \langle \mathcal{F}_i, v_{i,z_i} \rangle$, the fuzzy set v_{i,z_i} associated with p, and whether or not $\mathcal{F}_i$ appears in a rule.
- **Deep Structure**. It is a more specific description which expands the surface structure with the definitions of the membership functions. Optimization only affects the membership functions of the fuzzy sets. Using a metaheuristic algorithm we can reduce the explainability of the system in exchange for greater accuracy. We control this by preserving the shape of the fuzzy partitions, i.e., triangular Ruspini partitions as explained in Sect. 2.

Configuration. Each configuration C of the optimization process represents an explanation theory E, shown graphically in Fig. 2. For this purpose, we will use a four-part configuration $(C_F + C_R + C_H + C_U)$ as follows:

- C_F is the encoding of the fuzzy variables. We assume that the minimum and maximum values of $dom(\mathcal{X}_i)$ are known. As an example, in Fig. 1a we know $k_i = 3$, $min = 15$ and $max = 40$, and so we only have a free value (25) to codify the three fuzzy sets: $\{(15, 15, 25); (15, 25, 40); (25, 40, 40)\}$. Just by changing the value 25 to, e.g. 20, we modify the fuzzy semantics of the variable, obtaining a new partition: $\{(15, 15, 20); (15, 20, 40); (20, 40, 40)\}$. Thus, C_F has a length of $(\sum_{i=1}^{n_{cont}} k_i - 2)$, all of them being real numbers.
- C_R is the encoding of the rules. It has a length of $n \cdot |E|$ elements, where $|E|$ is the number of rules in the explanation theory, i.e., in the FRBS. Each n consecutive elements codify a rule with an ordinal encoding from the set $\{0, \ldots, k_i\}$, where 0 represents that the *i-th* variable does not appear in the rule and 1 to k_i identify each fuzzy set or value of the variable $\mathcal{F}_i$, depending on whether $\mathcal{X}_i$ is numerical or categorical.
- C_H is the encoding of the linguistic hedges. It has a length of $n_{cont} \cdot |E|$ elements, where each element belongs to the set $\{-1, 0, 1\}$ representing no linguistic hedge (-1), *very* (0) or *slightly* (1), for that particular continuous (fuzzy) variable.
- C_U is the encoding of the used rules. $|E|$ elements-long, encodes whether a rule is used in the final FRBS (1) or not (0).

3.3 Global Explanation Theory Generation

In order to generate the global explanation theory E_G, we exploit the encoding illustrated in the previous section to create the chromosomes of a genetic optimization process. Since the optimization process aims to simultaneously *(i)* accurately mimic the black box $b()$, and to *(ii)* have a compact FRBS, we designed an objective function that takes into account both aspects. In particular, we measure the first goal as the Area Under the ROC curve *(AUC)* to correctly handle imbalanced datasets, while we measure the second goal as the number of rules used in the system. Specifically, the objective function $f(C)$ that we maximize in our experimentation is defined as follows:

$$f(C) = \alpha \cdot (1 - \frac{\sum_{i=0}^{|C_U|-1} C_U[i]}{|C_U|}) + (1 - \alpha) \cdot AUC$$

with α used to balance the two values optimized. Note that this is an implementation of the objective function, but others may be used.

FLocalX can employ any metaheuristic algorithm as optimizer in order to obtain the global explanation theory. For this work, we adopted a genetic algorithm [12]. Inspired by evolutionary adaptation, genetic algorithms encode solutions in a chromosome space, and sequentially evolve them, each generation selecting, merging, and improving on the previous one. In our case, merging is encoded by *(i)* a *crossover* operation, which generates new solutions by blending existing ones, and *(ii)* a *mutation* operation, which randomly alters a subset of the current solutions. In genetic fashion, a *selection* operation picks the best solutions (according to an objective function) which will be carried on to the next generation. Next, we detail these crucial aspects of the genetic algorithm:

- **Initial Population.** The initial population of size $\rho + 1$ for the genetic algorithm is generated in an informed manner, i.e., by altering a known configuration (C^{Ec}) rather than generating all elements at random. Given an initial configuration representing of a FRBS, each part C_F, C_R, C_H and C_U, generates $\lceil \rho/4 \rceil$ chromosomes by applying the mutation operator to that part (see below). The original configuration is also included in the initial population.
- **Crossover.** It selects pairs of chromosomes and crosses them with a probability p_{cross}. Due to the encoding adopted, the chromosome is divided in two, and different crossovers are applied:
  - First, a min-max-arithmetic crossover [10] is applied in the C_F part, generating four children.
  - Second, a six-point crossover is applied in the remaining chromosome, choosing two points for each part (i.e. two for C_R, two for C_H, and two for C_U). This generates two children.

  After recombining both parts, eight children are generated. The two best children are selected in order to keep the same population size.
- **Mutation.** It selects chromosomes and mutates them with a probability p_{mut}. The mutation over each part of the chromosome is performed applying an operation to a single bit $C[i]$ of each part of the chromosome as follows:
  - For C_F, the bit is randomly generated by sampling a real number from a uniform distribution in the range of the continuous variable.
  - For C_R, the bit is randomly chosen in the set $\{0, \cdots, k_i\} \setminus C[i]$.
  - For C_H, the bit is randomly chosen in the set $\{-1, 0, 1\} \setminus C[i]$.
  - C_U is generated as $1 - C[i]$, i.e., altering the bit.
- **Selection.** A rank-based selection with respect to the fitness is used.
- **Replacement.** A replacement with elitism is performed, i.e., the best configuration from the previous population is kept.
- **Stop Criterion.** The genetic algorithm stops when the fitness of the best individual does not offer enough improvement beyond a threshold ϵ over a consecutive period of κ iterations.

These operations provide great flexibility and generality, and allow to directly *learn*, rather than *define*, the evolution of fuzzy rules. A challenging task such as the Local to Global one, which is not directly differentiable, requires flexible algorithms able to explore vast non-differentiable solution spaces, and adapt to a wide variety of users, and thus objectives. Optimizing global explanations to both be understandable by a user, as well as comprehensive enough to mimic a complex black-box classifier, is thus a perfect fit for our purpose.

4 Experiments

We evaluated FLocalX on three widely used multi-class datasets, i.e., *Iris*[3], *Wine*[4], and *Beer*[5]. The decision to use small datasets is driven by the main objective of developing and showcasing a framework to extract global explanations, rather than focusing on a specific metaheuristic (in this case, the genetic algorithm). As metaheuristic algorithms are resource-intensive and time-intensive processes, they often require specific optimizations made for each case and algorithm in order to tackle different problems. By employing simpler datasets, we can shift our focus towards illustrating the capability of the framework of working with different types of local explanations, as well as how it can seamlessly mimic a variety of black box algorithms. This is a first step in the line of work of a more complex experimentation where multiple metaheuristic algorithms are used and optimized with this framework in order to tackle much more complex problems. The implementation of FLocalX is available on Github[6]. **Experimental Setting.** We adopted the following metrics to evaluate the performance of FLocalX and the other classifiers used as baselines.

- *Accuracy*. It measures how close is the global explainer to the ground truth. We measure the accuracy of the black box (Acc-B), of the explanation theory formed by the union of the local explanations (Acc-U), and of the global explanation theory after applying FLocalX (Acc-F).
- *Fidelity*. It measures how well the global explainer mimics the black box classifier. We measure the fidelity of the explanation theory formed by the union of the local explanations (Fid-U), and the fidelity of the global explanation theory after applying FLocalX (Fid-F).
- *Number of Rules*. The total number of rules in the system. More rules indicate a more complex system and so a less interpretable system. We measure the number of rules before (#R) and after applying FLocalX (#R-F).
- *Number of Premises*. The number of premises in the antecedent of the rules. More premises are sometimes (falsely) perceived as being more helpful [14],

---

[3] https://archive.ics.uci.edu/dataset/53/iris.

[4] https://archive.ics.uci.edu/dataset/109/wine.

[5] https://gitlab.citius.usc.es/ilia.stepin/fcfexpgen/-/tree/master/all_datasets/BEER_exp1.

[6] GitHub: https://github.com/Kaysera/flocalx. FLocalX was programmed in Python 3.10, using libraries such as numpy and scikit-learn to properly manage the data structures and efficiently generate the explanations. To guarantee reproducibility, all the experiments are also published in a separate public Github repository https://github.com/Kaysera/ida2024-experiments.

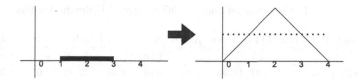

Fig. 3. LORE interval transformation to fuzzy set.

so shortening the rule together with a proper communication of attribute importance is a good practice. We measure the number of premises before (#P) and after applying FLocalX (#P-F).

We used a train-validation-test (60%-30%-10%) split for the experimentation. The training split was used to train the black box classifiers using default hyper-parameters. The validation partition was used to fit the hyperparameters of the local explanation methods, as well as to extract the local explanations (E_L). The test partition was used to measure the accuracy score for all algorithms. The genetic algorithm was repeated 20 times, altering the random seed and averaging the result between them. The parameters were chosen empirically[7] as follows: population size (ρ) = 128, size pressure (α) = 0.1, # iterations (κ) = 20, threshold (ϵ) = 0.01, # fuzzy sets (k_i) = 5, p_{mut}= 0.15 and p_{cross}= 0.8. The fuzzy sets for Iris and Wine were obtained using equal-width partitions, while the fuzzy sets for Beer were obtained from [24].

We experiment with FLocalX with a set of different alternatives:

- *Black-Box Models*: We used *SVM, Neural Network (NN)* and *Random Forest (RF)* as baseline classifiers as implemented by *scikit-learn* [17].
- *Rule-Based Models*: Algorithms from which a ruleset that can be used for both prediction and explanation can be extracted. They are used as global explanation systems. The algorithms used are *Fuzzy Decision Tree (FDT)* [21], *LORE* [8] and *FLARE*.
- *Local to Global Approaches*: They set local explanations and merge them into a global explanation theory that is able to predict and explain instances of the dataset. We considered:
  - *FLocalX + LORE*: We used LORE to extract local explanations and then applied FLocalX. As FLocalX takes fuzzy rules, the intervals were expanded into fuzzy sets as if they were an α-cut of 0.5 of the corresponding fuzzy set. For example, the interval [1, 3] would become the fuzzy set (0, 2, 4) as shown in Fig. 3.
  - *FLocalX + FLARE*: We used FLARE to extract local explanations and then FLocalX was applied.

---

[7] With these datasets, a large population size which is a power of 4 shows better results, and a small size pressure allows for faster convergence with a high accuracy. The rest of the parameters are standard for genetic tuning.

Table 1. Performance and Explainability of Different Models

	Method	Black Box	Fid-U	Fid-F	Acc-B	Acc-U	Acc-F	#R	#R-F	#P	#P-F
Iris	FDT	–	–	–	–	1.00	–	12.00	–	1.42	–
	FLARE	NN	**0.97**	0.93	1.00	0.95	0.91	32.00	5.05	1.31	**1.27**
		RF	0.94	0.91	0.93	0.94	0.91	26.00	**4.16**	1.81	1.70
		SVM	0.93	**0.95**	1.00	0.95	0.92	19.00	4.47	1.26	1.32
	LORE	NN	0.93	0.94	1.00	0.94	**0.93**	45.00	4.79	1.96	1.55
		RF	**0.97**	0.93	0.93	**0.97**	**0.93**	45.00	**4.16**	2.07	1.85
		SVM	0.92	**0.95**	1.00	0.94	0.92	45.00	4.37	1.58	1.36
Wine	FDT	–	–	–	–	0.94	–	36.00	–	3.00	–
	FLARE	NN	0.86	0.76	0.89	0.82	0.77	48.00	8.68	1.42	1.54
		RF	0.61	0.61	1.00	0.61	0.61	41.00	**2.89**	1.39	**1.23**
		SVM	**0.99**	0.73	0.67	0.71	0.76	17.00	4.95	**1.00**	1.32
	LORE	NN	0.77	0.78	0.89	0.76	0.76	54.00	4.79	2.52	1.94
		RF	0.90	**0.88**	1.00	**0.90**	**0.88**	54.00	6.21	3.19	3.02
		SVM	0.93	0.76	0.67	0.68	0.71	52.00	5.05	1.02	1.56
Beer	FDT	–	–	–	–	1.00	–	69.00	–	2.42	–
	FLARE	NN	0.69	0.71	0.80	0.67	0.79	128.00	20.42	1.85	**1.78**
		RF	0.87	0.88	1.00	0.87	0.88	129.00	26.68	2.34	2.18
		SVM	0.86	0.77	0.85	0.85	0.82	99.00	15.21	1.96	1.93
	LORE	NN	0.74	0.76	0.80	0.70	0.80	119.00	**13.42**	2.01	1.98
		RF	**0.92**	*0.89*	1.00	**0.92**	*0.88*	119.00	14.63	2.54	2.60
		SVM	0.78	0.82	0.85	0.67	0.86	119.00	15.58	2.02	2.07

Results. We compare the results of FLocalX for two different local explanation methods, using the union of the local explanations as a global explainer and studying how much improvement our framework provides. We also use a rule-based white box method (i.e., FDT) as baseline. Table 1 reports both the performance of the global explainers, as well as its level of complexity.

As one objective of the optimization process is to minimize the size of the rule-based system, testing the impact on performance is necessary. We can observe that problems where Acc-U is really high (i.e. >0.9), Acc-F is lower than Acc-U, likely because most rules are necessary to achieve that degree of accuracy. However, that decrease in accuracy is not so much as to lose trust in the explainer. On the other hand, in more complex problems where the starting point is not as good (the Beer dataset with FLARE and NN, or LORE and SVM are examples of this), the optimization process can even improve the starting point's accuracy. This suggests that a metaheuristic approach, while time-consuming, benefits hard-to-solve problems. Finally, it is worth mentioning that LORE rules tend to be a better starting point for FLocalX than FLARE rules. This finding might suggest that either crisp rules are better than fuzzy rules as a starting point, or that more premises provide a better starting point. More experiments will be done to explore the cause.

Turning to explanation complexity, the most relevant part is in the reduction of rules from the union of local explanations (#R) to after FLocalX is applied (#R-F). We can observe that we need around $10\% - 15\%$ of the number of rules from which FLocalX starts. Beer shows the largest explanation theories (at around 20 rules for FLARE and 14 for LORE), which are still readable for humans. Moreover, there is a great reduction from the baseline white-box classifiers, needing around 40% of the rules in simpler datasets and around $20\% - 30\%$ of the rules in more complex problems. The number of rules generated by the FDT increases with the complexity of the problems, which makes it unfit as a global explainer for difficult problems, i.e., valid for Iris and Wine but unreasonably long at 70 rules for Beer. On the other hand, looking at the number of premises, most rules have around 1–3 premises, also manageable for a human reader. #P-F is only marginally smaller than #P because $f(C)$ does not consider the length of the rule. Finally, we can see that LORE global explanations usually have fewer rules than FLARE, with some more premises per rule.

The results of this preliminary experimentation, with a single optimization algorithm (i.e., a genetic algorithm) and smaller datasets, showcase the flexibility of the framework, which can generate compact and performant global explanation theories that can be useful to a human reader.

5 Conclusions and Future Work

This work introduces FLocalX, a model agnostic local to global explanation framework based on fuzzy logic that leverages the power of evolutionary computing to obtain a global explanation of a black-box model. FLocalX uses local explanations formed as fuzzy rules as the starting point from which it builds a global fuzzy explanation that summarizes the model underneath. Using a genetic algorithm as the optimization method, the experimentation carried out in this paper shows that FLocalX is able to generate a short and accurate global explanation theory, improving upon the trivial union of local explanations, as well as upon the used baseline white box model. As future research directions, we intend to perform a comprehensive study on the different hyperparameters, as well as different operators and objective functions for the genetic tuning process of FLocalX. Moreover, we would like to study the usage of a different metaheuristic algorithm to replace the genetic procedure. Finally, the difference in performance shown between using FLARE to generate the local explanation theory and using LORE motivates the need to experiment with other local explainers.

Acknowledgements. This work has been funded by the following projects: SBPLY/21/180225/000062 (Government of Castilla-La Mancha and ERDF funds); PID2019–106758GB–C33, and FPU19/02930 (MCIN/AEI/10.13039/501100011033 and ERDF Next Generation EU); and 2022-GRIN-34437 (Universidad de Castilla-La Mancha and ERDF funds), EU NextGenerationEU programme PNRR-PE-AI FAIR (Future Artificial Intelligence Research), PNRR-SoBigData.it - Prot. IR0000013, H2020-INFRAIA-2019-1: Res. Infr. G.A. 871042 *SoBigData++*, ERC-2018-ADG G.A. 834756 *XAI*, and CHIST-ERA-19-XAI-010 SAI.

References

1. Alonso, J.M., et al.: Explainable fuzzy systems: paving the way from interpretable fuzzy systems to explainable AI systems. In: SCI (2021)
2. Angelov, P.P., et al.: Explainable artificial intelligence: an analytical review. WIREs Data Mining Knowl. Discov. **11**(5), e1424 (2021)
3. Arrieta, A.B., et al.: Explainable artificial intelligence (XAI): concepts, taxonomies, opportunities and challenges toward responsible AI. Inf. Fusion **58**, 82–115 (2020)
4. Casillas, J., et al.: Genetic tuning of fuzzy rule deep structures preserving interpretability and its interaction with fuzzy rule set. IEEE TFS **13**(1), 13–29 (2005)
5. Chen, T., et al.: Xgboost: extreme gradient boosting. R package **1**(4), 1–4 (2015)
6. Dai, Z., et al.: Coatnet: marrying convolution and attention for all data sizes. Adv. Neural. Inf. Process. Syst. **34**, 3965–3977 (2021)
7. Fernández, G., et al.: Factual and counterfactual explanations in fuzzy classification trees. IEEE Trans. Fuzzy Syst. **30**(12), 5484–5495 (2022)
8. Guidotti, R., et al.: Factual and counterfactual explanations for black box decision making. IEEE Intell. Syst. **34**(6), 14–23 (2019)
9. Guidotti, R., et al.: A survey of methods for explaining black box models. ACM Comput. Surv. **51**(5), 93:1–93:42 (2019)
10. Herrera, F., et al.: Fuzzy connectives based crossover operators to model genetic algorithms population diversity. Fuzzy Sets Syst. **92**(1), 21–30 (1997)
11. Hiabu, M., et al.: Unifying local and global model explanations by functional decomposition. In: AISTATS, vol. 206, pp. 7040–7060. PMLR (2023)
12. Holland, J.H.: Adaptation in Natural and Artificial Systems: an Introductory Analysis with Applications to Biology, Control, and AI. MIT Press, Cambridge (1992)
13. Huber, T., et al.: Local and global explanations of agent behavior: integrating strategy summaries with saliency maps. Artif. Intell. **301**, 103571 (2021)
14. Kliegr, T., et al.: A review of possible effects of cognitive biases on interpretation of rule-based machine learning models. Artif. Intell. **295**, 103458 (2021)
15. Lundberg, S.M., et al.: From local explanations to global understanding with explainable AI for trees. Nat. Mach. Intell. **2**(1), 56–67 (2020)
16. Maria, A.J., et al.: Explainable fuzzy systems: Paving the way from interpretable fuzzy systems to explainable AI systems. SCI **970** (2021)
17. Pedregosa, F., et al.: Scikit-learn: ML in python. JMLR **12**, 2825–2830 (2011)
18. Regulation, G.D.P.: General data protection regulation (GDPR). Intersoft Consulting, Accessed Oct 24 **1** (2018)
19. Schrouff, J., et al.: Best of both worlds: local and global explanations with human-understandable concepts. CoRR (2021)
20. Schrouff, J., et al.: Best of both worlds: local and global explanations with human-understandable concepts. CoRR **abs/2106.08641** (2021)
21. Segatori, A., et al.: On distributed fuzzy decision trees for big data. IEEE Trans. Fuzzy Syst. **26**(1), 174–192 (2017)
22. Setzu, M., et al.: Glocalx-from local to global explanations of black box AI models. Artif. Intell. **294**, 103457 (2021)
23. Stepin, I., et al.: Generation and evaluation of explanations for decision trees and fuzzy rule-based classifiers. In: FUZZ, pp. 1–8. IEEE (2020)
24. Stepin, I., Catala, A., Pereira-Fariña, M., Alonso, J.M.: Factual and counterfactual explanation of fuzzy information granules. In: Pedrycz, W., Chen, S.-M. (eds.) Interpretable Artificial Intelligence: A Perspective of Granular Computing. SCI, vol. 937, pp. 153–185. Springer, Cham (2021). https://doi.org/10.1007/978-3-030-64949-4_6

25. Varshney, A.K., et al.: Literature review of the recent trends and applications in various fuzzy rule-based systems. In: IJFS, pp. 1–24 (2023)
26. Zhang, S., et al.: The diversified ensemble neural network. Adv. Neural. Inf. Process. Syst. **33**, 16001–16011 (2020)

Interpretable Quantile Regression
by Optimal Decision Trees

Valentin Lemaire[1]([✉])(iD), Gaël Aglin[2](iD), and Siegfried Nijssen[2](iD)

[1] Euranova, Rue Emile Francqui, Mont-Saint-Guibert, Belgium
valentin.lemaire@euranova.eu
[2] Université Catholique de Louvain, Louvain-la-Neuve, Belgium

Abstract. The field of machine learning is subject to an increasing interest in models that are not only accurate but also interpretable and robust, thus allowing their end users to understand and trust AI systems. This paper presents a novel method for learning a set of optimal quantile regression trees. The advantages of this method are that (1) it provides predictions about the complete conditional distribution of a target variable without prior assumptions on this distribution; (2) it provides predictions that are interpretable; (3) it learns a set of optimal quantile regression trees without compromising algorithmic efficiency compared to learning a single tree.

Keywords: Interpretability · Robustness · Quantile regression · Optimal decision trees

1 Introduction

Recently, many studies have focused on rendering complex machine learning systems interpretable from a human perspective [5,6,15]. Indeed, in many domains, the explanation is of equal interest as the accuracy of the prediction; especially in medical settings, business strategy settings, etc., both because one wants to understand why the prediction is made, but also to gain insights into the data by inspecting the model's structure. However, most post-hoc techniques' results remain complex to understand and often fail to explain the whole decision process of the AI system [7]. Therefore, work has been conducted in the direction of inherently interpretable models [9,14]. Within this family of inherently interpretable models falls the family of decision trees. Indeed, given a tree of reasonable depth, a human can quickly analyze it and see how and why the model predicts a particular class or value.

Traditionally, decision trees are learned top-down by training them with a Mean Squared Error (MSE) heuristic, and by putting a single prediction in each leaf of the tree, corresponding to the mean of the target variable over the training examples. In this paper, we study *quantile regression*. A *quantile regression tree* does not predict a mean; for a given quantile parameter q, a quantile regression tree for that choice of q would predict a value such that q percent of the

I. Miliou et al. (Eds.): IDA 2024, LNCS 14642, pp. 210–222, 2024.
https://doi.org/10.1007/978-3-031-58553-1_17

observed values are below the predicted value. Hence, for $q = 10\%$ such a tree would underestimate, while for $q = 90\%$ the tree would overestimate. This can be important in applications where well-motivated under- or overestimation is important, for instance, when using a tree to predict demand in retail, where the retailer would prefer to have a larger stock than the demand predicted by a standard regression tree. However, there is no consensus on an efficient heuristic for quantile regression, which is why we interest ourselves in optimal decision trees, such as DL8 [11], DL8.5 [1] and MurTree [4]. Advantages of these methods include that (1) for trees constrained in depth these algorithms manage to find more accurate trees; (2) these algorithms can be used to learn trees without requiring the prior development of good heuristics for top-down tree induction.

This latter characteristic makes optimal decision trees very suitable to quantile regression. However, an important weakness of learning a single quantile regression tree is that one would have to choose a single parameter q. Moreover, a single regression tree only provides limited insight in the complete conditional distribution of the target variable. In this paper we argue that it is often desirable to model the whole target distribution without making one choice for q or without making prior assumptions with respect to the shape of this distribution: this increases trust in an AI system, provides insight in this target distribution, and allows to provide well-motivated under- or over-estimations of the target variable. We propose to do so by learning a set of decision trees, each corresponding to a *quantile regression* tree. Indeed, predicting quantiles rather than the most likely value renders the models more robust to outliers [12] and doing this for many different quantiles gives information about the whole distribution. We introduce Quantile DL8.5 (QDL8.5). This method efficiently learns a *set* of optimal, shallow, and explainable decision trees for multiple quantiles, achieving high accuracy and interpretability. QDL8.5 addresses the complexity of choosing the right quantile in Quantile Regression by learning multiple trees for different quantiles. We show that this can be done with virtually no computational overhead compared to learning a single tree. The contributions of this paper are twofold: (i) We first propose an extension of the DL8.5 algorithm that enables it to perform quantile regression by one optimal tree per quantile while exploring the tree space only once, thus limiting drastically the time increase of learning many trees instead of one and (ii), we provide a robust assessment in terms of accuracy, execution time performance and interpretability.

2   Related Work

Little work has gone towards performing quantile regression with interpretable decision trees. However, there are some notable works to mention. A very popular way to perform conditional quantile regression is Quantile Linear Regression (QLR) [8]. Indeed, it is possible to derive a linear transform on features that finds the best linear fit to optimize quantile loss. While easily interpretable, this method is limited in terms of expressiveness. Quantile Regression Forests (QRF) [10] represent a notable advancement in interpretable decision trees, sharing architecture with Random Forests [2]. Unlike traditional methods, QRF records samples associated with each leaf and predicts the empirical quantile based on the specified quantile during inference [10].

Cousins and Riondato introduced CaDET [3], a model that constructs interpretable decision trees or random forests predicting density functions within their leaves. They employ a selected statistical family (e.g., Gaussian, Pareto) instantiated in leaves and use empirical cross-entropy as the impurity measure for relevant splits. Despite fast training times due to heuristic-based tree growth, models like CaDET [3] and QRF [10] require more trees for sufficient expressiveness. Efficient heuristics exist for optimizing cross-entropy or mean squared error, but consensus is lacking for heuristics optimizing quantile loss in decision tree growth. Some other works have used more expressive but less interpretable methods to estimate conditional quantiles. Wang et al. [17] propose to use different standard regression models (random forests, support vector regressor and gradient-boosted decision trees) and combine them linearly to optimize quantile loss. Finally, they use Kernel Density Estimators (KDE) on the quantiles to predict a conditional probability density function. Similarly, Zhang et al. [18] combined linearly different methods that already perform quantile regression including QLR and QRF. They also combine those quantile results into pdfs using KDE. Both of these methods combine models that are complicated to interpret, making them uninterpretable.

Another approach might be to use model trees like those introduced by Quinlan et al. [13], where, rather than having a single prediction in the leaf, a simple model (e.g. QLR) is fitted on the mapped samples. While this is purposeful for point predictions, it increases complexity of the models and it does not help to give a fuller picture of a sample's distribution without prior assumptions on said distribution. These will therefore not be included in the experimental evaluation.

3   Background

As a first step into technical background, let us introduce quantiles.

Definition 1. *Given a probability density function (pdf) f describing a distribution, from which can be derived a cumulative density function (cdf) F, and given a quantile q, the corresponding quantile value y_q is such that the probability that a realization $y \in \mathcal{Y}$ of the random variable Y is lower than y_q is q.*

Empirically, given a set of realizations $\boldsymbol{y}$, it was shown [10] that the quantile value for a given quantile q is the value $\hat{y}_q$ that minimizes the quantile loss:

$$QL_q(\boldsymbol{y}, \hat{y}_q) = \sum_{i=1}^{|\boldsymbol{y}|} \max\left\{ q(\hat{y}_q - y_i), (1-q)(y_i - \hat{y}_q) \right\}. \tag{1}$$

Quantile Regression is identical to standard Regression but the loss to optimize is the loss in Eq. 1. Another main building block of this work is optimal decision trees, and more specifically, the DL8 and DL8.5 implementations of these. DL8 [11] and its newer, more efficient version DL8.5 [1] are algorithms that enable the learning of the best tree that optimizes any additive loss function under constraints of maximum depth, minimum support, etc. DL8.5 works as follows. Given a binary featured dataset, it goes through each feature following a dynamic programming principle in a branch-and-bound manner.

In these models, data is represented as itemsets, i.e. a collection of positive f or negative $\neg f$ items for each feature f. When a decision tree is built, it splits each feature into its positive and negative branches, thus building larger itemsets as it goes down in the trees. For the remainder of this paper, we will consider our datasets to be binary as any dataset can be binarized (categorical features turned into one-hot encoded features and continuous features turned into binary bins). Each leaf of a decision tree will map to a certain subset of the samples that all contain the corresponding itemset. For example, if a leaf contains the itemset $\{a, \neg d, g\}$, all samples having ones for features a and g and a zero for feature d will be mapped to that itemset (and leaf).

Algorithm 1 shows how DL8.5 works. At each level of the search in DL8.5, a split is performed on each feature (line 13) and the overall error associated with each split is computed, going down recursively in the tree, always keeping track of the path that leads to the lowest error. This is an exhaustive search, but the search space is pruned by efficient lower and upper bounding. Indeed, the best errors of the branches at the same level act as upper bounds for the following negative branches (line 18). The upper bound used for the positive branches that are explored after the negative ones is the difference between the previous upper bound and the error of the negative branch, as errors are additive (line 19). To summarize in a sentence: *at each step of the exploration, a new feature is chosen to be added to the itemset, the corresponding error is found for the two branches (positive and negative) and this feature is saved if it yields the best error so far. This is done by pruning as many unnecessary branches as possible.*

DL8.5 also makes use of a cache (line 8 and 30). Indeed, each node in the search tree can be seen as an itemset of positive or negative features that must be present in the samples mapped to that node. However, an itemset is an unordered collection of positive or negative features, meaning that it is possible to encounter the same itemset in different parts of the search tree. Indeed, exploring a after $\neg b$ yields the same itemset as exploring $\neg b$ after a. The cache allows us to save the best trees for these itemsets and therefore avoid performing the same computation twice. If, for a particular itemset, the search yielded no result, this is also saved with the upper bound that was used, as a later search might ask for the same itemset with an equal or lower upper bound. This search can be avoided as it is already known it will give no result.

Demirović et al. [4] proposed some optimisations to further the pruning by a better lower bounding and using a special computation for depth-two trees. Some of these optimisations are transferable to regression and have been included in the implementation but not shown in the algorithm as they are not principal for this paper. This dynamic programming approach combined with efficient branching and bounding allows the algorithm to ensure finding the best decision

---

Algorithm 1: DL8.5($maxdepth$, $minsup$)

1: **struct** $BestTree\{lb : float, tree : Tree, error : float\}$
2: $cache \leftarrow HashMap < Itemset, BestTree >$
3: $best\_solution \leftarrow$ DL8-RECURSE($\emptyset, +\infty, 0$)
4: **return** $best\_solution.tree$
5: **Procedure** DL8.5-RECURSE(I, ub)
6:     **if** $leaf\_error(I) = 0$ or $|I| = maxdepth$ or time-out is reached **then**
7:         **return** $BestTree(ub, make\_leaf(I), leaf\_error(I))$
8:     $solution \leftarrow cache.get(sort(I))$
9:     **if** $solution$ was found **then**
10:         **if** solution.tree $\neq$ NO_TREE or $ub \leq solution.lb$ **then**
11:             **return** $solution$
12:     $(\tau, b, left\_ub) \leftarrow$ (NO_TREE, $+\infty, ub$)
13:     **for** all attributes i in a well-chosen order
14:         **if** $cover(I \cup \{i\}) \geq minsup$ and $cover(I \cup \{\neg i\}) \geq minsup$ **then**
15:             $sol_1 \leftarrow$ DL8.5-RECURSE($I \cup \{\neg i\}, ub$)
16:             **if** $sol_1.tree =$ NO_TREE **then**
17:                 **continue**
18:             **if** $sol_1.error < left\_ub$ **then**
19:                 $sol_2 \leftarrow$ DL8.5-RECURSE($I \cup \{i\}, left\_ub - sol_1.error$)
20:                 **if** $sol_2.tree =$ NO_TREE **then**
21:                     **continue**
22:                 $feature\_error \leftarrow sol_1.error + sol_2.error$
23:                 **if** $feature\_error < left\_ub$ **then**
24:                     $\tau \leftarrow make\_tree(i, sol_1.tree, sol_2.tree)$
25:                     $b \leftarrow feature\_error$
26:                     $left\_ub \leftarrow b$
27:                 **if** $feature\_error = solution.lb$ **then**
28:                     **break**
29:     $solution \leftarrow BestTree(ub, \tau, b)$
30:     $cache.store(sort(I), solution)$
31:     **return** $solution$

---

tree under the given constraints. This formulation of the DL8.5 algorithm is already able to find the optimal decision tree optimizing quantile loss as it is an additive function. The only change that has to be brought is the implementation of the quantile loss as the leaf error function.

4   Quantile DL8.5

In this section, we will show how to extend the DL8.5 learning algorithm to perform *simultaneous* quantile regression[1]. This change in the algorithm will allow it to learn, for each given quantile, an optimal decision tree while only exploring the search space once, thus utilizing for all quantiles the common parts of the search trees. This will be specially marked if the obtained decision trees are very similar as they will have resulted from similar searches among possible decision trees. The output of the algorithm is, for each sample, an array of quantile values, each corresponding to a different quantile. Given enough quantiles, they describe most of the conditional distribution of a sample. The more quantiles, the more this full distribution will be precise[2].

---

[1] Source code is available at https://github.com/valentinlemaire/pydl8.5.

[2] We however recommend setting the number of quantiles to be lower than the minimum support in each leaf to avoid skewed estimations.

4.1   Simultaneous Tree Learning

The quantile loss in equation (1) is parametric, meaning that its value is dependent on the corresponding quantile. This also means that a tree that is optimal for one quantile is not guaranteed to be optimal for another. For this reason, we need to learn an optimal tree for each quantile. However, it is reasonable to assume that these trees will not be very different from each other, especially for close quantiles as they will describe close parts of the conditional distribution. Therefore, running independent searches for the optimal trees would be inefficient as the searches will go through many common itemsets for the different quantiles. For this reason, we have changed the DL8.5 algorithm to enable it to learn the optimal decision tree for different quantiles while only exploring each itemset once, no matter how many trees have to be learnt.

To enable DL8.5 to learn many trees at once a few changes had to be brought to the algorithm. Algorithm 2 shows these changes. This algorithm is essentially the same as Algorithm 1 except that all conditions of pruning change and some computations need to be performed once for each quantile. Indeed, in QDL8.5, a branch can only be pruned if all the searches relating to all the different quantiles do not give any results. Thus, line 10 shows a call to a *can_return* function. This function returns true if, *for all quantiles*, there is either no solution, either the upper bound is lower than the lower bound or the leaf error has attained its lower bound.

During the search, when considering an itemset, for each quantile we consider each possible feature. However, if we consider a particular itemset for a particular quantile, we compute the quantile values and quantile loss for all the quantiles. Therefore, if we get back to that itemset for a different quantile later in the search, the quantile value and loss will already be saved to the cache and therefore no additional computation will be needed. Section 4.2 shows how we can compute many quantile values and losses without additional computational cost. It can also be seen that for each quantile and for each itemset explored, there are lower and upper bounds and associated errors.

4.2   Efficient Quantile Loss Computation

The most costly operation in the DL8.5 (and QDL8.5) method is traversing the data to compute the predictions and the errors for each node (itemset) explored. It is therefore primordial to make that operation as efficient as possible. The naive way would be to apply equation (1) which can be computed in $\mathcal{O}(N)$ time for each quantile, leading to a calculation time of $\mathcal{O}(|q|N)$, where N is the number of samples mapped to that itemset, i.e. the cover of that itemset, and $|q|$ is the number of quantiles. As a first step towards computational efficiency, the complete data can be sorted according to the y values before starting the tree search. Indeed, the empirical estimation of a quantile is defined as

$$\hat{y}_q = y_{\lfloor h \rfloor} + (h - \lfloor h \rfloor)(y_{\lceil h \rceil} - y_{\lfloor h \rfloor}) \tag{2}$$

Algorithm 2: QDL8.5(*maxdepth, minsup*)

1: **struct**
    BestTree{*lbs* : *vector* < *float* >, *trees* : *vector* < *Tree* >, *errors* : *vector* < *float* >}
2: *cache* ← *HashMap* < *Itemset, BestTree* >
3: *best_solution* ← QDL8.5-RECURSE($\emptyset$, $+\infty^{|q|}$)
4: **return** *best_solution.tree*
5: **Procedure** QDL8.5-RECURSE*I, ubs*
6:     *solution* ← *cache.insertOrGet*(*sort*(*I*))
7:     *leaf_errors* ← *quantile_errors*(*I, q*)
8:     **if** $|I|$ = *maxdepth or time-out is reached* **then**
9:         **return** *BestTree*(*solution.lbs, make_leafs*(*I*), *leaf_errors*)
10:     **if** *can_return*(*solution, ubs, leaf_errors*) **then**
11:         **return** *solution*
12:     **for** all attributes F in a well-chosen order
13:         **if** *cover*($I \cup \{f\}$) ≥ *minsup* and *cover*($I \cup \{\neg f\}$) ≥ *minsup* **then**
14:             *sol₁* ← QDL8.5-RECURSE($I \cup \{\neg i\}$, *ubs*)
15:             **if** all *sol₁.trees* are NO_TREE **then** **continue**
16:             **if** $\exists i \in \{1, 2, \ldots, |q|\}$: *sol₁.errors$_i$* < *solution.errors$_i$* **then**
17:                 *sol₂* ← QDL8.5-RECURSE($I \cup \{i\}$, *ubs* − *sol₁.errors*)
18:                 **if** all *sol₂.trees* are NO_TREE **then** **continue**
19:                 **for** $i \in \{1, 2, \ldots, |q|\}$
20:                     *feature_errors$_i$* ← *sol₁.errors$_i$* + *sol₂.errors$_i$*
21:                     **if** *feature_errors$_i$* < *solution.errors$_i$* **then**
22:                         *solution.trees$_i$* ← *build_tree*(*F, sol₁.trees$_i$, sol₂.trees$_i$*)
23:                         *solution.errors$_i$* ← *feature_errors$_i$*
24:                         *ubs$_i$* ← *feature_errors$_i$*
25:                 **if** all *feature_errors* = *solution.lbs* **then** **break**
26:     **for** $i \in \{1, 2, \ldots, |q|\}$
27:         *solution.lbs$_i$* ← *ubs$_i$*
28:     **return** *solution*

---

with $h = q(N - 1) + 1$. This can be computed in $\mathcal{O}(1)$ time if the array of values $\boldsymbol{y}$ is sorted. This is also true for any subset of values as a subset of a sorted array is itself sorted. Using this sorted array we can compute the quantile values corresponding to all the quantiles $\boldsymbol{q}$ in $\mathcal{O}(|\boldsymbol{q}|)$ time. We can also notice that the quantile loss formulated in equation (1) can be rewritten as

$$QL_q(\hat{y}_q, \boldsymbol{y}) = (1 - q) \sum_{i: y_i \leq \hat{y}_q} y_i - (1 - q) \sum_{i: y_i \leq \hat{y}_q} \hat{y}_q + q \sum_{i: y_i > \hat{y}_q} \hat{y}_q - q \sum_{i: y_i > \hat{y}_q} y_i \quad (3)$$

Notice there is no index on $\hat{y}_q$ as the prediction is the same for all samples mapped to a leaf. Using this formulation we can see that the only need for the data is to store, for each quantile, the sum of the y values that are above the prediction and the sum of those under the prediction. This can be used to compute the errors for all quantiles by only traversing the data once. Indeed, it is possible to go through the samples in increasing order (since the data is sorted) and bin the different samples according to the quantile values. Then, with a loop over the quantiles, it is possible to compute the sum of elements above and below each quantile value, thus allowing us to compute the quantile loss for each quantile efficiently. Using this technique, we get a temporal complexity of $\mathcal{O}(N + |\boldsymbol{q}|)$. However, it makes little sense to compute more quantiles than there are samples, so we always recommend setting $|\boldsymbol{q}| \leq minsup$, with $minsup \leq N$, thus making the temporal complexity of this $\mathcal{O}(N)$.

Probability Density Estimates. With this implementation of the algorithm, the output at prediction time is an array of conditional quantile values, each corresponding to a different quantile. Given enough quantiles, they describe most of the conditional distribution of a sample. For a more visual interpretation, we can combine those quantiles in a distribution function using Kernel Density Estimators in the same way as in other works [17,18]. This adds two new parameters, the kernel and the width of these kernels. In the remainder of this paper, we'll use Scott's rule for kernel widths, which was shown to be optimal when the underlying distribution is Gaussian [16]. Even though we cannot make this assumption, we will use this method to estimate kernel widths. Correspondingly, we'll use Gaussian kernels.

5 Experiments

For our experiments, we consider three different aspects of the performance of our model: (i) the accuracy by measuring how well the outputs of the model (estimated quantiles and pdfs) describe the actual data, (ii) the efficiency in terms of execution time and (iii) a study of how interpretable the obtained models are. We consider our main competitors to be CaDET [3] and Quantile Random Forests [10] as they are both ensemble methods using decision trees and outputting distribution information. For our experiments, we have chosen 4 datasets; one synthetic and 3 real-world. The synthetic dataset is generated as follows: using 9 binary features we created 15 categories represented by a combination of those features. Each category has its associated Gaussian distribution, each with a different mean and standard deviation, from which target values were drawn. This dataset was created to have a dataset with a known conditional probability distribution. In addition to it, we measured quality on three real-world datasets: Air Quality, Solar Flares, and Stock Portfolio Performance as they are widely used benchmark datasets of varying sizes, with low dimensionality that have categorical features (thus losing less information in binarization). All our experiments were performed on a machine running Intel(R) Xeon(R) Gold 6134 CPU@3.20GHz processor with 32 physical cores and 128GB RAM with Ubuntu 18.04.6 operating system.

5.1 Metrics

To ensure the quality of the predictions we use different metrics. First is Mean Integrated Squared Error (MISE), which measures the integral of the squared difference between the true and predicted distributions. This can only be done for the synthetic dataset of which we know the true distribution. We also use Mean Quantile Error (MQE), which is the mean of all the different quantile errors over the different quantiles. We also measured the Negative Log Likelihood (NLL) of the samples with respect to the predicted distribution function and finally, the Continuous Ranked Probability Score (CRPS), which measures the integral of the squared difference between the predicted cdf for a sample and the step function for the actual realization of that sample.

Regarding interpretability, we wish to demonstrate that while we learn many trees which can impede on overall interpretability, the resulting trees are mostly similar and analyzing just a few trees would give sufficient insights to the analyst. To this effect, we evaluated the partitions of the training dataset generated by each tree and measured a Jaccard index on all pairs of these partitions. With this metric, we would expect to see block matrices attesting that trees corresponding to close quantiles are indeed similar, thus attesting to overall interpretability.

5.2   Results

This section will illustrate our experiments and related observations. *Quality* Our first experiment concerns the quality of the obtained regressors. In Table 1, we show, for each dataset and each model, the different quality metrics. All three methods get good results. QDL8.5 performs either as the best model or second best model on all metrics and datasets, often being close to the best result when not achieving it. From these observations, we can conclude that QDL8.5 is competitive with existing methods and matches the state of the art in decision trees that predict (some form of) conditional distributions.

Table 1. Quality metrics on all 4 datasets. For all metrics, lower is better. **Bold** means best, underlined means second best. We performed hyperparameter tuning with Bayesian search (20 trials) for each dataset and method.

Datasets	Metrics	Models		
		Quantile RF	CaDET RF	Quantile DL8.5
Synthetic dataset	MISE	0.121	0.122	**0.120**
	NLL	2.16	**2.15**	**2.15**
	MQE	0.0700	0.0616	**0.0603**
	CRPS	**1.22**	1.25	1.24
	n. trees/depth	*100/5*	*50/5*	*100/4*
Air Quality	NLL	1.49	1.48	**1.46**
	MQE	0.0235	0.0312	**0.0223**
	CRPS	**0.802**	0.832	0.818
	n. trees/depth	*100/5*	*100/5*	*100/4*
Solar Flares	NLL	1.10	**−0.972**	−0.0234
	MQE	**0.00185**	0.00597	0.00367
	CRPS	0.209	0.212	**0.195**
	n. trees/depth	*25/7*	*50/6*	*100/4*
Stock Performance	NLL	−0.870	−0.578	**−1.04**
	MQE	textbf0.00124	0.00324	**0.00124**
	CRPS	0.0824	0.0881	**0.0777**
	n. trees/depth	*50/4*	*25/6*	*100/3*

Efficiency. This work showed an algorithm modification that enables the DL8.5 algorithm to learn many trees at once, each optimizing a loss function with a different quantile parameter, while only exploring the tree space once. In this experiment, we measured, for different numbers of trees to learn, the execution time of QDL8.5 compared to the naive version that learns each tree by starting a new search every time. Figure 1 shows the results of this experiment. It shows that the naive version's execution time is linear with the number

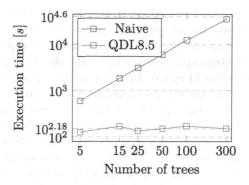

Fig. 1. Running times for naive and efficient versions of QDL8.5. Both axes are in *log* scale.

of trees. It also shows that the execution time of QDL8.5 is virtually independent of the number of trees. This shows that QDL8.5 can learn arbitrarily many trees, each describing a different point in the distribution range, at almost no additional computational cost compared to learning a single tree. We can also see that for 5 trees, the speedup of QDL8.5 is 4.74, which is close to the optimal speedup we could expect by learning the trees jointly. The same observation can be made for other numbers of trees.

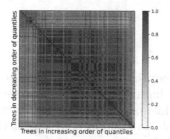

Fig. 2. Similarity matrix of trees learned by QDL8.5 for the Air Quality dataset.

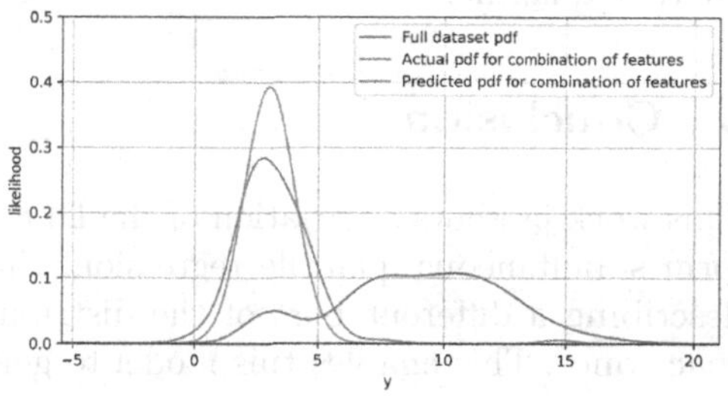

Fig. 3. Distribution plots for a category within the synthetic dataset.

Interpretability. Using optimal decisions is a way to provide inherently interpretable models. However, while we do generate shallow trees, we produce many of them, which may increase interpretability difficulty. However, if those trees happen to be similar, then inspecting only a few of them at different parts of the distribution range would reflect the majority of the information of the quantile regression ensemble. In this experiment, we have analyzed, for each tree, how it partitions the training dataset and performed a Jaccard index on those partitions to see how similar the trees are. The result of this experiment is shown in Fig. 2. Based on this plot, a few observations can be made. First, generated trees are all quite similar, as the lowest Jaccard index in this matrix is ~ 0.4. Secondly, we can see that trees corresponding to close quantiles partition the feature space similarly, as attested by high Jaccard index values on blocks close to the diagonal. In this example, if we define *tree zones* as being delimited by a change of 10% in Jaccard index in successive trees, we would end up with 5 zones, and thus by picking only one among those zones, we can interpret *whole* the distribution of the *whole* data with only 5 trees of depth 4.

Finally, we can see that the tails of the distribution (low and high quantiles) are described by trees that differ from the centre of the distribution (large darker blocks in the centre and diagonal corners of the matrix), thus justifying the use of quantile regression to describe those parts of the distribution. Another way to interpret the results of the QDL8.5 algorithm is to plot the predicted conditional pdf. Figure 3 shows this for a particular category in the synthetic dataset. This figure shows that the predicted pdf is quite close to the actual pdf and allows a human to understand how the target variable behaves for a sample.

Another point where the interpretability of our method is improved compared to other tree ensemble methods like Cadet RF and QRF is that in our case, each tree is linked with an interpretable value. Analyzing the tree corresponding to quantile 0.1 will give information about that specific part of the distribution for the whole dataset. For the other methods, all trees have to be analyzed to understand trends in the data, and their aggregation is not trivial. In addition, the optimality criteria ensures that the partitioning of the data is the best one for that quantile.

6   Conclusion

This work presents a variation of the DL8.5 [1] algorithm that enables it to perform simultaneous quantile regression, thus learning many trees at once, each describing a different part of the distribution while only exploring the search space once. This enables this model to generate shallow and interpretable decision trees that provide robust predictions via quantile regression and information about the complete conditional distribution of the data. Experiments have shown that this model achieves good quality matching or surpassing the state-of-the-art in this domain, that learning many trees comes at a small additional computing cost compared to learning only one, and finally, that the obtained trees are highly interpretable because they correspond to an interpretable parameter and because they only differ incrementally with the different quantiles.

References

1. Aglin, G., Nijssen, S., Schaus, P.: Learning optimal decision trees using caching branch-and-bound search. In: Proceedings of the AAAI Conference on Artificial Intelligence. vol. 34, pp. 3146–3153 (2020). https://doi.org/10.1609/aaai.v34i04.5711
2. Breiman, L.: Random forests. Mach. Learn. **45**, 5–32 (2001). https://doi.org/10.1023/A:1010933404324
3. Cousins, C., Riondato, M.: CaDET: interpretable parametric conditional density estimation with decision trees and forests. Mach. Learn. **108**(8), 1613–1634 (2019). https://doi.org/10.1007/s10994-019-05820-3
4. Demirović, E., et al.: MurTree: optimal decision trees via dynamic programming and search. J. Mach. Learn. Res. **23**(1), 1–47 (2022). https://doi.org/10.5555/3586589.3586615
5. Du, M., Liu, N., Hu, X.: Techniques for interpretable machine learning. Commun. ACM **63**(1), 68–77 (2019). https://doi.org/10.1145/3359786
6. Fong, R.C., Vedaldi, A.: Interpretable explanations of black boxes by meaningful perturbation. In: Proceedings of the IEEE International Conference on Computer Vision, pp. 3429–3437 (2017). https://doi.org/10.1109/ICCV.2017.371
7. Gilpin, L.H., Bau, D., Yuan, B.Z., Bajwa, A., Specter, M., Kagal, L.: Explaining explanations: an overview of interpretability of machine learning. In: 2018 IEEE 5th International Conference on Data Science and Advanced Analytics (DSAA), pp. 80–89 (2018). https://doi.org/10.1109/DSAA.2018.00018
8. Koenker, R., Hallock, K.F.: Quantile regression. J. Econ. Perspect. **15**(4), 143–156 (2001). https://doi.org/10.1257/jep.15.4.143
9. Letham, B., Rudin, C., McCormick, T.H., Madigan, D.: Interpretable classifiers using rules and bayesian analysis: Building a better stroke prediction model. Ann. Appl. Stat. **9**(3), 1350–1371 (2015). https://doi.org/10.1214/15-aoas848
10. Meinshausen, N.: Quantile regression forests. J. Mach. Learn. Res. **7**, 983–999 (2006). https://doi.org/10.5555/1248547.1248582
11. Nijssen, S., Fromont, É.: Mining optimal decision trees from itemset lattices. In: Knowledge Discovery and Data Mining (2007). https://doi.org/10.1145/1281192.1281250
12. John, O.O.: Robustness of quantile regression to outliers. Am. J. Appl. Math. Stat. **3**(2), 86–88 (2015). https://doi.org/10.12691/ajams-3-2-8
13. Quinlan, J.R., et al.: Learning with continuous classes. In: 5th Australian Joint Conference on Artificial Intelligence. vol. 92, pp. 343–348. World Scientific (1992). https://doi.org/10.1142/9789814536271
14. Rudin, C.: Stop explaining black box machine learning models for high stakes decisions and use interpretable models instead. Nat. Mach. Intell. **1**(5), 206–215 (2019). https://doi.org/10.1038/s42256-019-0048-x
15. Selvaraju, R.R., Cogswell, M., Das, A., Vedantam, R., Parikh, D., Batra, D.: Grad-CAM: visual explanations from deep networks via gradient-based localization. Int. J. Comput. Vis. **128**(2), 336–359 (2019). https://doi.org/10.1007/s11263-019-01228-7
16. Terrell, G.R., Scott, D.W.: Variable Kernel Density estimation. Ann. Stat. **20**(3), 1236–1265 (1992). https://doi.org/10.1214/aos/1176348768

17. Wang, S., Wang, S., Wang, D.: Combined probability density model for medium term load forecasting based on quantile regression and kernel density estimation. Energy Procedia **158**, 6446–6451 (2019). https://doi.org/10.1016/j.egypro.2019.01.169, innovative Solutions for Energy Transitions
18. Zhang, S., Wang, Y., Zhang, Y., Wang, D., Zhang, N.: Load probability density forecasting by transforming and combining quantile forecasts. Appl. Energy **277**, 115600 (2020). https://doi.org/10.1016/j.apenergy.2020.115600

SLIPMAP: Fast and Robust Manifold Visualisation for Explainable AI

Anton Björklund$^{(\boxtimes)}$ ⓘ, Lauri Seppäläinen ⓘ, and Kai Puolamäki ⓘ

University of Helsinki, Helsinki, Finland
{anton.bjorklund,lauri.seppalainen,kai.puolamaki}@helsinki.fi

Abstract. We propose a new supervised manifold visualisation method, SLIPMAP, that finds local explanations for complex black-box supervised learning methods and creates a two-dimensional embedding of the data items such that data items with similar local explanations are embedded nearby. This work extends and improves our earlier algorithm and addresses its shortcomings: poor scalability, inability to make predictions, and a tendency to find patterns in noise. We present our visualisation problem and provide an efficient GPU-optimised library to solve it. We experimentally verify that SLIPMAP is fast and robust to noise, provides explanations that are on the level or better than the other local explanation methods, and are usable in practice.

Keywords: Manifold visualisation · Explainable AI · Local approximation

1 Introduction

The goal of manifold visualisation is to find a low-dimensional visualisation of high-dimensional data. We recently introduced a method that combines manifold visualisation with *explainable artificial intelligence* (XAI), called SLISEMAP [6,7]. SLISEMAP creates an embedding of data points such that points nearby in the embedding have similar explanations. (for a given black box machine learning model). Figure 1 shows an example of an embedding (left) and explanations in the form of linear coefficients (right). SLISEMAP has already been used in studying physical systems [29], for studying molecular properties [4], and to reduce data dimensionality in manufacturing [27].

The practical application of SLISEMAP is hindered by four shortcomings: (i) *Speed.* SLISEMAP scales quadratically with the amount of data, so it is impractical to visualise large datasets (larger than $\sim 10^4$ points). The solution in [7] is subsampling: train on a subset of the data and, if necessary, add the remaining points to the trained SLISEMAP post-hoc. (ii) *New data.* However, adding new data is only possible if the value of the target variable is known [7]. (iii) *No predictive model.* Since there is no principled way of adding points to the embedding, SLISEMAP cannot predict the values of the target variable. (iv) *Behaviour*

© The Author(s) 2024
I. Miliou et al. (Eds.): IDA 2024, LNCS 14642, pp. 223–235, 2024.
https://doi.org/10.1007/978-3-031-58553-1_18

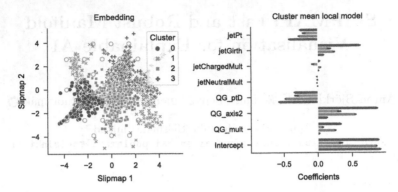

Fig. 1. SLIPMAP embedding of the *Jets* dataset used in a classification task described in Sect. 4 is shown on the left. The local models explaining the black box classifier have been clustered, and the mean coefficients for each cluster are shown on the right.

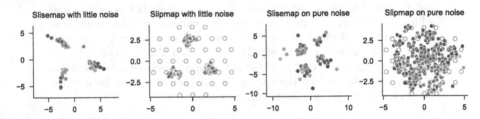

Fig. 2. Both SLISEMAP (left) and SLIPMAP (2nd from the left) correctly find the three modes for a toy data of 500 points constructed as in Fig. 1 of [7] ($y = \max(x_1, x_2, x_3) + \mathcal{N}(0, 0.01)$ and $x \sim N(0,1)^4 \in \mathbb{R}^4$), each of the three visual clusters corresponding to a linear model $f_j(x) \approx x_j$, where $j = \arg\max_{i \in \{1,2,3\}} x_i$. However, SLISEMAP outputs visual clusters, even when the target variable y is Gaussian noise (2nd from the right). In contrast, SLIPMAP (right) overfits less due to the equally spaced prototypes and Gaussian kernel (see Sect. 2.1), leading to fewer misleading visual structures; for SLIPMAP, noise looks like noise.

for noisy data. SLISEMAP works well for low-noise data, but in the presence of noise, it tends to cluster data in random clusters, as shown in Fig. 2.

The contributions of this paper are: we introduce a new prototype-based variant, coined SLIPMAP, that solves the scalability issues, define the computational problem, and present a simple modification to SLISEMAP that allows it to be a generative model that makes predictions (and is, therefore, an actual interpretable model). We show that SLIPMAP is fast, the modification results in a predictive model having good fidelity, and the explanations are stable even in the presence of noise, and valuable in practice.

Related Work. Starting at the introduction of ISOMAP in 2000 [30], countless *manifold visualisation* methods have been developed, of which t-SNE [22] and UMAP [23] are currently commonly used with several variants proposed (e.g.,

[15,16]). Manifold visualisations present high-dimensional data in a typically two-dimensional embedding such that neighbouring points in the embedding are similar by some pre-defined criteria. Unlike SLIPMAP and SLISEMAP, none of the prior methods defines the neighbourhood in terms of local explanations. Manifold visualisations are an indispensable tool in various disciplines where understanding of complex datasets is necessary, from genetics [11,18] to astronomy [3] and linguistics [19].

XAI is essential due to the increasing complexity and widespread use of black-box machine learning models. The primary objective in XAI is to understand and explore *black box* supervised learning algorithms [14]. Explanation methods can be divided into *model specific* and *model agnostic*, the latter of which can be applied to any supervised learning model.

XAI methods can further be split into *global* and *local*. Global methods try to explain the global behaviour of the supervised learning model for all data points. The obvious drawback of this approach is that if the black box model is too complicated, it is impossible to find a simple explanation that approximates it with sufficient fidelity. On the other hand, local explanations methods such as LIME [28], SHAP [21], and SLISE [5] produce an explanation that is valid only for individual data items. In this categorisation, SLIPMAP falls into the class of model-agnostic methods, which provide local explanations for all data points. However, combined with the embedding, the local explanations effectively produce a global explanation of the black-box model.

A common approach for local, model-agnostic explanation methods is to locally approximate the black box model with an interpretable model [5,7,21,28]. However, most other methods rely on randomly sampling new data points [21,28]. In contrast, SLIPMAP only uses the training data. As a result, SLIPMAP is especially useful for explaining models where random sampling of new data is not straightforward; e.g., with scientific data, generating random data that obeys all physical constraints is often challenging.

2 Problem Definition

In this section, we define the computational problem we want to solve in Sect. 2.1, and how we can get interpretable predictions for new data items, in Sect. 2.2.

2.1 SLIPMAP

The main difference between SLIPMAP and SLISEMAP is the introductions of "prototypes" in the embedding (the regular grid of circles in Fig. 1). In SLIPMAP, only the prototypes have local models instead of every data item, making the algorithm faster as we only need to optimise the parameters for a smaller number of prototypes, yielding a linear computational complexity (Sect. 3.1).

SLIPMAP also uses a Gaussian kernel instead of an exponential kernel (distances in the exponent are squared). The squared distances and the fixed spacing of the prototypes reduce the tendency of SLIPMAP to form clusters with random data. SLIPMAP solves the following optimisation problem:

Problem 1. (SLIPMAP) Assume you are given a dataset $\{(\boldsymbol{x}_i, \boldsymbol{y}_i)\}_{i \in [n]}$,[1] prototype vectors $\{\boldsymbol{c}_j\}_{j \in [p]}$, embedding dimensionality $d \in \mathbb{N}$ (typically $d = 2$), and a radius $r \in \mathbb{R}_{>0}$, where $\boldsymbol{x}_i \in \mathbb{R}^m$ are the vectors of features, $\boldsymbol{y}_i \in \mathbb{R}^o$ are the targets, and $\boldsymbol{c}_j \in \mathbb{R}^d$ are embedding coordinates. Find the embedding $\boldsymbol{z}_i \in \mathbb{R}^d$ and the local models $f_j : \mathbb{R}^m \rightarrow \mathbb{R}^o$, where $i \in [n]$ and $j \in [p]$, that minimise

$$\mathcal{L}_0 = \sum_{i=1}^{n} \sum_{j=1}^{p} \frac{e^{-\|\boldsymbol{z}_i - \boldsymbol{c}_j\|_2^2}}{\sum_{k=1}^{n} e^{-\|\boldsymbol{z}_k - \boldsymbol{c}_j\|_2^2}} l(f_j(\boldsymbol{x}_i), \boldsymbol{y}_i), \tag{1}$$

where $\|\cdot\|_2$ is the Euclidean distance and $l(\cdot, \cdot)$ is a loss function for the local models under the constraint that

$$\text{radius}(\boldsymbol{Z}) = \left(\sum_{i=1}^{n} \sum_{k=1}^{d} z_{ik}^2 / n \right)^{1/2} = r. \tag{2}$$

We use the following matrices: $\boldsymbol{X}_{i.} = \boldsymbol{x}_i$, $\boldsymbol{Y}_{i.} = \boldsymbol{y}_i$, and $\boldsymbol{Z}_{i.} = \boldsymbol{z}_i$ for $i \in [n]$ and $\boldsymbol{C}_{j.} = \boldsymbol{c}_j$ for $j \in [p]$. The rows $\boldsymbol{B}_{j.}$ of matrix $\mathbf{B} \in \mathbb{R}^{p \times q}$ contain the parameters for the local models f_j, where q is the number of parameters in the local models. The loss function in Eq. (1) can be augmented with regularisation terms,

$$\mathcal{L} = \mathcal{L}_0 + \sum_{j=1}^{p} \sum_{k=1}^{q} (\lambda_{\text{lasso}} |\mathbf{B}_{jk}| + \lambda_{\text{ridge}} \mathbf{B}_{jk}^2), \tag{3}$$

where $\lambda_{\text{lasso}} \in \mathbb{R}_{\geq 0}$ and $\lambda_{\text{ridge}} \in \mathbb{R}_{\geq 0}$ are the parameters for Lasso and Ridge regularisation, respectively.

As local, interpretable models, we use linear models for regression problems and multi-variate logistic regression for classification problems. The loss functions are a quadratic loss for regression and Hellinger loss for classification; see [7] for details and discussion.

2.2   Mapping from Covariates to the Target Variable

Next, we define a mapping from the covariates to the embedding coordinates and the local models. In principle, these mappings could be arbitrary functions. Here, we have chosen the 1-nearest neighbour regression model as the mapping for simplicity and computational efficiency. The simplicity also makes the whole prediction procedure very transparent since we *"use an interpretable model that works well for similar data items"*.

The implied predictive model $f : \mathbb{R}^m \rightarrow \mathbb{R}^o$ for SLIPMAP is then the distance-weighted average over the local models in the embedding:

$$f(\boldsymbol{x}) = \sum_{j=1}^{p} \frac{e^{-\|\boldsymbol{z}_i - \boldsymbol{c}_j\|_2^2}}{\sum_{k=1}^{p} e^{-\|\boldsymbol{z}_i - \boldsymbol{c}_k\|_2^2}} f_j(\boldsymbol{x}), \tag{4}$$

where $i = \arg\min_{i \in [n]} \|\boldsymbol{x} - \boldsymbol{x}_i\|_2$. We can define an equivalent mapping for SLISEMAP by replacing p with n and $\boldsymbol{c}_j$ by $\boldsymbol{z}_j$ in Eq. (4).

---

[1] We use shorthand notation $[n] = \{1, \ldots, n\}$.

Algorithm 1: The SLIPMAP algorithm, where $\mathcal{L}$ is given in Eq. (1). See the text for discussion.

---

1 **Function** Slipmap(X, Y, C, r, d)
2     $Z \leftarrow \text{PCA}(\mathbf{X})_{.,1:d}$ // Initialise the embedding
3     $Z \leftarrow Z \cdot r/\text{radius}(Z)$ // Normalise the embedding
4     $B \leftarrow \arg\min_B[\mathcal{L}(X,Y,Z,C,B,r,d)]$ // Initialise the local models
5   **do**
6        $Z \leftarrow \text{Escape}(X,Y,C,B,r)$
7        $Z, B \leftarrow \arg\min_{Z,B} \mathcal{L}(X,Y,Z \cdot r/\text{radius}(Z),C,B,r,d)$
8   **while** *not converged*
    Result: Z, B

9 **Function** Escape(X, Y, C, B, r)
10     $W_{jk} \leftarrow e^{-\|c_j - c_k\|_2^2}/\sum_{l=1}^p e^{-\|c_j - c_l\|_2^2}$ for all $j,k \in [p]$
11     $L_{ij} \leftarrow l(f_j(X_{i\cdot}), Y_{i\cdot})$ for all $i \in [n]$ and $j \in [p]$
12     $Z_{i\cdot} \leftarrow C_{k\cdot}$ where $k = \arg\min_k (LW)_{ik}$ for all $i \in [n]$
    Result: $Z \cdot r/\text{radius}(Z)$

---

3 Algorithm

This section discusses how we implement and solve Prob. 1, including the computational complexity in Sect. 3.1.

To optimise Eq. (1), we use the gradient-based quasi-Newton LBFGS optimiser [20]. We combine the optimiser with a heuristic for escaping local optima, just as with SLISEMAP [7]. The pseudocode can be seen in Alg. 1.

The algorithm starts by initialising the embedding for the data items and the local models for the prototypes (lines 2–4 in Alg. 1). Then, it alternates between the escape heuristic and the optimisation until no better solution is found (lines 6–8). The escape heuristic consists of greedily assigning each item the embedding of the prototype that minimises the weighted loss (lines 10–12).

SLIPMAP is implemented using PyTorch [26], which enables GPU acceleration. The source code for our implementation and experiments (Sect. 5) is available under an open-source licence at https://github.com/edahelsinki/slisemap.

3.1 Computational Complexity

The time complexity of Eq. (1) is $\mathcal{O}(np)$, not counting the time for evaluating a local model on one data item. For many simple models and loss functions, including linear and logistic regression, the time complexity increases by a factor of $\mathcal{O}(m+q+o)$. The optimisation contributes an unknown number of iterations, depending on the convergence difficulty. The memory complexity of Eq. (1) is $\mathcal{O}(npo + nm + pq)$, and the LBFGS optimisation only adds a constant factor for the history. The complexities are empirically evaluated in Sect. 5.4.

4   Datasets

We use the following datasets in the experiments (Sect. 5).

Air Quality [25] contains 7355 hourly instances of 12 different air quality measurements. One of the measurements is chosen as a dependent regression variable, and the others are used as covariates.

Covertype [8] is a classification dataset of forest cover types containing over half a million instances with 54 features and seven classes. The instances are various cartographic variables of natural forests.

Gas Turbine [1, 17] is a regression dataset with 36,733 instances of 9 sensor measurements from a gas turbine to study gas emissions.

HIGGS [31] is a two-class classification dataset consisting of signal processes that produce Higgs bosons or are background. The dataset contains nearly 100,000 instances with 28 features.

Jets [10] contains simulated LHC proton-proton collisions. The collisions produce quarks and gluons that decay into cascades of stable particles called jets. The classification task is to distinguish between jets generated by quarks and gluons. The dataset consists of 266,421 instances with seven features.

QM9 [9] is a regression dataset comprising 133,766 small organic molecules. As the dependent variable, we use HOMO energies obtained from [12], and create interpretable features with the Mordred molecular description calculator [24].

5   Experiments

In this section, we empirically evaluate SLIPMAP by first comparing predictions on unseen data in Sect. 5.1. Then, we verify the embedding quality in Sect. 5.2 and local explanations in Sect. 5.3. Finally, we validate the claims about improved scaling in Sect. 5.4.

All experiments use normalised data (zero mean and unit variance). The density of the prototype grid is one prototype per unit square, and the regularisation coefficients λ_{lasso} and λ_{ridge} and the radius r have been optimised using Bayesian hyperparameter optimisation. All experiments have been run ten times with different seeds and randomly subsampled datasets. Since SLIPMAP is implemented with PyTorch [26], we run the experiments with GPU acceleration, except for the experiments measuring time.

5.1   Predictions

In this experiment, we measure the predictive performance of SLISEMAP and SLIPMAP, using Eq. (4). We also compare the predictions against the nearest neighbours to verify that the local models improve the predictions. As target values, we try both predictions from various black box models and the ground truth labels, with increasing subsamples of the training data.

In Fig. 3, we see how the losses from the SLIPMAP predictions on unseen test data approach that of the predictions from the black box models. In some cases,

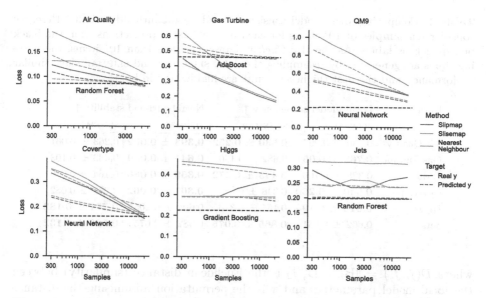

Fig. 3. Loss curves for SLIPMAP, SLISEMAP, and nearest neighbour models trained on predicted y:s and ground truth y:s compared to various black box models. The loss for regression (top row) is mean squared error; for classification (bottom row), the loss is Hellinger loss. Lower is better.

such as the *Jets* dataset, only very little data is needed. Predictions from black box models provide smoothing, especially for discrete class labels. However, with sufficient data, SLIPMAP trained on ground truth labels often converge to similar losses. The AdaBoost regressor is non-optimal for the *Gas Turbine* dataset since SLIPMAP and SLISEMAP, trained directly on the ground truth, actually provide better predictions. Generally, SLIPMAP performs slightly better than SLISEMAP and clearly better than the nearest neighbour.

5.2 Robustness

Explanations are only helpful if they are consistent. If, for example, slightly changing the training dataset causes a significant explanation shift, the explanations are less trustworthy.

Local model consistency [29] measures how stable the set of local models is with respect to resampling the data. If the local models are inconsistent, the local models are not trustworthy as explanations. To measure local model consistency, we train two models on subsamples taken from a dataset such that there is no overlap between the samples. This yields two sets of local models $\{f_1, f_2, ... f_p\}$ and $\{f_1', f_2', ..., f_p'\}$. We then match each local model to its most similar counterpart and calculate the average distance between the models:

$$\mathcal{M}_B = 1 - \min_{\pi} \frac{\sum_{i=1}^{p} D(f_i, f_{\pi(i)}')}{\frac{1}{n}\sum_{i=1}^{p}\sum_{j=1}^{p} D(f_i, f_j')}, \tag{5}$$

Table 1. Comparing local model consistency and neighbourhood stability. Here, we consider ten samples of 10^4 items for each dataset, using predictions from the black box models as labels. As the *Air Quality* dataset has less than 10^4 items, the missing items are generated by resampling the data. SLISEMAP and SLIPMAP show similar performance, and the best (highest) results are highlighted in bold.

Data	Local model consistency ↑		Neighbourhood stability ↑	
	SLIPMAP	SLISEMAP	SLIPMAP	SLISEMAP
Air Quality	0.460 ± 0.097	$\mathbf{0.530 \pm 0.252}$	$\mathbf{0.393 \pm 0.062}$	0.263 ± 0.061
Gas Turbine	$\mathbf{0.762 \pm 0.051}$	0.682 ± 0.190	$\mathbf{0.641 \pm 0.039}$	0.433 ± 0.103
QM9	0.328 ± 0.106	$\mathbf{0.443 \pm 0.272}$	$\mathbf{0.369 \pm 0.086}$	0.164 ± 0.036
Covertype	$\mathbf{0.540 \pm 0.260}$	0.348 ± 0.380	$\mathbf{0.301 \pm 0.062}$	0.276 ± 0.082
Higgs	$\mathbf{0.515 \pm 0.193}$	0.167 ± 0.376	0.604 ± 0.206	$\mathbf{0.771 \pm 0.183}$
Jets	0.662 ± 0.061	$\mathbf{0.865 \pm 0.075}$	0.382 ± 0.075	$\mathbf{0.523 \pm 0.132}$

where $D(f_i, f'_j) = \|\mathbf{B}_i. - \mathbf{B}'_j.\|_2$ is the Euclidean distance (similarity) between the local model parameters and π is the permutation minimising the distance between the local models.

Neighbourhood stability measures the stability of the embedding with respect to resampling. It measures how well models trained on partly overlapping data retain the relative locations of the data items in the embedding, i.e., whether or not the neighbouring relations between the items are preserved. To measure neighbourhood stability, we train models on datasets sampled such that half of the items overlap. Let $\mathcal{S}$ be the set of overlapping points. Then, for each shared item, we form the set of neighbours in both learned embeddings (denoted as $N(i) = \{j \in \mathcal{S} | \|z_i - z_j\|_2 < 1\}$ and $N'(i) = \{j \in \mathcal{S} | \|z'_i - z'_j\|_2 < 1\}$) and calculate the Jaccard similarity between the neighbour sets:

$$\mathcal{M}_{neighbourhood} = |\mathcal{S}|^{-1} \sum_{i \in \mathcal{S}} |N(i) \cap N'(i)| / |N(i) \cup N'(i)| \tag{6}$$

Table 1 shows a comparison between the explanation robustness of SLIPMAP and SLISEMAP. SLIPMAP performs comparably to SLISEMAP with respect to local model concistency and neighbourhood stability. As discussed in [7], local explanations have inherent ambiguity; a given data item can have multiple local explanations with comparable performance. The neighbourhood stability results show SLIPMAP also exhibits this behaviour.

5.3   Local Explanation Comparison

In this section we quantitatively compare the local models from SLIPMAP with other model-agnostic, local explanation methods: LIME [28] (with and without discretisation), (partition) SHAP [21], SLISE [5], and SLISEMAP [7]. These all provide explanations in the form of local, linear approximations. As metrics, we consider the following:

Time. How long does it take to get one explanation (dividing the setup time between the data items)?

Table 2. Comparing local explanation methods. We measure how well the approximation predicts the selected data item (local loss), the five nearest neighbours (stability), and the number of other data items with a loss less than a threshold. All explanations are based on 5,000 data items, and the best results are highlighted in bold.

Data	Method	Time (s) ↓	Local loss ↓	Stability ↓	Coverage ↑
Air Quality	LIME	3.648 ± 0.02	0.147 ± 0.07	0.180 ± 0.05	0.079 ± 0.00
	LIME (nd)	0.062 ± 0.02	0.041 ± 0.01	0.046 ± 0.01	0.464 ± 0.04
	SHAP	0.723 ± 0.21	**0.000 ± 0.00**	0.049 ± 0.02	0.217 ± 0.01
	SLISE	13.723 ± 2.36	**0.000 ± 0.00**	**0.019 ± 0.01**	**0.853 ± 0.01**
	SLIPMAP	**0.005 ± 0.00**	0.004 ± 0.00	**0.021 ± 0.01**	0.366 ± 0.01
	SLISEMAP	0.366 ± 0.14	0.001 ± 0.00	**0.018 ± 0.00**	0.759 ± 0.02
Gas Turbine	LIME	2.982 ± 0.02	0.205 ± 0.07	0.259 ± 0.07	0.219 ± 0.02
	LIME (nd)	0.030 ± 0.00	0.186 ± 0.07	0.180 ± 0.07	0.299 ± 0.04
	SHAP	0.693 ± 0.10	**0.000 ± 0.00**	0.116 ± 0.05	0.333 ± 0.03
	SLISE	26.577 ± 4.90	**0.000 ± 0.00**	0.056 ± 0.02	**0.407 ± 0.04**
	SLIPMAP	**0.007 ± 0.00**	0.004 ± 0.00	**0.052 ± 0.02**	0.325 ± 0.03
	SLISEMAP	0.482 ± 0.19	0.007 ± 0.00	**0.048 ± 0.01**	0.270 ± 0.04
QM9	LIME	6.256 ± 0.09	0.773 ± 0.23	0.778 ± 0.25	0.153 ± 0.02
	LIME (nd)	0.029 ± 0.00	0.323 ± 0.05	0.366 ± 0.04	0.179 ± 0.01
	SHAP	1.326 ± 0.79	**0.000 ± 0.00**	0.299 ± 0.07	0.207 ± 0.01
	SLISE	28.176 ± 3.71	**0.000 ± 0.00**	0.218 ± 0.04	**0.393 ± 0.01**
	SLIPMAP	**0.013 ± 0.00**	0.011 ± 0.00	**0.158 ± 0.03**	0.303 ± 0.02
	SLISEMAP	0.737 ± 0.22	0.016 ± 0.00	**0.160 ± 0.04**	0.292 ± 0.01
Higgs	LIME	8.425 ± 0.13	0.025 ± 0.00	**0.033 ± 0.00**	0.349 ± 0.01
	LIME (nd)	0.032 ± 0.00	0.034 ± 0.00	0.037 ± 0.00	0.333 ± 0.01
	SHAP	0.561 ± 0.04	**0.000 ± 0.00**	0.038 ± 0.00	0.315 ± 0.00
	SLISE	24.785 ± 2.35	**0.000 ± 0.00**	0.042 ± 0.00	**0.445 ± 0.01**
	SLIPMAP	**0.003 ± 0.00**	0.023 ± 0.01	0.040 ± 0.00	0.337 ± 0.03
	SLISEMAP	1.103 ± 0.35	0.034 ± 0.00	0.041 ± 0.00	0.302 ± 0.00
Jets	LIME	2.346 ± 0.02	0.011 ± 0.00	0.016 ± 0.00	0.106 ± 0.01
	LIME (nd)	0.063 ± 0.01	0.013 ± 0.00	0.014 ± 0.00	0.179 ± 0.01
	SHAP	0.246 ± 0.01	**0.000 ± 0.00**	0.007 ± 0.00	0.163 ± 0.01
	SLISE	10.357 ± 1.34	**0.000 ± 0.00**	**0.006 ± 0.00**	**0.431 ± 0.02**
	SLIPMAP	**0.018 ± 0.01**	0.000 ± 0.00	**0.006 ± 0.00**	0.357 ± 0.02
	SLISEMAP	1.956 ± 0.57	0.000 ± 0.00	**0.006 ± 0.00**	0.308 ± 0.02

Local loss. [5, 21] How well does the approximation match the black box model for the selected data item using the losses mentioned in Sect. 2?

Stability. [2, 5] Does a slight change in the input need a significant change in the explanation? Measured by calculating the mean loss of the local models on the five nearest neighbours.

Coverage. [7, 13] Does the explanations generalise to other data items? Measured by counting the number of data items with a loss less than a threshold. The threshold is the 0.3 quantile of the losses from a global linear approximation.

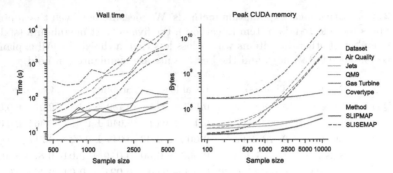

Fig. 4. Comparison of time and memory scaling between SLISEMAP and SLIPMAP. SLIPMAP is consistently faster as sample size increases and needs radically less memory in all six datasets (notice the logarithmic scale). Lower is better.

The results can be seen in Table 2 where SLIPMAP performs comparably or better than SLISEMAP. SLIPMAP does not guarantee a zero local loss like SLISE and SHAP, but they are generally quite small (whereas LIME sometimes have a smaller loss for the synthetic neighbourhood than the item being explained [5]).

5.4 Scaling

This experiment shows that SLIPMAP scales better with the number of data items than SLISEMAP, both in time and in memory. We measure the time on a CPU (to avoid the overhead of GPU communication on small data sizes) and the memory on a GPU, since that is usually the limiting factor SLISEMAP. As the left panel of Fig. 4 demonstrates, for each dataset, SLIPMAP converges faster than SLISEMAP by at least an order of magnitude. The difference is even more dramatic when considering memory complexity (Fig. 4 right panel), as SLIPMAP scales linearly (Sect. 3.1) compared to the quadratic scaling of SLISEMAP [7].

6   Conclusions and Future Work

We propose SLIPMAP, a novel model-agnostic method for explainable AI. SLIPMAP finds all local explanations for a complex black-box regression or classification model and produces an informative embedding where data items with similar explanations (local models) are embedded nearby. We substantially improved our earlier work by making our algorithm fast and robust to noise, leading to fewer false patterns in the embedding (Fig. 2). We have shown that the local explanations produced by SLIPMAP have high fidelity, good stability, and coverage. When trained on the predictions of the black-box model (instead of raw target values), SLIPMAP is, in our use cases, always able to mimic the black-box model with almost perfect fidelity.

Also, even though SLIPMAP is not meant to replace purpose-built classification and regression models, it performs similarly to the state-of-the-art models in real-world use cases.

SLIPMAP allows adding data items to the embedding and making predictions, even when the target variable is unknown, unlike the original SLISEMAP, extending the usage of SLIPMAP from a pure XAI method (which requires a pre-trained regression or classification model to work) to a more general supervised data exploration tool (which finds an interpretable predictive model for the data). In the future, we can still improve on SLIPMAP, e.g., by replacing the simple nearest neighbour model and kernel density estimate for making the predictions with a more general model, such as Gaussian Processes. The improved scalability, especially the GPU memory requirements, also unlocks applications with larger datasets.

Acknowledgement. We acknowledge funding from the Research Council of Finland (decisions 346376 and 345704) and the University of Helsinki.

References

1. Gas Turbine CO and NOx Emission Data Set (2019). https://doi.org/10.24432/C5WC95
2. Alvarez-Melis, D., Jaakkola, T.S.: On the Robustness of Interpretability Methods (2018). https://doi.org/10.48550/arXiv.1806.08049
3. Anders, F., et al.: Dissecting stellar chemical abundance space with t-SNE. Astron. Astrophys. **619**, A125 (2018). https://doi.org/10.1051/0004-6361/201833099
4. Besel, V., Todorović, M., Kurtén, T., Rinke, P., Vehkamäki, H.: Curation of high-level molecular atmospheric data for machine learning purposes. Tech. Rep. (2023). https://doi.org/10.5194/egusphere-egu23-1135
5. Björklund, A., Henelius, A., Oikarinen, E., Kallonen, K., Puolamäki, K.: Explaining any black box model using real data. Front. Comput. Sci. **5**, 1143904 (2023). https://doi.org/10.3389/fcomp.2023.1143904
6. Björklund, A., Mäkelä, J., Puolamäki, K.: SLISEMAP: combining supervised dimensionality reduction with local explanations. In: ECML PKDD, vol. 13718, pp. 612–616 (2023). https://doi.org/10.1007/978-3-031-26422-1_41
7. Björklund, A., Mäkelä, J., Puolamäki, K.: SLISEMAP: supervised dimensionality reduction through local explanations. Mach. Learn. **112**(1), 1–43 (2023). https://doi.org/10.1007/s10994-022-06261-1
8. Blackard, J.: Covertype (1998). https://doi.org/10.24432/C50K5N
9. Blum, L.C., Reymond, J.L.: 970 million druglike small molecules for virtual screening in the chemical universe database GDB-13. J. Am. Chem. Soc. **131**, 8732 (2009)
10. CMS Collaboration: Simulated dataset QCD_Pt-15to3000_TuneZ2star_Flat_8TeV_pythia6 in AODSIM format for 2012 collision data (2017). https://doi.org/10.7483/OPENDATA.CMS.7Y4S.93A0
11. Diaz-Papkovich, A., Anderson-Trocmé, L., Gravel, S.: A review of UMAP in population genetics. J. Hum. Gene. **66**(1), 85–91 (2021)
12. Ghosh, K.: MBTR_QM9 (2020). https://doi.org/10.5281/zenodo.4035918
13. Guidotti, R., Monreale, A., Ruggieri, S., Pedreschi, D., Turini, F., Giannotti, F.: Local Rule-Based Explanations of Black Box Decision Systems (2018). https://doi.org/10.48550/ARXIV.1805.10820
14. Guidotti, R., Monreale, A., Ruggieri, S., Turini, F., Giannotti, F., Pedreschi, D.: A survey of methods for explaining black box models. ACM Comput. Surv. **51**(5), 1–42 (2019). https://doi.org/10.1145/3236009

15. Heiter, E., Kang, B., Seurinck, R., Lijffijt, J.: Revised conditional t-SNE: looking beyond the nearest neighbors. In: IDA, vol. 13876, pp. 169–181 (2023)
16. Kang, B., García García, D., Lijffijt, J., Santos-Rodríguez, R., De Bie, T.: Conditional t-SNE: more informative t-SNE embeddings. Mach. Learn. 110(10), 2905–2940 (2021). https://doi.org/10.1007/s10994-020-05917-0
17. Kaya, H., Tüfekci, P., Uzun, E.: Predicting CO and NOxemissions from gas turbines: novel data and a benchmark PEMS. Turk. J. Elec. Eng. Comp. Sci. 27(6), 4783–4796 (2019). https://doi.org/10.3906/elk-1807-87
18. Kobak, D., Berens, P.: The art of using t-SNE for single-cell transcriptomics. Nat. Commun. 10(1), 5416 (2019). https://doi.org/10.1038/s41467-019-13056-x
19. Levine, Y., et al.: SenseBERT: driving some sense into BERT. In: ACL, pp. 4656–4667 (2020). https://doi.org/10.18653/v1/2020.acl-main.423
20. Liu, D.C., Nocedal, J.: On the limited memory BFGS method for large scale optimization. Math. Program. 45(1–3), 503–528 (1989)
21. Lundberg, S.M., Lee, S.I.: A unified approach to interpreting model predictions. In: NeurIPS. vol. 30 (2017)
22. van der Maaten, L., Hinton, G.: Visualizing data using t-SNE. J. Mach. Learn. Res. 9(86), 2579–2605 (2008). http://jmlr.org/papers/v9/vandermaaten08a.html
23. McInnes, L., Healy, J., Saul, N., Großberger, L.: UMAP: uniform manifold approximation and projection. J. Open Source Softw. 3(29), 861 (2018)
24. Moriwaki, H., Tian, Y.S., Kawashita, N., Takagi, T.: Mordred: a molecular descriptor calculator. J. Cheminform. 10(1), 4 (2018)
25. Oikarinen, E., Tiittanen, H., Henelius, A., Puolamäki, K.: Detecting virtual concept drift of regressors without ground truth values. Data Min. Knowl. Discov. 35(3), 726–747 (2021). https://doi.org/10.1007/s10618-021-00739-7
26. Paszke, A., Gross, S., Massa, F., et al.: PyTorch: an imperative style, high-performance deep learning library. In: NeurIPS. vol. 32 (2019)
27. Peng, G., Cheng, Y., Zhang, Y., Shao, J., Wang, H., Shen, W.: Industrial big data-driven mechanical performance prediction for hot-rolling steel using lower upper bound estimation method. J. Manuf. Syst. 65, 104–114 (2022)
28. Ribeiro, M.T., Singh, S., Guestrin, C.: "Why Should I Trust You?": explaining the predictions of any classifier. In: ACM SIGKDD, pp. 1135–1144 (2016)
29. Seppäläinen, L., Björklund, A., Besel, V., Puolamäki, K.: Using slisemap to interpret physical data. PLoS ONE 19(1), e0297714 (2024). https://doi.org/10.1371/journal.pone.0297714
30. Tenenbaum, J.B., de Silva, V., Langford, J.C.: A global geometric framework for nonlinear dimensionality reduction. Science 290(5500), 2319–2323 (2000)
31. Whiteson, D.: HIGGS (2014). https://doi.org/10.24432/C5V312

A Frank System for Co-Evolutionary Hybrid Decision-Making

Federico Mazzoni[1]([✉])[iD], Riccardo Guidotti[1][iD], and Alessio Malizia[1,2][iD]

[1] University of Pisa, Pisa, Italy
federico.mazzoni@phd.unipi.it, riccardo.guidotti@unipi.it
[2] Molde University College, Molde, Norway
alessio.malizia@unipi.it

Abstract. We introduce FRANK, a human-in-the-loop system for co-evolutionary hybrid decision-making aiding the user to label records from an un-labeled dataset. FRANK employs incremental learning to "evolve" in parallel with the user's decisions, by training an interpretable machine learning model on the records labeled by the user. Furthermore, advances state-of-the-art approaches by offering inconsistency controls, explanations, fairness checks, and bad-faith safeguards simultaneously. We evaluate our proposal by simulating the users' behavior with various levels of expertise and reliance on FRANK's suggestions. The experiments show that FRANK's intervention leads to improvements in the accuracy and the fairness of the decisions.

Keywords: Human-Centered AI · Hybrid Decision Maker · Skeptical Learning · Incremental Learning · Explainable AI · Fairness Checking

1 Introduction

Automated decision-makers based on Machine Learning (ML) are still not widely adopted for high-stakes decisions such as medical diagnoses or court decisions [22]. In these fields, humans are aided but not replaced by Artificial Intelligence (AI), resulting in Hybrid Decision-Makers (HDM) [15]. While HDM literature is flourishing, certain key aspects have not yet been considered, preventing HDM systems from covering possible use cases. HDM systems promote the collaboration between human and AI decision-makers, resulting in a final set of "hybrid" decisions (some taken by the human, others by the machine). In *Learning-to-Defer* [10] systems, the machine plays the primary role, deferring decisions on records with a high degree of uncertainty to an external human supervisor. In [22], a rule-based AI model with inferred rules suggests replacing some user's decisions to maximize fairness, whereas in [9], the model mediates between a user and their supervisor if it is not confident in the user's decisions. On the other hand, in the Skeptical Learning (SL) paradigm, an ML model learns "in parallel" to the decisions taken by a human and queries them if it is "skeptical" of the human decision [4,19,23,24]. SL aims to help the user remain consistent with their past decisions, still giving them veto power against the model's suggestions. SL has been extensively applied to personal context recognition [4,24] and image classifications [19]. In [19], SL suggestions are also supported by *contrastive explanations*. Our system employs and extends traditional SL, by taking into account simultaneously fairness aspects, explainable suggestions, and the involvement of the user's supervisor. In line with [4], our proposal is powered by a *Incremental Learning* (IL) model. IL, also known as Continual Learning, is an ML paradigm where the model is continuously trained on small data batches, potentially including only one data point, instead of the entirety of the training set [12,21].

The *eXplainable AI (XAI)* research field aims to create humanly interpretable proxies of "black-box" ML models used for decision-making. An explanation is *global* if it unveils the whole model logic, or *local* if it justifies the decision of a specific record [7]. A global explanation can be achieved by approximating black-box models with interpretable-by-design ones, such as a decision tree, which also offers local explanations as decision rules [3]. Also, instance-based explanations make use of examples and counter-examples, i.e., similar records with the same/different decision by the AI system [6]. Our proposal offers both a model approximation, employing an interpretable decision tree, and instance-based (counter-)examples to explain the model's suggestions to the user.

Finally, we also account for the *fairness* of the decisions. Two major approaches have been proposed to quantify a dataset's fairness [2]. For *individual fairness*, similar individuals should receive similar treatment, while for *group fairness*, each group should receive a similar treatment [16]. The discriminatory feature to be monitored (e.g., *Race, Gender*) is often defined *sensitive* or *protected attribute* [20]. Given a sensitive attribute, our proposal checks both individual and group fairness, helping the user avoid discriminating behavior.

We propose FRANK, a HDM system overcoming the current limitations of SL related to explainability, fairness, consistency, and bad-faith users. As in SL, if the user's label is inconsistent with FRANK's prediction, the user is warned of possible contradictions with their past behavior and suggested to modify their decision. Besides, provides explanations that become increasingly detailed as the model learns more from the user, who can, in turn, learn more about their behavior. Also, can prevent bad-faith behavior and discriminating decisions. Ultimately, and the human have a symbiotic co-evolutionary relationship, with FRANK's model able to predict the user's behavior, thus aiding them, and the human feeding FRANK's model with new data. Experimental results show that pairingwith less reliable users provides noticeable improvements in terms of accuracy and fairness, and that the usage of explanations increases the number of acceptance for suggestions in case of skepticism.

2   Setting the Stage

We keep the paper self-contained by reporting in the following a brief overview of concepts necessary to understand our proposal. We indicate with X, Y a dataset where $X = \{x_1, \ldots, x_n\} \in \mathcal{X}^{(m)}$ is a set of n records described by m attributes (features), i.e., $x_i = \{(a_1, v_1), \ldots, (a_m, v_m)\}$, where a_i is the attribute name and v_i is the corresponding value, and $\mathcal{X}^{(m)}$ is the feature space consisting of m input features, while $Y = \{y_1, \ldots, y_n\} \in \mathcal{Y}$ is the set of the target variable in the target space $\mathcal{Y}$. With $A = \{a_1, \ldots, a_m\}$ we indicate the set of feature names, and for an instance $x \in X$, we write $x[a_k]$ to refer to the value v_k of attribute a_k. For classification problems, $y_i \in \{1, \ldots, l\} = L$ where L is the set of different class labels and l is the number of the classes, while when dealing with regression problems, $y_i \in \mathbb{R}$. Without losing in generality, we consider $l = 2$, i.e., binary classification problems. We indicate a trained decision-making model with a function $f : \mathcal{X}^{(m)} \to \mathcal{Y}$ that maps data instances x from the feature space $\mathcal{X}^{(m)}$ to the target space $\mathcal{Y}$. We write $f(x) = y$ to denote the decision y taken by f, and $f(X) = Y$ as a shorthand for $\{f(x_i) \mid x_i \in X\} = Y$.

Skeptical Learning. Given a ML model f and a dataset X, the user is tasked to assign a label y_i to each record $x_i \in X$. In SL, the user assigns to x_i the label $\hat{y}_i$, according to their own belief and background and, independently from them, f assigns the label $\tilde{y}_i$, i.e., $\tilde{y}_i = f(x_i)$. The ML model implementing f can be pre-trained on a small training set. If $\hat{y}_i \neq \tilde{y}_i$ and f is *skeptical* (see below), the user is asked if they want to accept $\tilde{y}_i$ as y_i. If they do, y_i takes the value $\tilde{y}_i$. If the user refuses, if $\hat{y}_i = \tilde{y}_i$ or if the model is not skeptical, y_i is assigned $\hat{y}_i$. The ML model is then incrementally trained on x_i and y_i.

The definition of the model's *skepticality* varies in the literature [19]. However, skepticism is always related to model's *epistemic uncertainty*, which is independent of the notion of *confidence* score towards a certain decision, i.e., the prediction probability[1]. Epistemic uncertainty is the model's *ignorance*, and given enough data, it should be minimized [8]. Only a limited number of ML model offers by-design access to epistemic uncertainty, e.g., Naive Bayes, Gaussian Process [4,8]. In the context of SL, it has been approximated by the *empirical accuracy* of past predictions both of the user and the model, i.e., the ratio between the number of times a label has been proposed by the user or predicted by the model, and the times it has been accepted as y [23]. Thus, given x_i and the prediction $\tilde{y}_i$, the skepticism towards the user's $\hat{y}_i$ is:

$$skpt(x_i, \tilde{y}_i, \hat{y}_i, Y, \tilde{Y}, \hat{Y}) = c(f, x_i, \tilde{y}_i) \cdot ea(\tilde{y}_i, Y, \tilde{Y}) - c(f, x_i, \hat{y}_i) \cdot ea(\hat{y}_i, Y, \hat{Y}) \quad (1)$$

where $c(f, x_i, \tilde{y}_i)$ and $c(f, x_i, \hat{y}_i)$ are the model confidence score towards $\tilde{y}_i$ and $\hat{y}_i$. The function ea computes the empirical accuracy of either the model or the user toward their respective label. The empirical accuracy is computed as the cardinality of the intersection between the subset of all their past decisions with label either $\hat{y}_i$ or $\tilde{y}_i$ and the corresponding subset in Y, i.e., the final decision, over the subset of all their past decisions with either $\hat{y}_i$ or $\tilde{y}_i$. Therefore, each possible label $l \in L$ has two accuracy values – following the user's and the model's track record. In [23], the user's accuracy values are initialized with 1, and the model's with 0 (therefore, the model is not skeptical of earlier decisions).

---

[1] Note that there's a general lack of normativity w.r.t. these terms; e.g., [23] uses the term *confidence* to refer to the epistemic uncertainty.

Incremental Decision Tree. We employ Extremely Fast Decision Tree (EFDT) [13], a variant of Hoeffding Tree, which offers performance on par with the non-incremental counterpart [1,5]. EFDT splits a node as soon as the split is deemed useful, with the possibility of later revisiting the decision [13]. Being a decision tree, EFDT can also be exploited to provide explanations to the user [7].

Preferential Sampling. We include an interactive variant of Preferential Sampling (PS), an algorithm increasing group fairness [11]. PS assumes that in the set of class labels L we can recognize a favorable $+$ and an unfavorable $-$ decision, i.e., $L = \{+, -\}$, while among A we can denote a binary sensitive attribute $sa \in A$, e.g., *Sex*. The possible values $\{v, \bar{v}\}$ of sa refers to a discriminated group v and privileged group $\bar{v}$, e.g., *Female* and *Male*. The algorithm identifies the size of the groups of *D*iscriminated records with a *P*ositive (DP) or *N*egative label (DN), and of *P*rivileged records with a *P*ositive (PP) or *N*egative label (PN). Given X, it computes the *dataset discrimination score* as:

$$disc(X, sa, v) = |PP|/|PP \cup PN| - |DP|/|DP \cup DN| \qquad (2)$$

Then, it computes how many records from PP and DN should be removed, and how many from DP and PN should be duplicated to reach $disc \approx 0$. Records are selected w.r.t. the prediction probability of a classifier trained on X. A variant supporting non-binary sensitive attributes, and where the user does not need to know *a priori* the discriminated group(s), is presented in [14].

3   A Frank System

FRANK is a system for HDM, learning from the decisions of the human decision-maker (typically identified as the "user"), continuously evolving with them, and aiding the human to remain consistent by offering suggestions and explanations. FRANK is named after its frank behavior – it interacts with the user as soon as something "unexpected" happens. Other thanand the user, in line with [9], we also suppose a third agent, i.e., the user's *supervisor*. Depending on the context, the supervisor could be someone enforcing company policies to the user's decisions, e.g., making sure they are not biased by personal beliefs, or someone with higher expertise than the user, e.g., a senior doctor.

The pseudocode ofis reported in Algorithm 1. FRANK requires a set of records to label X, which are received one by one, a set of rules R provided by the user's supervisor, a sensitive attribute sa, a skepticality threshold s, the number of iterations k after which a group fairness check is performed on the records and decisions analyzed so far, and a stopping condition stp. At this stage, we are very general about the stopping condition stp as it might be implemented as reaching a certain number of labeled records, or an accuracy higher than a threshold[2] for f. The initialization of $X', Y', \tilde{Y}, \hat{Y}, \ddot{Y}$ in line 1 can rely on empty sets for

---

[2] In our experiments, we consider as stp a certain number of instances to be analyzed, leaving for future work the study of measures automatically unveiling when to stop the training.

Algorithm 1: FRANK

Input : X - records to label, R - supervisor rule set, sa - sensitive attribute,
s - skepticality thr, k - nbr of iter. for GFC, stp - stopping condition,

1:   $X', Y', \tilde{Y}, \hat{Y}, \ddot{Y}, f \leftarrow initialize;$ // sets initialization

2: **while** $stop \neq True$ **do**   // until a stop condition is met

3:     $x_i \leftarrow receive\_record(X);$ // receive a new un-label record

4:     $\hat{y}_i \leftarrow user\_decision(x_i);$ // get user decision

5:     $\tilde{y}_i \leftarrow f(x_i);$ // get model prediction

6:     **if** $ideal\_rule_R(x_i)$ **then**   // if x_i covered by expert rule

7:       $\bar{y}_i \leftarrow rule\_label_R(x_i);$ // get $\bar{y}_i$ from rule

8:       $y_i \leftarrow \bar{y}_i;$ // $\bar{y}_i$ is compulsorily accepted

9:     **else if** $individual\_fairness_{sa}(x_i, X')$ **then**   // if x_i is similar to past records

10:       $y'_p \leftarrow get\_similar\_past\_label(x_i, X', Y');$ // get y'_p from past records

11:       **if** $\hat{y}_i \neq y'_p$ **then**   // conflict with a past decision

12:         $\langle y_i, Y' \rangle \leftarrow solve\_conflict(x_i, y'_p, \hat{y}_i, Y');$ // solve conflict & update Y'

13:       **else** $y_i \leftarrow \hat{y}_i;$ // otherwise, user decision $\hat{y}_i$ is accepted

14:     **else if** $\hat{y}_i \neq \tilde{y}_i \wedge skept_s(f, x_i, \tilde{y}_i, \hat{y}_i, \ddot{Y}, \tilde{Y}, \hat{Y})$ **then** // if clash & skepticism

15:       **if** $is\_expl\_desired(x_i, \tilde{y}_i)$ **then**   // if an explanation for $\tilde{y}_i$ is desired

16:         $e_i \leftarrow get\_and\_show\_expl(x_i, \tilde{y}_i, f, X');$ // return explanation e_i

17:       **if** $accept\_label\_change(x_i, \tilde{y}_i)$ **then** $y_i \leftarrow \tilde{y}_i;$ // $\tilde{y}_i$ is accepted

18:       **else** $y_i \leftarrow \hat{y}_i;$ // $\tilde{y}_i$ is refused

19:     **else** $y_i \leftarrow \hat{y}_i;$ // otherwise $\hat{y}_i$ is accepted

20:     $X' \leftarrow X' \cup \{x_i\}; Y' \leftarrow Y' \cup \{y_i\}; \ddot{Y} \leftarrow \ddot{Y} \cup \{y_i\};$ // update sets

21:     $\hat{Y} \leftarrow \hat{Y} \cup \{\hat{y}_i\}; \tilde{Y} \leftarrow \tilde{Y} \cup \{\tilde{y}_i\}; f \leftarrow update(f, x_i, y_i);$ // update sets and model

22:     **if** $|Y'| \% k = 0$ **then**   // every k records

23:       $Y', f \leftarrow group\_fairness\_check_{sa}(X', Y', f);$ // run GFC

---

a cold start execution, or they might be initialized with records and decisions of previous runs. We use X' to collect the set of records analyzed so far, Y' for the set of final hybrid decisions taken on the records in X', $\tilde{Y}$ for the set of decisions of FRANK's EFDT model f alone, $\hat{Y}$ for the set of decisions proposed by the user alone, and $\ddot{Y}$ to store the decisions taken byand the user without re-labelling due to fairness corrections. Also, f might be completely untrained, pre-trained non-interactively on some records, or pre-trained in a past run of FRANK[3]. Until the stopping condition stp is met (line 2), receives a x_i from X (line 3). As in SL [19], the user assigns a label $\hat{y}_i$, and FRANK's model f a label $\tilde{y}_i$, i.e., the prediction (lines 4 and 5).

With *Ideal Rule Check* (IRC), checks if the record x_i is covered by a rule in the rule set R provided by the user's supervisor (line 7). If it is, then the decision $\bar{y}_i$ is derived from the rule and assigned to the final decision y_i (line 8). If none of the rules from R cover the record, with *Individual Fairness Check* (IFC), checks if the user's decision complies with the individual fairness condition by comparing $\hat{y}_i$ to the labels assigned to "similar" past records (lines 9–13). The definition of

---

[3] In our experiments, we consider the sets $X', Y', \tilde{Y}, \hat{Y}, \ddot{Y}$ initialized with empty sets and f pre-trained non-interactively on 500 records. Future works will investigate further these aspects.

similarity is further defined below. *Skeptical Learning Check* (SLC) is triggered if no similar records exist and the user's decision $\hat{y}_i$ and FRANK's prediction $\tilde{y}_i$ are not the same. Ifis skeptical of $\hat{y}_i$, the user is asked if they want an explanation for $\tilde{y}_i$ (line 15). If the user accepts, they are shown the explanation e_i (line 16). Regardless, the user is then asked if they accept $\tilde{y}_i$ as the final decision y_i (lines 17). If the user refuses (line 18), ifis not skeptical, or if it agrees with the user (line 19), the user's decision $\hat{y}_i$ is accepted as the final decision y_i. Regardless of the triggered checks, x_i and y_i are added respectively to X' and Y' (line 20), and are used to update FRANK's model f (line 21). Similarly, $\tilde{y}_i$ and $\hat{y}_i$ are added to $\tilde{Y}$ and $\hat{Y}$, respectively. Also, y_i is added to $\ddot{Y}$, which might differ from the set of labels Y' in the case of relabeling. Finally, every k records, performs *Group Fairness Check* (GFC, lines 24–25), asking the user if they want to change the label of some past records to reduce the dataset's discrimination as computed by Prefential Sampling [11]. FRANK prioritizes IRC to follow the guidelines of the supervisor, then IFC for fairness among similar records, and, finally, SLC. To avoid contradictions, once a final label y_i is set, checks with less priority are never triggered, and GFC cannot relabel records labeled by IRC or IFC. We stress that the user *has* to accommodate suggestions offered by IRC and IFC. On the other hand, the user is free to disregard suggestions by SLC and GFC. Depending on the use cases, certain checks might be turned off, e.g., IFC and GFC in health contexts. As some functions cycle the previously-seen records, FRANK's algorithmic complexity is $O(n)$ with $n = |X'|$. In the following, we provide a detailed explanation of FRANK's four checks.

Ideal Rule Check. Each rule $r \in R$ includes a set of conditions and a label $\bar{y}$. The *ideal_rule* function checks if x_i follows the conditions of one of the rules in R (line 6), and if it is, it provides the label $\bar{y}_i$ (line 7), which is selected as the final decision y_i, regardless of the user's label $\hat{y}_i$. In case of divergence between the user's decision and the supervisor's rule, the user is notified that their decision is not compliant. Since IRC leaves no freedom of choice, the rules R should only cover very limited, specific, and ideal cases, describing records which should *absolutely* receive a certain label. The supervisor should also make sure the rules R are mutually exclusive. Besides, to avoid conflicts with fairness-related functions, the rules' conditions should not rely on sensitive attributes.

Individual Fairness Check. IFC is meant to reduce the pairs of records violating individual fairness condition, i.e., similar individuals should be treated similarly, by assessing if records similar to x_i received a different label than $\hat{y}_i$. FRANK checks through the *individual_fairness* function (line 9) if there is at least one past record $x'_p \in X'$ identical or "similar" to the current record x_i. Given a binary sensitive attribute $sa \in A$, defines two records x_i and x'_p *similar* if $v_j = v'_j \forall a_j \in A - \{sa\}$, i.e., x_i and x'_p are similar if they are identical, save for the value of sa. More than one similar or identical record $x'_p \in X$ can be found, and, by construction, they have all the same past label $y'_p \in Y'$ (line 10). If there is a disagreement between the current decision and past decisions, i.e., $y'_p \neq \hat{y}_i$ (line 11), then in line 12 *solve_conflict* prompts the user either to change their decision to make it compliant with past records, i.e., to select y'_p as y_i, or to keep the decision but relabel past records with $\hat{y}_i$, i.e., modifying the labels in Y'^4. If the latter is chosen, f is also retrained, accounting for the modified labels. Otherwise, if $y'_p = \hat{y}_i$, x_i is assigned $\hat{y}_i$, i.e., the user's decision is accepted (line 13) as it is consistent with past records.

Skeptical Learning Check. If there is a disagreement between the decision of the user and f, i.e., $\hat{y}_i \neq \tilde{y}_i$, the *skept* function (line 14) computes FRANK's *skepticality* following Eq. 1. If it is higher than s, is skeptical. Empirical accuracy values are initialized as in Sect. 2. We emphasize that *skept* does not take as input Y', i.e., the set of decisions after possible re-labeling, but $\ddot{Y}$, i.e., the set of decisions made by the user after FRANK's checks for each record[5]. If skeptical, proposes $\tilde{y}_i$ for y_i, and asks the user if they want an explanation e_i (line 15). The user is then asked to accept $\tilde{y}_i$ (line 17). The user has the full veto power against FRANK, and if they reject $\tilde{y}_i$, the user label is collected as the final decision y_i (line 18). If the user accepts to see an explanation, runs the *get_and_show_expl* function and provides it to the user (line 16). FRANK can provide *Logic-based Explanations*, where a global representation of the EFDT is shown alongside the local decision rule followed for the record x_i and $\tilde{y}_i$ (such as *IF Years_of_Experience > 5 AND Attitude = True THEN Hire*), or *Instance-based Explanations*, i.e., records similar to x_i which can be either *real* or *synthetic*. These records are classified by f either with $\tilde{y}_i$, i.e., an *example* of FRANK's decision, or $\hat{y}_i$, i.e., a *counter-example*. FRANK's explanations are the result of a *co-evolutionary relationship* with the user, leading to more detailed justifications over time. Thus, the user should progressively trustmore.

Group Fairness Check. GFC checks if one of the value of a binary sensitive attribute $sa \in A$ are discriminating against the other group w.r.t. Y'. GFC is independent from the other checks, and it is always triggered every k records (see lines 22–23). FRANK computes *disc* and the DN, DP, PN, and PP groups of the set of records X' w.r.t. the labels Y', following [14]. Then, it orders the records from DN and PP following the prediction probability of f, and calculates how many of them should be removed. Finally, the records with higher probability are shown to the user, who can choose to change their label. The new labels replace the older ones in Y', and f is retrained from scratch. Thus, GFC is an interactive implementation of PS, where the user is made aware of their discriminating behavior and is asked to relabel past records to mitigate the discrimination.

4   Experiments

We evaluated FRANK[6] on three real-world datasets and, in line with [4,10], we employed simulated users to assess its impact in a variety of conditions.

Users. We employed five kinds of *simulated* users: the *Real Expert*, who always makes decisions following the ground truth (which is unknown in a real

---

[4] Note that $\ddot{Y}$ is not modified, nor taken into account by IFC.

[5] Y' and $\ddot{Y}$ coincide until the user relabel older records if prompted by IFC or GFC.

[6] The Python code is available here: https://github.com/FedericoMz/Frank.

scenario), the *Absent-Minded*, an easily-distracted expert who follows the ground truth 75% of times, the *Coin-Tosser*, who makes decisions by flipping a coin, and the *Bayesian* and *Similarity* experts, simulated by Naive Bayes and KNN [18]. For IFC, we suppose that all the users have conservative behavior w.r.t. their past decisions, with 80.00% of chance of changing the label assigned to the current record x_i, instead of re-labeling past records. For SLC, we set a threshold s of 0.05, increasing the timesis skeptical. We assumed that the users can always *accept* or *decline* FRANK's suggestions, or *randomly* choose. For Bayesian and Similarity experts, we also envisioned users who request *explanations*, i.e., five synthetic examples and counterexamples, monitoring their reaction[7]. If they agree with more than half, they accept FRANK's suggestions. For GFC, we suppose that the user selects to re-label the top 25% DN and PP records.

Datasets. The `Adult`, `COMPAS` and `HR` datasets[8] simulate classification tasks for granting credits, predicting recidivism, or giving a promotion, i.e., possible real use-cases for FRANK. In `HR`, only 8% of records belong to the positive class, compared to the 25% and the 50.00% in `Adult` and `COMPAS`, which are, however, highly discriminating [17]. In contrast, `HR` is fair w.r.t. *Sex*. After removing duplicated or incomplete records, we randomly selected 2,000 records to incrementally train FRANK, i.e., X. We set labeling all the records in X as our stopping condition *stp*. The Naive Bayes and KNN models were trained on an additional 500 records. Half of them were also used to pre-train FRANK's ML model f. Finally, a dataset X_T includes 500 records reserved to test f. For IRC, we set the following rules: for `Adult`, IF *capital_gain* > 9000 THEN $\bar{y} = +$; for `COMPAS`, IF *priors_count* > 0 THEN $\bar{y} = +$; for `HR`, IF *awards_won* = *True* THEN $\bar{y} = +$.

Evaluation Measures. We measured the *Co-evolutionary Accuracy* (*CA*) by comparing Y' with the ground truth Y, and the Model Accuracy (*MA*) by comparing the prediction of f on X_T with its ground truth Y_T. Likewise, we measured the *Co-evolutionary Discrimination* (*CD*) and the Model Discrimination (*MD*). The *disc* score was computed towards *Female* for all datasets[9]. Finally, we counted the number of Unfair Couples (*UC*), i.e., similar records violating individual fairness by receiving a different label. Ideal values are 1 for *CA* and *MA*, 0 for the others. Each experiment was repeated 10 times. The tables report the average results, standard deviations are very low and not reported.

Results. As an ablation study of FRANK's structure, in Table 1, we report the results when *None* of FRANK's functions are enabled, and when only IRC, IFC, or GFC are enabled (*oIRC, oIFC, oGFC*). The impact of IRC is minimal on `Adult` and `HR`, whereas it negatively affects all the experts except for the *Coin-Tosser* in `COMPAS`. This is probably due to the selected rules, either too narrow in scope or inaccurate. These results highlight the importance of selecting good rules for FRANK. On the other hand, comparing the *oIFC* and *oGFC* columns to

---

[7] As synthetic records lack a ground truth, this option cannot be implemented with the other users.

[8] kaggle.com/datasets/.

[9] Note that a negative *disc* implies that *Male* is discriminated.

Table 1. Ablation study of FRANK's checks.

		None CA MA CD MD UC	oIRC CA MA CD MD UC	oIFC CA MA CD MD UC	oGFC CA MA CD MD UC
Adult	Real	1.0 .83 .21 .18 7.0	.96 .82 .23 .15 7.0	1.0 .84 .22 .17 0.0	.84 .75 −.02 .01 6.0
	Abs.	.75 .77 .10 .09 5.3	.74 .76 .13 .09 5.3	.75 .77 .11 .12 0.0	.78 .76 .01 .04 4.2
	Coin	.50 .56 .00 .02 5.6	.52 .51 .03 −.01 5.6	.50 .52 .00 .00 0.0	.55 .55 .03 .04 5.3
	Bayes	.80 .77 .12 .07 0.0	.79 .76 .11 .09 0.0	.80 .77 .12 .07 0.0	.80 .77 .09 .09 0.0
	Sim.	.79 .76 .20 .24 1.0	.79 .76 .20 .24 1.0	.79 .76 .20 .24 0.0	.80 .77 .03 .17 0.0
COMPAS	Real	1.0 .69 −.14 −.21 42.	.65 .61 −.15 −.21 18.	.98 .68 −.14 −.24 0.0	.75 .64 −.06 −.15 17
	Abs.	.75 .63 −.07 −.19 50.	.60 .61 −.12 −.21 24.	.74 .64 −.08 −.19 0.0	.64 .62 −.03 −.17 24
	Coin	.50 .57 .00 −.17 56.	.54 .55 −.09 −.08 27.	.50 .52 −0.0 −.09 0.0	.49 .48 .01 −.01 32
	Bayes	.63 .63 −.20 −.19 0.0	.59 .62 −.18 −.25 0.0	.63 .63 −.20 −.19 0.0	.61 .63 −.15 −.18 0.0
	Sim.	.63 .66 −.31 −.18 25.	.58 .62 −.21 −.25 15.	.62 .66 −.28 −.17 0.0	.63 .66 −.01 −.21 17
HR	Real	1.0 .93 −.01 .00 39.	.99 .89 −.02 0.02 39.	.98 .93 −.01 .00 .99	.94 .93 .00 .00 31
	Abs.	.75 .93 −.01 .00 24.	.74 .92 −.01 −0.01 24.	.74 .93 .00 .00 0.0	.85 .93 .00 .00 20
	Coin	.50 .93 −.01 .00 21.	.50 .83 −.02 −.06 21.	.50 .93 .01 .00 0.0	.62 .65 .01 .06 19
	Bayes	.89 .92 .00 −.02 0.0	.89 .92 .00 −.02 0.0	.89 .92 .00 −.02 0.0	.90 .93 .00 .00 0.0
	Sim.	.89 .93 .00 .00 0.0	.89 .91 .00 −.02 0.0	.89 .93 .00 .00 0.0	.89 .93 .00 .00 0.0

None, we can see a significant improvement in terms of fairness. IFC always successfully minimizes *UC* with no side effects, whereas GFC consistently reduces both *CD* and *MD*. For Adult and COMPAS and with the *Real Expert*, this is at the expense of *CA* and *MA*. However, we should stress that the "accuracy" of very biased datasets does not necessarily mirror "right" decisions. In fact, on the already balanced HR, the impact on *CA* and *MA* with the *Real Expert* is minimal. Additionally, with Adult and HR, GFC improves the accuracy of *Absent-Minded* and *Coin-Tosser* experts without negatively impacting the model-based ones.

Table 2 compares traditional SL [19] withwith everything enabled, except for IRC in COMPAS. As mentioned for IRC, consistently minimizes UC. In Adult, provides each expert better CA and MA if they always accept the suggestions, whereas CD and sometimes MD is slightly better with SL. By declining the suggestions or randomizing the choices with SL, the *Real Expert* gets better CA and MA, but worse CD and MD. With other experts, is better than, or very close to, SL for CA and MA, while consistently improving CD and MD. In COMPAS, always has a better CD, and often a better MD. When the *Real Expert* and the *Absent-Minded* randomize or decline, this is at the expense of CA and, to a lesser extent, MA, with a strong fairness-performance trade-off. In other cases, performs a bit better or on par with SL. As for HR, the two methods are very close for the *Real, Bayesian,* and *Similarity* experts, with SL slightly better. With the *Absent-Minded* and the *Coin-Tosser*, declining or randomizing decision greatly enhances the CA. In fact, the randomizing *Coin-Tosser* reaches a CA comparable to the *Absent-Minded*'s. Also, with the same example we can notice a lower MA than SL's. This might be due to the fact that IRC, IFC, and GFC are not triggered when f makes decisions on X_T.

Figure 1 shows CD and CA over time for different experts, randomly accepting the suggestions fromand SL. Plots are in log scale along the x-axis. At first, for each userand SL follow a similar pattern, both in terms of CA and CD. Their

Table 2. FRANK vs traditional SL. Best scorer in bold, parity in italics.

			Real Expert		Absent-Minded		Coin-Tosser		Bayesian		Similarity	
			SL	FRANK	SL	FRANK	SL	FRANK	SL	FRANK	SL	FRANK
Adult	accept	CA	0.74	**0.78**	0.73	**0.78**	0.73	**0.78**	0.73	**0.78**	0.74	**0.77**
		MA	0.66	**0.75**	0.66	**0.75**	0.65	**0.75**	0.64	**0.75**	0.74	**0.75**
		CD	**0.02**	0.05	**0.03**	0.05	**0.04**	0.05	**0.04**	0.05	**0.00**	0.01
		MD	−0.10	**0.05**	−0.06	**0.05**	−0.01	**0.05**	**0.00**	0.05	−0.02	**0.01**
		UC	1.00	**0.00**	1.00	**0.00**	1.00	**0.00**	1.00	**0.00**	*0.00*	*0.00*
	decline	CA	**1.00**	0.83	0.75	**0.77**	0.50	**0.57**	**0.80**	0.79	0.79	**0.80**
		MA	**0.83**	0.75	**0.77**	0.75	0.56	**0.58**	**0.77**	0.76	0.76	**0.77**
		CD	0.21	**0.03**	0.11	**0.01**	*−0.01*	*−0.01*	0.12	**0.11**	0.20	**0.03**
		MD	0.18	**0.05**	0.12	**0.05**	−0.04	**0.05**	0.07	**0.09**	0.24	**0.17**
		UC	7.00	**0.00**	6.10	**0.00**	5.60	**0.00**	0.00	0.00	1.00	**0.00**
	random	CA	**0.89**	0.80	0.74	**0.77**	0.55	**0.57**	*0.79*	*0.79*	*0.77*	*0.77*
		MA	**0.76**	0.75	0.71	**0.75**	*0.58*	*0.58*	*0.76*	*0.76*	0.73	**0.75**
		CD	0.14	**0.03**	0.09	**0.01**	0.04	**0.01**	0.10	**0.08**	0.14	**0.00**
		MD	0.09	**0.05**	0.04	0.05	0.01	**−0.01**	0.07	0.08	0.14	**0.02**
		UC	4.60	**0.00**	3.10	**0.00**	4.20	**0.00**	0.10	**0.00**	0.60	**0.00**
COMPAS	accept	CA	*0.52*	*0.52*	0.51	**0.53**	0.52	**0.54**	*0.52*	*0.52*	0.55	**0.58**
		MA	**0.52**	0.51	0.49	**0.54**	0.49	**0.55**	**0.52**	0.51	0.56	**0.62**
		CD	−0.02	**0.00**	−0.01	**0.00**	−0.01	**0.00**	−0.02	**0.00**	−0.09	**−0.05**
		MD	**−0.02**	−0.04	0.01	**−0.07**	0.01	**−0.08**	**−0.02**	−0.04	−0.03	**−0.11**
		UC	8.00	**0.00**	17.60	**0.00**	17.60	**0.00**	8.00	**0.00**	1.00	**0.00**
	decline	CA	**1.00**	0.77	**0.75**	0.64	*0.50*	*0.50*	**0.63**	0.61	**0.63**	0.62
		MA	**0.69**	0.66	**0.65**	0.62	**0.53**	0.52	*0.63*	*0.63*	**0.66**	0.65
		CD	**−0.14**	0.00	−0.06	**−0.01**	0.01	**0.00**	−0.20	**−0.15**	−0.31	**−0.04**
		MD	**−0.21**	−0.19	−0.21	**−0.13**	**−0.07**	−0.10	−0.19	**−0.18**	*−0.18*	*−0.18*
		UC	42.00	**0.00**	50.00	**0.00**	51.10	**0.00**	*0.00*	*0.00*	25.00	**0.00**
	random	CA	**0.80**	0.66	**0.65**	0.58	0.50	**0.54**	**0.62**	0.60	*0.62*	*0.62*
		MA	**0.64**	0.61	**0.57**	0.56	0.48	**0.57**	**0.63**	0.60	0.65	0.65
		CD	**−0.15**	−0.01	−0.10	**0.00**	−0.02	**−0.01**	−0.18	**−0.08**	−0.23	**−0.05**
		MD	**−0.16**	−0.12	−0.17	**−0.11**	**0.00**	−0.08	−0.18	**−0.15**	−0.17	**−0.16**
		UC	34.00	**0.00**	45.20	**0.00**	43.30	**0.00**	3.20	**0.00**	23.70	**0.00**
HR	accept	CA	**0.90**	0.89	**0.90**	0.86	**0.90**	0.88	**0.90**	0.89	**0.90**	0.89
		MA	**0.93**	0.92	**0.93**	0.89	**0.93**	0.91	**0.93**	0.92	**0.93**	0.92
		CD	*0.00*	0.01	*0.00*	0.01	*0.00*	*0.00*	*0.00*	*0.00*	*0.00*	*0.00*
		MD	**0.00**	−0.02	**0.00**	−0.02	**0.00**	−0.02	**0.00**	−0.02	**0.00**	−0.02
		UC	*0.00*	*0.00*	*0.00*	*0.00*	*0.00*	*0.00*	*0.00*	*0.00*	*0.00*	*0.00*
	decline	CA	**1.00**	0.91	0.75	**0.83**	0.50	**0.66**	*0.89*	*0.89*	*0.89*	*0.89*
		MA	**0.93**	0.92	**0.93**	0.92	**0.90**	0.71	*0.92*	*0.92*	**0.93**	0.92
		CD	*−0.01*	*−0.01*	−0.01	**0.00**	*−0.01*	*−0.01*	*0.00*	*0.00*	*0.00*	*0.00*
		MD	**0.00**	−0.02	−0.01	**0.00**	−0.01	**0.05**	*−0.02*	*−0.02*	**0.00**	−0.02
		UC	39.00	**0.00**	26.9	**0.00**	22.6	**0.00**	*0.00*	*0.00*	*0.00*	*0.00*
	random	CA	**0.95**	0.91	0.82	**0.86**	0.70	**0.82**	*0.89*	*0.89*	*0.89*	*0.89*
		MA	**0.93**	0.92	**0.93**	0.90	**0.93**	0.86	**0.93**	0.92	**0.93**	0.92
		CD	−0.01	**0.00**	*0.00*	*0.00*	**0.00**	0.01	*0.00*	*0.00*	*0.00*	*0.00*
		MD	**0.00**	−0.02	**0.00**	−0.02	**0.00**	0.03	**0.00**	−0.02	**0.00**	−0.02
		UC	18.10	**0.00**	17.40	**0.00**	16.20	**0.00**	*0.00*	*0.00*	*0.00*	*0.00*

lines then diverge due to fairness interventions. In Adult, this results in a drop of CA for the *Real Expert*, and in COMPAS also for the *Absent-Minded*. In HR, the *Real Expert* is far less affected, as the dataset is less biased. On the other hand, the *Absent-Minded* and *Coin-Tosser* receive a noticeable boost in terms of CA. In Adult and COMPAS, the *Real* and the *Similarity* experts make biased decisions while paired with SL, whereas their CD withis near 0. FRANK's CD lines tend to converge to 0 for all the datasets.

Table 3 compares the impact of having users accepting FRANK's suggestions randomly (RND) against users deciding on top of FRANK's explanations (XAI).

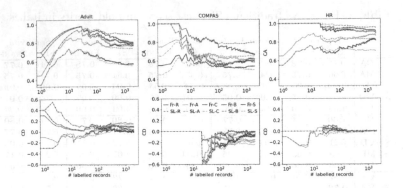

Fig. 1. CA and CD evolution over time with different experts.

Table 3. Users accepting randomly (RND) or w.r.t. explanations (XAI).

	Adult				COMPAS				HR			
	Bayesian		Similarity		Bayesian		Similarity		Bayesian		Similarity	
	RND	XAI	RND	XAI	RND	XAI	RND	XAI	RND	XAI	RND	XAI
Agr %	96.38	88.72	77.14	77.37	89.64	74.53	68.83	60.42	100.00	100.00	99.12	99.16
Ske %	3.49	11.20	22.47	22.47	10.32	25.41	31.04	39.45	0.00	0.00	0.79	0.73
Dis %	0.11	0.05	0.37	0.15	0.03	0.05	0.12	0.13	0.00	0.00	0.08	0.10
Acc %	51.99	94.03	49.61	37.58	50.49	93.94	50.37	74.02	N/A	N/A	54.30	0.00
Dec %	48.01	5.97	50.39	62.42	49.51	6.06	49.63	25.98	N/A	N/A	45.70	100.00
CA	0.79	0.77	0.77	0.76	0.60	0.53	0.62	0.59	0.89	0.89	0.89	0.89
CD	0.08	0.03	0.0	0.01	−0.08	−0.01	−0.05	−0.02	0.00	0.00	0.00	0.00

The first three rows report the percentage of Agreements, Skepticism, and Disagreement between the user and FRANK. We notice that they tend to agree, and the disagreement almost always leads to skepticism. The fourth and fifth rows show the percentage of the Accepted and Declined FRANK's suggestions. When XAI is used, we observe a lower agreement rate (Agr) in Adult and COMPAS, but ultimately, looking at the acceptance rate (Acc), these users rely onmore than their randomizing counterparts, also resulting in a better CD at the expense of CA. This confirms thatis able to provide satisfying explanations to the *Bayesian* and *Similarity* users. We underline that the *Similarity* expert on Adult is the exception, as they tend to decline. Finally, in HR, SLC was never triggered by the *Bayesian*, and only 14 times by the *Similarity* expert (who then declined the 14 suggestions, hence the anomalous percentage).

5   Conclusion

We have presented FRANK, a system based on Skeptical Learning that evolves with the user. Compared to traditional SL,checks the fairness of the decisions, if they are compliant with external rules, and provides explanations for the suggestions. Through these additional functions,successfully improves the fairness of the datasets and of the model, often outperforming SL in terms of accuracy, especially with less-skilled users. Moreover, we noticed that our simulated users accept FRANK's explanations most of the time. However, at the moment, is limited to tabular data and better suitable to those of low dimensionality. Future works might extendto other data types and decision models, explore alternative stopping conditions, and focus on the FRANK-user relationships. For example, could build trust or distrust towards the user, and react accordingly. Finally, after being trained in the co-evolutionary process, FRANK's model f could be used within a Learning-to-Defer system, withmaking decisions and asking the user when uncertain.

Acknowledgment. . This work is partially supported by the EU NextGenerationEU programme under the funding schemes PNRR-PE-AI FAIR, PNRR-SoBigData.it - Prot. IR0000013, H2020-INFRAIA-2019-1: Res. Infr. G.A. 871042 *SoBigData++*, TANGO G.A. 101120763, and ERC-2018-ADG G.A. 834756 *XAI*. We thank Andrea Bontempelli for his insights on Skeptical Learning.

References

1. Bifet, A., Gavaldà, R.: Adaptive learning from evolving data streams. In: Adams, N.M., Robardet, C., Siebes, A., Boulicaut, J.-F. (eds.) IDA 2009. LNCS, vol. 5772, pp. 249–260. Springer, Heidelberg (2009). https://doi.org/10.1007/978-3-642-03915-7_22
2. Binns, R.: On the apparent conflict between individual and group fairness. In: FAT*, pp. 514–524. ACM (2020)
3. Blanco-Justicia, A., Domingo-Ferrer, J.: Machine learning explainability through comprehensible decision trees. In: Holzinger, A., Kieseberg, P., Tjoa, A.M., Weippl, E. (eds.) CD-MAKE 2019. LNCS, vol. 11713, pp. 15–26. Springer, Cham (2019). https://doi.org/10.1007/978-3-030-29726-8_2
4. Bontempelli, A.,Teso, S., et al.: Learning in the wild with incremental skeptical gaussian processes. In: IJCAI, pp. 2886–2892 (2020). ijcai.org
5. Domingos, P., et al.: Mining high-speed data streams. In: KDD, p. 71. ACM (2000)
6. Guidotti, R.: Counterfactual explanations and how to find them: literature review and benchmarking. Data Min. Knowl. Disc. 1–55 (2022). https://doi.org/10.1007/s10618-022-00831-6
7. Guidotti, R., Monreale, A., Ruggieri, S., et al.: A survey of methods for explaining black box models. ACM CSUR **51**(5), 93:1–93:42 (2019)
8. Hüllermeier, E., Waegeman, W.: Aleatoric and epistemic uncertainty in machine learning: an introduction to concepts and methods. ML **110**(3), 457–506 (2021)
9. Jarrett, D., et al.: Online decision mediation. In: NeurIPS (2022)
10. Joshi, S., Parbhoo, S., Doshi-Velez, F.: Pre-emptive learning-to-defer for sequential medical decision-making under uncertainty. *CoRR*, abs/2109.06312 (2021)
11. Kamiran, F., Calders, T.: Classification with no discrimination by preferential sampling. In: BNAIC, vol. 1. Citeseer (2010)
12. Lange, M.D., et al.: A continual learning survey: defying forgetting in classification tasks. IEEE Trans. Pattern Anal. Mach. Intell. **44**(7), 3366–3385 (2022)

13. Manapragada, C., Webb, G.I., Salehi, M.: Extremely fast decision tree. In: KDD, pp. 1953–1962. ACM (2018)

14. Mazzoni, F., Manerba, M.M., Cinquini, M., Guidotti, R., Ruggieri, S.: GenFair: A genetic fairness-enhancing data generation framework. In: Bifet, A., Lorena, A.C., Ribeiro, R.P., Gama, J., Abreu, P.H. (eds.) Discovery Science, DS 2023, LNCS, vol. 14276, pp. 356–371. Springer, Cham (2023). https://doi.org/10.1007/978-3-031-45275-8_24

15. Mosier, K.L., et al.: Human decision makers and automated decision aids: made for each other? In: Automation and Human Performance, pp. 201–220. CRC (2018)

16. Pessach, D., Shmueli, E.: A review on fairness in machine learning. ACM CSUR 55(3), 51:1–51:44 (2023)

17. Quy, T.L., et al.: A survey on datasets for fairness-aware machine learning. WIREs Data Mining Knowl. Discov. 12(3), e1452 (2022)

18. Tan, P.-N., et al.: Introduction to Data Mining. Pearson Education India, Chennai (2016)

19. Teso, S., Bontempelli, A., Giunchiglia, F., Passerini, A.: Interactive label cleaning with example-based explanations. In: NeurIPS, pp. 12966–12977 (2021)

20. Verma, S., et al.: Fairness definitions explained. In: ICSE, pp. 1–7. ACM (2018)

21. Wang, L., et al.: A comprehensive survey of continual learning: theory, method and application. CoRR, abs/2302.00487 (2023)

22. Wang, T., Saar-Tsechansky, M.: Augmented fairness: an interpretable model augmenting decision-makers' fairness. CoRR, abs/2011.08398 (2020)

23. Zeni, M., et al.: Fixing mislabeling by human annotators leveraging conflict resolution and prior knowledge. ACM IMWUT 3(1), 32:1–32:23 (2019)

24. Zhang, W.: Personal context recognition via skeptical learning. In: IJCAI, pp. 6482–6483 (2019). ijcai.org

Industrial Challenge

Predicting the Failure of Component X in the Scania Dataset with Graph Neural Networks

Maurizio Parton[1]([✉]) [iD], Andrea Fois[2] [iD], Michelangelo Vegliò[1], Carlo Metta[3] [iD], and Marco Gregnanin[4,5] [iD]

[1] University of Chieti-Pescara, Chieti, Italy
maurizio.parton@gmail.com
[2] University of Parma, Parma, Italy
[3] ISTI-CNR, Pisa, Italy
[4] IMT School for Advanced Studies Lucca, Lucca, Italy
[5] KU Leuven, Leuven, Belgium

Abstract. We use Graph Neural Networks on signature-augmented graphs derived from time series for Predictive Maintenance. With this technique, we propose a solution to the Intelligent Data Analysis Industrial Challenge 2024 on the newly released SCANIA Component X dataset. We describe an Exploratory Data Analysis and preprocessing of the dataset, proposing improvements for its description in the SCANIA paper.

Keywords: SCANIA Component X · Predictive Maintenance · Graph Neural Networks · Visibility Graphs

1 Introduction

The growing sophistication and accessibility of machine learning and deep learning models have significantly impacted the automotive industry, particularly in maintenance practices. Leveraging big data and data-driven models has revolutionized vehicle maintenance by enabling the implementation of Predictive Maintenance (PdM) strategies. PdM aims to optimize vehicle performance and prevent component failure, thereby minimizing maintenance costs. However, the scarcity of real-world PdM datasets presents a substantial challenge. Most existing datasets rely on synthetic data, which introduce biases into the models.

This paper addresses this challenge by investigating the performance of two models on a real-world multivariate timeseries dataset, "SCANIA Component

C. Metta—EU Horizon 2020: G.A. 871042 SoBig-Data++, NextGenEU - PNRR-PEAI (M4C2, investment 1.3) FAIR and "SoBigData.it".

M. Parton—Funded by GNSAGA INdAM group.

M. Parton, A. Fois, M. Vegliò, C. Metta, M. Gregnanin—Computational resources provided by CLAI laboratory, Chieti-Pescara, Italy.

I. Miliou et al. (Eds.): IDA 2024, LNCS 14642, pp. 251–259, 2024.
https://doi.org/10.1007/978-3-031-58553-1_20

X", for predicting maintenance needs in a vehicle fleet. One of the novelties introduced in this paper is the application of Graph Neural Networks (GNNs) [2] to PdM. GNNs demonstrate exceptional capabilities in processing data structured as graphs, making them well-suited to capture the complex and intricate relationships present within sensor data collected from vehicles. We introduce a framework that utilizes both time series signatures and visibility graphs for effective PdM analysis. Timeseries signatures, a concept derived from path theory, serve as a potent tool for identifying the most relevant variables crucial for constructing an informative graph representation of the multivariate timeseries data. Subsequently, these identified key variables are used to construct the graph using the visibility graph algorithm, which is able to transform a time series in a graph preserving its temporal characteristics such as trends [3,10].

Section 2 details the methods. Sections 3 and 4 focuses on data analysis and preprocessing. Section 5 describe hyperparameters and results, and Sect. 6 provides potential avenues for future research.

2   Methods

Geometric Deep Learning. Graph Neural Networks (GNNs) represent a specific deep learning model tailored to data structured as a graph. When dealing with multivariate time series, determining the graph structure becomes imperative, as no predefined structure exists as in social networks, chemistry, or traffic forecasting problems [4]. Two primary approaches emerge: constructing a graph representation from a similarity matrix, typically utilizing the correlation matrix, or selecting a variable and computing the graph using the visibility graph method [3]. In our analysis, we opt for the latter approach due to its simplicity and effectiveness. Utilizing the correlation matrix necessitates filtering it using a threshold method to eliminate noise and irrelevant connections in the graph. Not filtering creates a complete graph with all nodes interconnected, making it complex and less useful.

Signature. The concept of the signature originates from the domain of path theory, offering a structured and comprehensive depiction of the temporal evolution within a time series. Its efficacy lies in capturing both temporal and geometric patterns inherent in the time series. Temporal patterns encompass long-term dependencies and recurrent trends over time, while geometric patterns involve the shape of time series trajectories, including intricate behaviors such as loops and self-intersections [6]. The signature structure offers a hierarchical interpretation, with lower-order components capturing broad path attributes and higher-order terms revealing intricate characteristics, including higher-order moments and local geometric features. Importantly, the signature remains invariant under reparameterization, preserving integral values despite time transformations, and adheres to translation invariance and concatenation properties [7].

Furthermore, when representing a stochastic process as a cumulative process and computing its signature, referred to as the "cumulative signature," we can effectively capture and portray temporal patterns. Specifically, the cumulative signature aggregates information over time, thereby encapsulating the temporal evolution of the data. Additionally, the final cumulative signature is influenced by the entire sequence of signatures, reflecting the order in which information accumulates over time. Lastly, it mitigates noise and fluctuations in the original data, offering a more reliable representation of the underlying patterns [6].

Consider a discrete time series $S = \{s_0, s_1, \ldots, s_T\}$, and let $T((\mathbb{R}^d))$ be the tensor algebra $\bigoplus_{k=0}^{\infty} (\mathbb{R}^d)^{\otimes k}$, encompassing the signatures of $\mathbb{R}^d$-valued paths. As the signature method is applicable solely to continuous processes, discrete time series must undergo a transformation into a continuous form. This transformation can be achieved through methods such as the lead-lag transformation or the time-join transformation [8]. We opt to focus on continuous functions mapping from a compact time interval $J := [a, b]$ to $\mathbb{R}^d$ with finite p-variation, denoted as $C_0^p(J, \mathbb{R}^d)$. Thus, we choose the lead-lag transformation, denoted by L, to convert the discrete time series S into a continuous form due to its ability to extract path volatility. Consequently, we define the signature, denoted as $\mathcal{S}$, and the truncated signature at level M, denoted as $\mathcal{S}_M$, as outlined in [9]. The rationale behind utilizing the signature of a time series to compare two or more stochastic processes stems from the "Expected Signature" theorem [12]. This theorem asserts that if the signatures computed on two distinct stochastic processes are equal, then the processes themselves are equal in distribution. The expected signature theorem enables the comparison of different stochastic processes because, under suitable assumptions, the expected signature uniquely determines the distribution of the stochastic process, analogous to the role of the moment generating function [13].

For a more comprehensive and precise mathematical definition, please refer to [6,14,15].

Visibility Graph. In this study, we employ the visibility graph algorithm proposed in [3] to derive the graph representation. The visibility graph algorithm operates specifically on univariate time series data. To identify the time series suitable for graph representation, we opt to compute the truncated signature for each time series. Subsequently, leveraging the "Expected signature" theorem, we construct a distance matrix wherein the entries correspond to the Euclidean Distance between the signature of each time series and all others. We then transform this distance matrix into a similarity matrix using a strictly monotone decreasing function, specifically $F(x) = \frac{1}{a+x}$ with $a = 1$. In order to ascertain the most significant time series, we conduct Principal Component Analysis (PCA) [18] on the similarity matrix. This analysis enables us to identify the most significant eigenvalue and its associated time series. We generate the graph associated to the selected time series using visibility graph algorithm.

The concept underlying the visibility graph entails representing time series values as vertical bins, wherein each bin connects to others if an unobstructed line of sight exists from one bin's top to another's. Therefore, given a discrete time series $S = \{s_0, s_1, \ldots, s_T\}$ with its associated set of discrete timestamps $\mathcal{T} = \{t_0, t_1, \ldots, t_T\}$, the visibility condition can be defined using the following equation: $s_k < s_j + (s_j - s_i)\frac{t_j - t_k}{t_j - t_i}$. Meaning that any two data points (s_i, t_i) and

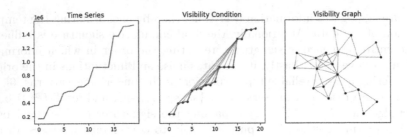

Fig. 1. The visibility graph technique is applied to feature 171 of a randomly selected truck from the training dataset.

(s_j, t_j) are connected in the graph if any other data point (s_k, t_k) satisfies this condition. This approach yields a connected graph, with the number of nodes corresponding to the observations within the time series. Importantly, the visibility graph is chosen for its ability to preserve both the temporal and structural characteristics of the time series. Specifically, periodic time series result in regular graphs, random time series yield random graphs, and fractal time series produce random graphs. Additionally, the visibility graph remains invariant under affine transformations of the time series. The type of visibility condition applied determines whether the resulting graph is undirected or directed. Computation of the visibility graph is facilitated using the Python package "Time series to visibility graphs" (ts2vg) [19]. Figure 1 illustrates the functioning of the visibility graph using a time series extracted from the training set of a randomly selected truck. The first plot displays the path trajectories of the time series, the middle plot represents the visibility condition, and the last plot depicts the associated undirected graph. For a more comprehensive understanding, including a formal definition and visual representations, refer to [3].

3 Exploratory Data Analysis

Initially, we conducted an exploratory analysis of the features. This step was crucial due to the lack of information about the origin of these features, caused by the strong anonymization of the dataset. In this exploratory analysis description, we also highlight some issues mentioned in [11] that, in our opinion, should be corrected in a future version. Recall that the dataset describes 14 attributes measured at non-homogeneous timesteps for a fleet of vehicles. Six of these attributes are divided into bins related to a certain condition imposed on that attribute, and the columns represent the values of these bins. The remaining six attributes are numerical counters.

The first analysis aimed to verify that NaNs appear jointly in attributes, meaning that a NaN can appear at time t for vehicle i in attribute j if and only if all other bins for attribute j at time t for vehicle i are NaN. This is consistent with the fact that all bins of the same attribute measure the same type of data.

According to the paper, the features are cumulative, meaning the values in the columns should be increasing when restricted to individual vehicles. However, this property only holds for intervals of values without NaNs. In the paper, NaNs are explained in Sect. 2.1.1, where it is particularly mentioned that "the data collection counters could be reset, i.e., start again from the beginning", but in reality, the value restarts from zero only in 3.4% of cases. If the columns were increasing, or if the reason why they might not be was known, it would be possible to interpolate missing data. Instead, the lack of information makes imputation very difficult.

Another issue detected concerns the initial values of the features. If the timestep column measures "the duration in time_step that each vehicle has been utilizing Component X during its operational lifespan," it is reasonable to assume that at time 0, the features should have a value of 0. This is not the case, see for example vehicle 5.

The analysis also revealed that, only in train_operational, in 7 cells the non-decreasing property described above is not verified. This was interpreted as a transcription error or artificial noise added during the compilation and anonymization of the dataset, and it was corrected by linear interpolation. Note that in val_op and in test_op, this does not occur.

Multiple statistics on validation_operational_data and test_operational_data were evaluated and compared. These statistics showed that from all statistically evaluable perspectives, the data distribution is similar, which is expected from a dataset initially constructed for a challenge. This suggests that in future versions of the dataset, it might be useful to artificially introduce a slight distribution shift in the test, to better simulate real-world applications.

This analysis also allowed us to identify a significant outlier in the length of the time series (number of timesteps per vehicle) in test operational data. While in the validation, the maximum length is 209, and in the test, it is 262, there is only one vehicle in test with this length, all the others being at most 209 timesteps long. This observation was used to reduce the computational complexity of the training set during preprocessing.

4   Data Preprocessing

First, a new column, class_label, was created using the information contained in train_tte. For each row, class_label contains the ground truth for the vehicle and the corresponding timestep. If in_study_repair = 0, class_label = 0, otherwise class_label is 4, 3, 2, 1, 0 if length_of_study_time_step - timestep is in $[0, 6)$, $[6, 12)$, $[12, 24)$, $[24, 48)$, $[48, +\infty)$, respectively. This highlighted another issue in the data collection description: When in_study_repair = 1, what did the vehicle do from the last timestep to length_of_study_time_step? If Component X continued to work until that moment, the described ground truth is correct, but in that case, measurements at time length_of_study_time_step, which are probably the most important, are missing. Otherwise, it would seem reasonable to calculate the ground truth with the last timestep.

Given the significant scale difference between the values of a single vehicle as the timestep increases, it was necessary to choose a normalization procedure. The values of bin i of attribute j were divided by the average sum of the values of all bins of attribute j across the entire dataset. Given the semantics of the columns, this normalization represents a scaling change for attribute j.

Subsequently, the 8 columns taken from train_specifications were integrated, transformed into integers. Note that the statement that the specifications "can take categories in Cat0, Cat1, ..., Cat8" is a typo, as these columns take integer values up to 28.

NaNs were replaced with 0. Various other imputation methods were tried, but none provided satisfactory results, also due to the issues described above. The value 0, although rarely appearing as a true value in some measurements, was deemed distinctive enough to allow a model with good predictive capacity to use it.

Subsequently, for each vehicle with in_study_repair $= 0$, the final records were removed until the last timestep t satisfied length_of_study_time_step $-t \geq 48$. This is because the ground truth $= 0$ at a time $t \in [0, 48)$ might be wrong, and false negatives have a very high cost. 202454 out of 1122452 records were removed, but given the high proportion of class 0 compared to others, this did not pose problems in the subsequent choice of time series to use for training.

For training, a dataset balanced between class 0 and non-zero classes was chosen. Given the high cost of false negatives, it was preferred not to risk the model learning to distinguish 0 better than the other classes. In the visibility graph approach training was done by aggregating all positive classes as class 4.

To maintain a high number of training samples, data augmentation was performed on time series with class_label > 0, iteratively removing the last timestep until the class became 0. At the end of this augmentation, time series with class_label $= 0$ were randomly chosen and cut to a random length, sampled from the distribution of time series lengths in the test (which the EDA showed to be similar to the distribution of lengths in the validation). This data leakage should not be used for a model aiming to generalize in the real world, but it seemed reasonable to use it for a challenge. Graphs were augmented with the time-series signature as global feature.

For models that require time series of constant length, the option was to bring all series to the maximum length using zero-padding inserted at the initial timesteps. Semantically, this corresponds to performing some initial measurements when the vehicle is not yet operational, and seemed more reasonable than adding padding at the end. Thus, the maximum length should have been 262. However, since the EDA highlighted that this length is only reached by a single vehicle, for this single vehicle the option was to truncate the initial timesteps. This allowed for a significant reduction in one dimension of the training set.

Finally, a tensor X of shape (50511, 209, 114) containing the timestep (necessary because of the irregular sampling), all attribute bins (105), and specifications (8), and a vector y of shape (50511), with the ground truths were created.

CM	0	1	2	3	4
0	2292	0	0	0	2618
1	4	0	0	0	12
2	3	0	0	0	11
3	11	0	0	0	19
4	15	0	0	0	61

set	label	0	1	2	3	4
validation	#	2325	0	0	0	2721
test	#	2314	0	0	0	2731

Fig. 2. Confusion matrix on validation. True labels on rows.

Fig. 3. Number of predictions for each class on validation and test sets.

5   Experiments

Graph Neural Networks. We used the pytorch-geometric [16] GIN [17] graph neural network, with 114 input channels (the time-series features, time step, and truck specifications), $8 \times 114 = 912$ hidden channels with 15 layers and 2 output channels. We found that, due to the high cost involved in mispredicting a class with a lower one, it was worth using only class 0 and class 4 for training and classification. During training sequences that had a class in $\{1, 2, 3\}$ were treated as if they belonged to class 4. The undirected graphs on which the GNN operated were visibility graphs constructed using the 171_0 time series for the truck. The number of nodes was thus equal to the number of timesteps in the time series. Using a threshold of 0.6 for class 0, our model obtains a cost of 40109.0 on the validation set, where costs are given in [11]. In Fig. 2, the confusion matrix on validation, and totals on validation and test in Fig. 3.

ROCKET. In the realm of time series analysis, the classification model known as ROCKET (RandOm Convolutional KErnel Transform) stands out for its remarkable efficiency and accuracy. Stemming from [1], ROCKET [5] generates a vast feature space by applying thousands of random convolutions over the input time series. We used ROCKET in its sktime implementation, with 10000 random kernels over fixed-length time series, see preprocessing in Sect. 4. This resulted in an embedding space with 20000 dimensions, where we tested a Ridge classifier, and a fully connected model augmented with multi-bias and global connection as in [21, 22]. However, the final performance was below the baseline, and is not reported here. This is somewhat surprising, because ROCKET is considered state of the art for multivariate time series classification. While this is an indication of the complexity of the dataset, we think this low performance could be due to how the training set was built, and deserves future investigation.

6   Conclusion

We show that the Component X dataset presents a significant challenge for PdM and time-series analysis, marking a valuable contribution to the field. We anticipate that updates to the foundational paper [11] will incorporate our feedback. Our future work aims to expand the use of signature-augmented data beyond graph-based models. Moreover, we believe that adopting a multilayered graph approach, as outlined in [20,23], could enhance graph prediction techniques.

References

1. Guang-BinHuang, Q.-Y.Z., Siew, C.-K.: Extreme learning machine: theory and applications. Neurocomputing **70**(1–3), 489–501 (2006)
2. Scarselli, F., Gori, M., et al.: The graph neural network model. IEEE Trans. Neural Networks **20**(1), 61–80 (2008)
3. Lacasa, L., Luque, B., et al.: From time series to complex networks: the visibility graph. Proc. Nat. Acad. Sci. **105**(13), 4972–4975 (2008)
4. Wu, Z., Pan, S., et al.: A comprehensive survey on graph neural networks. IEEE Trans. Neural Networks Learn. Syst. **32**(1), 4–24 (2020)
5. Dempster, A., Petitjean, F., Webb, G.I.: ROCKET: exceptionally fast and accurate time series classification. arXiv:1910.13051 (2019)
6. Lyons, T.: Rough paths, signatures and the modelling of functions on streams. arXiv preprint arXiv:1405.4537 (2014)
7. Chen, K.T.: A faithful representation of paths by noncommutative formal power series. Trans. AMS **89**(2), 395–407 (1958)
8. Levin, D., Lyons, T., Ni, H.: Learning from the past, predicting the statistics for the future, learning an evolving system. arXiv:1309.0260 (2016)
9. Gregnanin, M., De Smedt, J., et al.: Signature-based community detection for time series. In: Cherifi, H., Rocha, L.M., Cherifi, C., Donduran, M. (eds.) Studies in Computational Intelligence, vol. 1142, pp. 146–158. Springer, Cham (2024). https://doi.org/10.1007/978-3-031-53499-7_12
10. Gregnanin, M., De Smedt, J., et al.: Stock Price Time Series Foresting using Dynamic Graph Neural Networks and Attention Mechanism in Recurrent Neural Networks. In: MIDAS - ECML-PKDD (2023, to appear)
11. Kharazian, Z., Lindgren, T., et al.: SCANIA component X dataset: a real-world multivariate time series dataset for predictive maintenance. arXiv:2401.15199 (2024)
12. Lyons, T., Ni, H.: Expected signature of brownian motion up to the first exit time. Ann. Probab. **43**(5), 2729–2762 (2015)
13. Chevyrev, I., Lyons, T.: Characteristic functions of measures on geometric rough paths. Ann. Probab. **44**(6), 4049–4082 (2016)
14. Levin, D., Lyons, T., Ni, H.: Learning from the past, predicting the statistics for the future, learning an evolving system. arXiv:1309.0260 (2013)
15. Chevyrev, I., Kormilitzin, A.: A primer on the signature method in machine learning. arXiv preprint arXiv:1603.03788 (2016)
16. Fey, M., Lenssen, J.E.: Fast graph representation learning with PyTorch geometric. In: ICLR Workshop on Representation Learning (2019)
17. Xu, K., Hu, W., Leskovec, J., Jegelka, S.: How powerful are graph neural networks? arXiv preprint arXiv:1810.00826 (2018)
18. Tipping, M.E., Bishop, C.M.: Probabilistic principal component analysis. J. Roy. Stat. Soc. Ser. B **61**(3), 611–622 (1999)
19. Time Series to Visibility Graphs (ts2vg) Python Packages. https://cbergillos.com/ts2vg
20. Lacasa, L., Nicosia, V., Latora, V.: Network structure of multivariate time series. Sci. Rep. **5**, 15508 (2015)

21. Metta, C., Fantozzi, M., et al.: Increasing biases can be more efficient than increasing weights. In: IEEE/CVF Winter Conference on Applications of Computer Vision WACV (2024)
22. Di Cecco, A., Metta, C., Fantozzi, M., Morandin, F., Parton, M.: GloNets: globally connected neural networks. In: Piatkowski, N., et al. (eds.) IDA 2024. LNCS, vol. 14641, pp. xx–yy. Springer, Cham (2024)
23. Freitas Silva, V., Eduarda Silva, M., et al.: MHVG2MTS: multilayer horizontal visibility graphs for multivariate time series. arXiv:2301.02333 (2023)

Towards Contextual, Cost-Efficient Predictive Maintenance in Heavy-Duty Trucks

Louis Carpentier[1,2]([⊠]), Arne De Temmerman[1,2], and Mathias Verbeke[1,2]

[1] Department of Computer Science, KU Leuven, Bruges, Belgium
{louis.carpentier,arne.detemmerman,mathias.verbeke}@kuleuven.be
[2] Flanders Make@KU Leuven, Leuven, Belgium

Abstract. Predictive maintenance is a crucial yet challenging task in many industrial applications. This work explores a large repository of existing techniques and approaches to process historical data and predict if an asset is at risk of failure. In particular, the operational condition and specification of Scania trucks in heavy-duty applications is considered as part of the IDA 2024 Industrial Challenge.

Keywords: Predictive maintenance · Time series analysis · Contextual decision-making · Feature extraction · Survival analysis

1 Introduction

Predictive maintenance, i.e., predicting whether an industrial asset is at risk of imminent failure, is a crucial yet challenging task in many industrial applications. Prediction of future failure allows maintenance to be planned before the fault happens, which in turn reduces the downtime [4]. Monitoring capabilities tremendously increased due to the paradigm shift towards Industry 4.0, enabling the development of automated decision-making processes.

Specifically, we look at predicting the failure of an anonymized component in Scania trucks in heavy-duty applications as part of the IDA 2024 Industrial Challenge [3]. The dataset includes variables capturing the operational condition of the trucks over time and the truck specification. In this work, we explore a wide range of techniques and methods to predict upcoming component failures for condition monitoring in general and for Scania trucks in particular. The main contributions of this work are as follows:

1. Transformation of observational readouts into history-aware feature vectors;
2. Application of inherently different models for predictive maintenance based on classification, regression, and survival analysis;
3. Open-source implementation of the methodology, including the source code, notebooks, and raw experimental results, to ensure reproducibility[1].

L. Carpentier and A. De Temmerman contributed equally to the paper.
[1] https://gitlab.kuleuven.be/u0158714/ida-industrial-challenge-2024.

Table 1. Size and amount of missing values in the train, validation, and test sets

	Train set	Validation set	Test set
#vehicles	23 550	5046	5045
%vehicles	70%	15%	15%
#readouts	1 122 452	196 227	198 140
#readouts per vehicle (average)	48	46	45
#missing values	354 634	60 339	66 403
%missing values	0.0030%	0.0029%	0.0032%

2   Problem Formulation

The train, validation, and test datasets consist of 33 651 vehicles in total with the following information:

1. **Operational Readouts.** Onboard sensors monitor crucial parameters regarding the vehicle's real-time condition and performance. To reduce storage requirements, the data is stored locally in histograms. In particular, multiple ranges or bins are defined for each variable, and each bin has one counter. Whenever a value is measured in a certain bin, the corresponding counter is incremented. The operational readouts are snapshots of these counters taken at irregular time intervals. Note that the operational readouts for each variable are monotonically increasing because the counters are only incremented. In total, 14 parameters were selected with varying numbers of bins, resulting in 105 attributes. Figure 1 plots the ten bins of feature 158 in function of the timestep for vehicle 100. Table 1 lists the total number of vehicles and readouts together with the average number of readouts per vehicle and their missing values. For each vehicle in the validation and test set, a random observation is selected, and only the earlier readouts are provided to simulate the usage of a prediction model in a real-world setting.
2. **Specification.** The specification of a truck corresponds to its model, including information such as engine type, weight capacities, and other technical details. There are eight categorical features with a variable number of categories, which totals in 11 181 240 possible specifications, of which only 3 607 occur in the training data.

The train set also contains a time-to-event table, indicating if and when a vehicle had a defect. The last observation of each vehicle in the validation set is assigned 1 of 5 classes, indicating when the vehicle will fail: class 0 instances will not fail in the next 48 time steps, class 1 instances will fail within 48 to 24 time steps, class 2 instances within 24 to 12 time steps, class 3 instances within 12 to 6 time steps, and class 4 instances will fail within 6 time steps. Thus, classes 1, 2, 3 or 4 indicate failure with increasing severity. The goal of the IDA challenge is to predict these labels for the last observation of each vehicle in the test set.

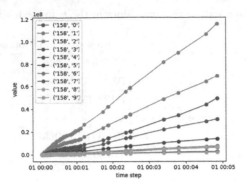

Fig. 1. Sample of the ten bins for feature 158 in operational readout data in function of the time step for vehicle 100

Table 2. Table of prediction costs

	Predicted: 0	Predicted: 1	Predicted: 2	Predicted: 3	Predicted: 4
Actual: 0	$C_{0,0} = 0$	$C_{0,1} = 7$	$C_{0,2} = 8$	$C_{0,3} = 9$	$C_{0,4} = 10$
Actual: 1	$C_{1,0} = 200$	$C_{1,1} = 0$	$C_{1,2} = 7$	$C_{1,3} = 8$	$C_{1,4} = 9$
Actual: 2	$C_{2,0} = 300$	$C_{2,1} = 200$	$C_{2,2} = 0$	$C_{2,3} = 7$	$C_{2,4} = 8$
Actual: 3	$C_{3,0} = 400$	$C_{3,1} = 300$	$C_{3,2} = 200$	$C_{3,3} = 0$	$C_{3,4} = 7$
Actual: 4	$C_{4,0} = 500$	$C_{4,1} = 400$	$C_{4,2} = 300$	$C_{4,3} = 400$	$C_{4,4} = 0$

The costs associated with predictive maintenance are highly unbalanced, caused by a larger cost to repatriate and repair a defective vehicle than to replace a non-defective component pre-emptively. Therefore, a cost function is defined by domain experts, which assigns a cost $C_{i,j}$ for predicting j if the actual label is i. These costs are shown in Table 2. In general, $C_{i,j}$ indicates the cost of a false positive if $i < j$, and the cost of a false negative if $i > j$. Consequently, false negatives have a much higher cost. For some model M, denote $N_{i,j}$ as the number of times that M predicted class i if the actual label was j. Then, the summarized cost of M is computed according to Eq. 1.

$$cost(M) = \sum_{i=0}^{4} \sum_{j=0}^{4} C_{i,j} \times N_{i,j} \tag{1}$$

3 Methodology

3.1 Data Transformation

The non-uniform time-series from the operational readouts of the train, validation, and test dataset were processed to a) handle the missing values, b) produce

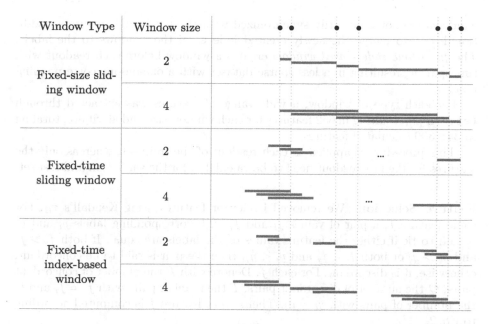

Fig. 2. The fixed-size sliding window, fixed-time sliding window, fixed-time index-based window for windows sizes of 2 and 4 on an illustrative data sample

rolling windows based on the time step, c) extract time-series features from these rolling windows, and d) reduce the number of features with feature importance and dimensionality reduction methods. The labels were transformed into the necessary format for model training and validation. Finally, the features and labels were resampled for a better representation of each defect class.

Feature Extraction. The operational readouts contain some missing values, because not all features are captured for each readout. The readout values increase linearly, as shown in Fig. 1. Therefore, linear interpolation was selected to fill in the missing values.

As these features are cumulatively increasing counters, the difference between the readouts is used to (a) focus on the change between consecutive readouts instead of the cumulative value and (b) reduce the scale of the values.

Further processing is done by applying several sliding windows to the time-series, using multiple, exponentially increasing window sizes $w \in \{4, 8, 16, 32, 64\}$, for improved temporal relevance of the extracted features. Several types of sliding windows were applied and compared. Figure 2 compares a fixed-size, fixed time-interval, and index-based sliding window. The *fixed-size sliding window* creates windows with exactly w observations. While this gives a constant number of observations in each window, there is a major variation in the recorded timespan due to non-uniformly sampled data. The *fixed-time sliding window* creates windows with a timespan w and stride 1. However, the windows are not necessarily synchronized with the labels and require an additional step to merge the newly created indexes of the windows to the labels. The *fixed-time index-based window* creates a window before each readout with timespan w, resulting in a less sparse dataset with a one-on-one corresponding label.

For each type of window, a wide range of features was extracted through tsfresh [2], resulting in 652 features for each window size and attribute, totaling to 342 300 calculated features.

This procedure is applied to each readout of the train set, whereas only the features for the last readout need to be calculated for the validation and test set.

Feature Selection. We removed irrelevant features using Kendall's τ_B. For some feature f, a pair of values f_i and f_j with corresponding labels y_i and y_j is said to tie if either the feature values or the labels are equal. If both $f_i > f_j$ and $y_i > y_j$ or both $f_i < f_j$ and $y_i < y_j$, then the pair is said to be concordant, otherwise, it is discordant. For each f, Denote with P the number of concordant pairs, Q the number of discordant pairs, T the number pairs with $f_i = f_j$ and U the number of pairs with $y_i = y_j$. Then, τ_B of feature f is computed according to Eq. 2.

The null hypothesis for Kendall's Tau states that there is no association between the two variables. The p-value represents the probability of observing a value of Tau as extreme as, or more extreme than, the one calculated from the sample data, under the assumption that the null hypothesis is true (Eq. 3). We varied the threshold on the resulting p-values between 0.0005, 0.001, 0.005, 0.01, and 0.05 to create multiple datasets with a variable number of relevant features. At the end, a threshold of 0.01 was used.

$$\tau_B = \frac{(P - Q)}{\sqrt{(P + Q + T) \times (P + Q + U)}} \tag{2}$$

$$p = P\left(|\tau_B| \geq |\tau_{\text{observed}}|\right) \tag{3}$$

Besides univariate feature selection, the Pearson correlation coefficient $\rho_{i,j}$ was used to remove highly collinear features f_i and f_j, and is computed as $\rho_{i,j} = \text{cov}(f_i, f_j)/(\sigma_{f_i} \cdot \sigma_{f_j})$. For two correlated features, we removed the feature with the smallest τ_k. Here, the threshold on $\rho_{i,j}$ was set to 0.5.

Label Transformation. Train labels can be inferred from the time-to-event table. If a vehicle was repaired at time t, all its readouts with a time step in $[t - 6, t]$ are of class 4, those in $[t - 12, t - 6]$ in class 3, and so on. Additionally, for a healthy vehicle that was studied until time t, we removed the readouts in the interval $[t - 48, t]$ because it might fail immediately after the monitored period.

Resampling for Improved Class Representation. The dataset is highly imbalanced because failures are rare. Specifically, there are 1 096 712 instances of class 0, 12 503 of class 1, 6 179 of class 2, 3 200 of class 3, and 3 858 of class 4. Training on this data introduces extreme bias in the models. To cope with this, we undersample class 0 such that there are an equal number of healthy instances as instances of class 1, the largest non-healthy class.

3.2 Predictive Model

Due to the openness of the training labels, multiple inherently different models were used to predict component failure. In particular, we used the implementations from `scikit-learn` [5] and `XGBoost` [1].

Classification Models. The data can be treated as a multiclass classification problem, with one healthy state and four failure states. We trained K-NN, Decision Tree, Random Forest, Logistic Regression, Support Vector Machine, and XGBoost classifiers to predict the class labels directly.

Regression Models. Classification models typically treat each class independently, while the classes in this dataset represent varying levels of severity. This knowledge is encoded by treating the discrete class labels as regression values. Then, we trained several regression models, including K-NN, Decision Tree, Random Forest, Linear, Support Vector Machine, and XGBoost regressors. The models' continuous predictions were rounded to derive discrete class labels. We also experimented with using a ceiling function to convert predictions into class labels, addressing the imbalanced cost function in Eq. 1, but this did not significantly impact the results.

Survival Analysis. The train data contains the failure time of a component, providing more fine-grained information than the five discrete labels. Therefore, we trained a survival analysis model that predicts the time until the component fails, which can subsequently be translated to class labels. In particular, we used XGBoosting with the Accelerated Failure Time model.

Including Truck Specification. Besides historical observations, the data also includes vehicle specifications, which the models incorporate as contextual information. Specifically, the distance between two vehicles is computed as a convex combination between the number of matches in the vehicle specification, on the one hand, and the Euclidean Distance of the min-max normalized trend of the operational readouts, on the other hand. Hierarchical clustering with average linkage is used to cluster the vehicles, which enables to train a model within each cluster or context. During prediction, we check which cluster the vehicle specification belongs to and use the corresponding model. Additionally, a global, non-contextual model is trained using all data, which predicts the label for vehicles that do not belong to any context.

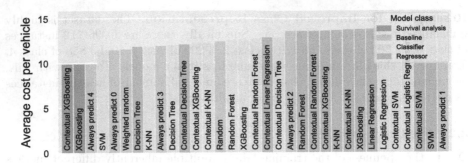

Fig. 3. Average prediction cost for the different models on the validation set

4   Evaluation

The validation set enables a quantitative comparison of the different models but does not indicate if the models generalize properly to unseen data. Therefore, we included several baselines:

1. **Always predict X.** A model that always predicts a single label, independent of the observational readouts or vehicle specification;
2. **Random.** Predict a random label for each vehicle. This is done 100 times, of which the average cost is taken to reduce variance;
3. **Weighted random.** Similar to random sampling, but a weight is assigned to each class label based on the number of occurrences of each class in the train data, thus taking class imbalance into account.

The prediction cost on the validation set, according to Eq. 1, over the number of vehicles in the validation set is shown in Fig. 3. Multiple conclusions can be drawn from this figure.

- Baseline *Always predict 4* gives a comparatively low average cost per vehicle. This is due to the highly imbalanced cost function. A model that overestimates the actual time until failure is heavily penalized, while a model that underestimates this time is penalized less. Although the healthy instances significantly outnumber the unhealthy instances, always predicting 4 results in a relative small cost on the validation set.
- Besides this baseline, the survival analysis models give the best results. Depending on the chosen sampling method for the train set, the model initially produces representative predictions. However, the predictions do not supersede *Always predict 4* and converge to this prediction towards the later boosting rounds. Including contextual information does not affect this.
- Classification models treat the different classes independently, while the regression models encode the severity levels. Nevertheless, in general, the classification models have a higher performance than the regression models.
- Generally speaking, including the specification as context does not increase the performance. This is especially true for the SVM classifier, for which

the average cost increases from 11.38 to 14.02 by the inclusion of context. However, for linear regression, the cost decreases by including context, from 13.70 to 12.80. This is because the full dataset is not linear, but data within a context follows a more linear trend.

5 Conclusion

This work explored a wide range of different models to predict if a component is at risk of imminent failure, specifically for Scania trucks in heavy-duty applications [3]. We showed that this task is non-trivial due to the highly imbalanced dataset and complex feature space. Therefore, we believe out-of-the-box models are insufficient for predictive maintenance, and dedicated and fine-tuned approaches must be developed for each application. Additionally, we illustrate that it is crucial to compare with baselines. By only comparing out-of-the-box models, one only measures the relative performance of different models without actually knowing if the model performs well.

Acknowledgement. This research is supported by Flanders Innovation & Entrepreneurship (VLAIO) through the AI-ICON project CONSCIOUS (HBC.2020.2795), the Flemish government under the Flanders AI Research Program, and by Internal Funds KU Leuven (STG/21/057).

References

1. Chen, T., Guestrin, C.: XGBoost: a scalable tree boosting system. In: Proceedings of the 22nd ACM SIGKDD International Conference on Knowledge Discovery and Data mining, pp. 785–794 (2016)
2. Christ, M., Braun, N., Neuffer, J., Kempa-Liehr, A.W.: Time series feature extraction on basis of scalable hypothesis tests (tsfresh-a python package). Neurocomputing **307**, 72–77 (2018)
3. Kharazian, Z., Lindgren, T., Magnússon, S., Steinert, O., Reyna, O.A.: SCANIA component x dataset: a real-world multivariate time series dataset for predictive maintenance. arXiv preprint arXiv:2401.15199 (2024). 10.48550/arXiv. 2401.15199
4. Li, Z., Wang, K., He, Y.: Industry 4.0 - potentials for predictive maintenance. In: Proceedings of the 6th International Workshop of Advanced Manufacturing and Automation, pp. 42–46. Atlantis Press (Nov 2016). https://doi.org/10.2991/iwama-16.2016.8
5. Pedregosa, F., et al.: Scikit-learn: machine learning in Python. J. Mach. Learn. Res. **12**, 2825–2830 (2011)

Implementing Deep Learning Models for Imminent Component X Failures Prediction in Heavy-Duty Scania Trucks

Jie Zhong and Zhenkan Wang[✉]

Volvo Group Trucks Technology (Volvo GTT), 417 15 Göteborg, Sweden
{jie.zhong,zhenkan.wang}@volvo.com

Abstract. This paper explores the application of predictive maintenance (PdM) in vehicle management, focusing on improving performance and reliability of critical truck components. By leveraging a newly acquired, comprehensive real-world dataset, the study aims to develop machine learning models for accurately predicting component failures. The dataset, sourced from the Symposium on Intelligent Data Analysis (IDA 2024), includes multivariate time series data from an anonymized engine component of a fleet of trucks, featuring operational data, repair records, and specifications. The research employs advanced deep learning techniques like Convolutional and Recurrent Neural Networks, including Long Short-Term Memory (LSTM) networks, to identify patterns indicative of potential failures. This initiative aims to optimize maintenance interventions, resource allocation, and fleet management by predicting the time or class of potential failures, thereby reducing downtime and maintenance costs.

Keywords: Predictive Maintenance · Machine Learning · Deep Learning · Time Series Data · LSTM

1 Background

1.1 Predictive Maintenance

Predicting component failures is crucial for maintaining vehicle reliability and performance, especially for trucks operating under diverse conditions [1]. The consequences of a truck breaking down mid-journey are significant, not only in terms of repair costs but also due to potential delays. These situations become even more challenging when repair facilities lack the necessary components, leading to extended downtime.

Predictive Maintenance (PdM) emerges as a proactive solution to these challenges, offering a proactive strategy to prevent failures before they occur [2, 3]. This enables the performance of timely and cost-effective maintenance interventions, informed by data-driven insights [4]. PdM with Machine Learning (ML) for automotive systems has been investigated over a decade [5–11]. This development not only promises to transform maintenance practices but also to ensure vehicles operate at peak efficiency while minimizing downtime and maintenance costs.

© The Author(s), under exclusive license to Springer Nature Switzerland AG 2024
I. Miliou et al. (Eds.): IDA 2024, LNCS 14642, pp. 268–276, 2024.
https://doi.org/10.1007/978-3-031-58553-1_22

1.2   Contest Background Description

The newly released real-world dataset from SCANIA is described in detailed in the reference [12], which holds significant potential for advancing the field of PdM. It is a comprehensive multivariate time series dataset, demonstrating gradual degradation in equipment in the form of time series readouts. The used dataset in our paper includes information gathered from Scania trucks used in heavy-duty applications and it is introduced for the ongoing Industrial Challenge 2024 at the 22nd International Symposium on Intelligent Data Analysis (IDA) [13] with the title of "Developing an Effective Predictive Model for Imminent Component X Failures in Heavy-Duty SCANIA Trucks" in Sweden.

This paper describes the approach for solving the challenge. The goal is to develop predictive models that can accurately predict whether a specific engine component in a vehicle is at risk of imminent failure. The remainder of the paper is organized as follows: flowchart of input data preparation is described in Sect. 2. Section 3 explains the modeling and results. Section 4 describes some reflection on the challenge. Finally, Sect. 5 concludes the paper and discusses future research directions.

2   Analysis Method for Input Data Preparation

This section describes the critical steps for the preparation of input data to ML models. Figure 1 demonstrates the flowchart of overall data preparation. The raw dataset underwent extensive processing with high data quality [12], e.g. relevant features selected, limited missing data rate and dataset divided into train, validation, and test subset, which reduce much workload on data processing.

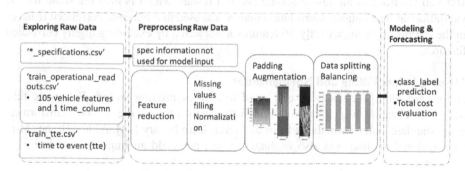

Fig. 1. Flowchart of the data preparation

2.1   Exploring Raw Data

Initially, an exploration of vehicle specification distributions within both the training and validation datasets was conducted based on '*_specifications.csv' files. From the observation of the distributions, for example, as shown in Fig. 2 for Spec_0, there are small differences in the distributions between normal and repaired vehicles. Due to page limitations, not all distributions of specifications are shown. However, observations across the datasets remain similar, leading to the decision not to utilize specification information for training the ML models.

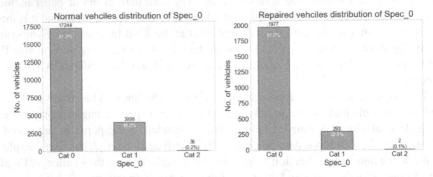

Fig. 2. Comparison of specification distributions example over Spec_0 between normal and repaired vehicles.

2.2 Preprocessing Raw Data

Feature Reduction
Subsequently, a feature reduction process was undertaken based on correlations identified within the dataset. The initial dataset 'train_operational_readouts.csv' contains 105 vehicle features alongside a time column feature which is also the same for the validation and test dataset. Correlation matrix was used to reduce the correlated features in the dataset and consequently 56 features are selected by excluding highly correlated features.

Missing Values Filling and Normalization
A forward filling method was employed to address missing values. Following this, data normalization was achieved through the application of the 'PowerTransformer' and 'StandardScaler()' methods from the Scikit-learn library [14]. Although the 'Min-MaxScaler()' method was also evaluated, it did not yield an improvement in model accuracy.

Padding and Augmentation
There are 56 features including the time unit column for each vehicle. A notable challenge arises from the varying number of data samples per vehicle, since the numbers of readout for each vehicle are different. To make the input data of each vehicle the same length or size, a fixed-length padding strategy was implemented. This technique ensures that each vehicle's data conforms to a consistent size.

Figure 3 shows an example of the padding process over one vehicle. The left plot shows the raw data of the vehicle with 106 features, after feature reduction and padding,

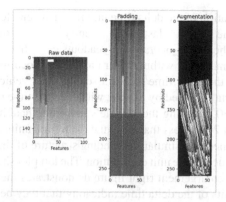

Fig. 3. Example of padding one vehicle's operational readouts to a fix length.

the processed data is shown in the middle plot. The right plot illustrates the processed data after augmentation.

Data Splitting and Balancing

The dataset contained in the "train_operational_readouts.csv" is segregated into training and validation dataset with 80-20 partition ratio. The validation dataset is used for validation of the models. Because the distribution of data classes is imbalanced, as depicted in Fig. 4, a weighted sampler function is implemented for creating a balance training dataset [15]. This approach ensures that each class is represented proportionally to its prevalence in the dataset, as shown in Fig. 4, contributing to a more equitable and robust model training procedure. Although weighted loss function method [16] was also evaluated, it did not yield an improvement in model accuracy.

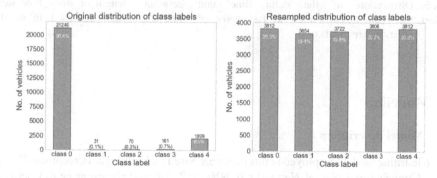

Fig. 4. Distribution of classes in training dataset before and after weighted sampler function.

2.3   Preprocessing Train Labels

To create the class_labels for the training dataset, i.e. all the vehicles with unique 'vehicle_id' in the 'train_tte.csv', the following steps were conducted.

Firstly, the computation of the delta time unit for each vehicle was accomplished by subtracting   the   last   entry   of   'time_step' value recorded in of the 'train_operational_readouts.csv', from the last readout time of 'length_of_study_time_step' within the 'train_tte.csv' document.

Secondly, binning the delta time to 5 classes for each vehicle. This classification was based on the criteria provided by Scania, where a delta time unit exceeding 48 time units denotes as Class 0, meaning the vehicle's in normal healthy status. A delta time unit that is greater than 24 but less than or equal to 48 time units falls into Class 1, with successive classes defined in a similar descending structure of time units [12].

Figure 5 shows the delta time unit distribution. The left plot shows the overall vehicle distribution, while the enlargement right figure demonstrates there are vehicles having longer than 48 time units of the delta time indicating that they belong to class 0.

As a result, the labeled data for all vehicles contained in "train_tte.csv" is similar to the content in "validation_labels.csv". In our study, the provided validation datasets ("validation_labels.csv" and "validation_operational_readouts.csv") were treated as test datasets to evaluate the final performance of the ML models.

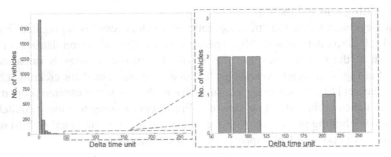

Fig. 5. Distribution of the delta time unit between 'length_of_study_time_step' within the 'train_tte.csv' and the last entry of 'time_step' value recorded in of the 'train_operational_readouts.csv'.

3   Modeling

3.1   Model Description

Deep learning models employed in this paper are based on Multilayer Perceptrons (MLP) [17], Convolutional Neural Networks (CNNs) [18, 19] and recurrent neural network (RNN) [20, 21]. Their hyperparameters are listed in Table 1.

MLP models can be a baseline for its computed relevance between features and output, i.e. the class of vehicles, while CNNs are adept at identifying the relationships among combined features. This implies that patterns formed by the values of multiple features could be associated with the output. In this work, we also employ transfer learning by utilizing a pretrained ResNet18 architecture and replacing the first and last layers to better align with the input data dimension and the desired output classes.

Table 1. Model description.

Deep Learning Architecture	Hyperparameters
MLP	Number of layers: 4 Layer 1 units: 1024 Layer 2 units: 256 Layer 3 units: 128 Activation function: ReLU
CNN	Number of layers: 2 Layer 1 units: 64 (Convolutional) Layer 2 units: 128 (Convolutional) Activation function: ReLU Fully Connected layers with 512 and 64 units
Bi-LSTM-attention	Number of layers: 4 Layer 1–4 units: 128

Since the readouts of each vehicle are time series data, we input a sequence, which consists a series of feature vectors across consecutive time steps, into a bidirectional long short-term memory (Bi-LSTM) model for a single vehicle. The model's output is a condensed summary of the sequence, which is used to classify the vehicle or engine. To accommodate sequences of varying lengths within the same batch, dynamic padding ensures uniform sequence length. Furthermore, an attention-based Bi-LSTM [22] model was implemented which can map a set of key-value pairs to the output. Such model leverages the attention mechanism to assign weights to different parts of the sequence enabling the model to focus on particular segments of the sequence instead of just considering the sequence's end.

Random search was conducted for each model to determine the optimal set of hyperparameters. All experimental results are averages over 5 runs with 5-fold cross validation. It is worth mentioning that the weighted sampler function has been applied separately in each fold.

3.2 Summary of Model Performance

Table 2 reports the best results obtained by hyperparameter optimization across various deep learning architectures discussed in this study. To assess model performance, we introduced a baseline scenario in which class predictions are deliberately set to false negatives.

The MLP model exhibits lower accuracy compared to the CNN and LSTM models but maintains one of the lowest costs, albeit with significant uncertainty. Whereas the CNN model achieves the highest accuracy but suffers a greater cost, potentially exceeding that of a model with random initial weights, as illustrated in Fig. 6. Despite the high accuracy, there's a risk of overfitting to accuracy due to the loss function as Cross-Entropy loss, where a small number of false negative predictions can significantly increase the cost.

Apparently, model enhancements through augmentation did not yield improvements in accuracy or cost, indicating the need for a better selection of augmentation. The performance of ResNet models is reduced by the limited number of trainable parameters.

RNN models achieve moderate accuracy and nearly the best cost. Therefore, it was chosen to generate the final submission file.

Table 2. Summary of model performance.

Model	Additional processing	Accuracy [%]	Cost
All false negative		0	50460
MLP		75 ± 3.6	42385 ± 2342
CNN		88.2 ± 1.2	50636 ± 1136
Resnet18		53.7 ± 10.7	47187 ± 1542
CNN	Augmentation	32.2 ± 7.4	61181 ± 2807
Resnet18	Augmentation	38.4 + 30.6	55369 ± 4011
Bi-LSTM		77.6 ± 5.6	41728 ± 1514
Bi-LSTM-attention		81 ± 4.2	43465 ± 1534
Bi-LSTM-attention	Dynamic padding	81 ± 4.8	43224 ± 831

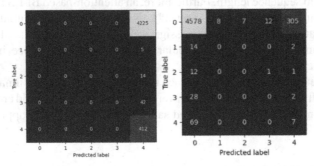

Fig. 6. Confusion matrix of the prediction and true label using CNNs. The left plot shows results from a randomly initialized model, while the right one displays outcomes following model training.

4   Reflection

In evaluating the performance of predictive maintenance models for this challenge, the primary objective was to minimize the 'Total_cost', a metric defined by a cost function [12] that aggregates 'Cost_n_m' across instances.

$$\text{Total\_cost} = \text{Cost\_n\_m} \times \text{No\_instances}$$

The function's design inherently places differing penalties on false positive and false negative predictions, with a significantly higher cost assigned to the latter. Upon reflection, it has been observed that models are not subjected to optimization or comprehensive training—thus inclined towards predicting a majority of instances in the 'Cost_0_4' category—could paradoxically achieve a lower 'Total_cost' compared to models that offer more precise predictions across all classes but incur higher total costs due to the severe penalties of false negatives, as shown in Fig. 6.

This phenomenon underscores a critical issue: the cost function, while effectively prioritizing the minimization of false negatives, may inadvertently encourage strategies that compromise the overall accuracy and reliability of PdM models. A more effective loss function needs to be developed that incorporates constraints aimed at optimizing both cost and accuracy simultaneously. On the other hand, incorporating metrics that specifically reward class accuracy could mitigate the risk of participants exploiting the cost function's bias towards lower penalties for certain types of errors.

5 Conclusion

In this study, various machine learning models, including MLP, CNN and RNN, were explored for analyzing a comprehensive multivariate time series dataset, specifically focusing on predictive maintenance (PdM) within the context of the Intelligent Data Analysis Industrial Challenge 2024. Although CNN models had the best accuracy, RNN models yielded the most favorable results according to the challenge's evaluation criteria. It's important to note that this final choice did not coincide with the highest accuracy model. This outcome highlights the needed balance between cost efficiency and prediction accuracy in the domain of PdM, and it opens opportunities for further research to refine predictive capabilities while also optimizing cost-related outcomes.

References

1. Ness Digital Engineering: Predictive Maintenance on Commercial Vehicle Fleets (2023). http://www.ness.com/enhancing-fleet-management-with-predictive-maintenance-for-com mercial-vehicles
2. Prytz, R., Nowaczyk, S., Rögnvaldsson, T., Byttner, S.: Analysis of truck compressor failures based on logged vehicle data. In: Proceedings of the 9th International Conference on Data Mining, CSREA Press, Las Vegas, NV, USA, 22–25 July (2013)
3. Jennions, I.K. (ed.): Integrated Vehicle Health Management: The Technology. SAE International, Warrendale (2013). https://doi.org/10.4271/R-429
4. Theissler, A., Pérez-Velázquez, J., Kettelgerdes, M., Elger, G.: Predictive maintenance enabled by machine learning: use cases and challenges in the automotive industry. Reliab. Eng. Syst. Saf. **215**, 107864 (2021). https://doi.org/10.1016/j.ress.2021.107864
5. Prytz, R., Nowaczyk, S., Rögnvaldsson, T., Byttner, S.: Predicting the need for vehicle compressor repairs using maintenance records and logged vehicle data. Eng. Appl. Artif. Intell. **41**, 139–150 (2015). https://doi.org/10.1016/j.engappai.2015.02.009
6. Fan, Y., Nowaczyk, S., Rögnvaldsson, T.: Evaluation of self-organized approach for predicting compressor faults in a city bus fleet. Procedia Comput. Sci. **53**, 447–456 (2015)

7. Costa, C.F., Nascimento, M.A.: Using machine learning for predicting failures. In: Boström, H., Knobbe, A., Soares, C., Papapetrou, P. (eds.) Advances in Intelligent Data Analysis XV, pp. 381–386. Springer, Cham (2016). https://doi.org/10.1007/978-3-319-46349-0

8. Gurung, B.R., Lindgren, T., Boström, H.: Predicting NOx sensor failure in heavy duty trucks using histogram-based random forests. Int. J. Prognost. Health Manage. **8**(1) (2017). https://doi.org/10.36001/ijphm.2017.v8i1.2535

9. Biteus, J., Lindgren, T.: Planning flexible maintenance for heavy trucks using machine learning models, constraint programming, and route optimization. SAE Int. J. Mater. Manuf. **10**(3), 306–315 (2017). https://doi.org/10.4271/2017-01-0237

10. Dhada, M., Parlikad, A.K., Steinert, O., Lindgren, T.: Weibull recurrent neural networks for failure prognosis using histogram data. Neural Comput. Appl. **35**(4), 3011–3024 (2023). https://doi.org/10.1007/s00521-022-07667-7

11. Fan, Y., Atoui, A., Nowaczyk, S., Rognvaldsson, T.: Evaluation of multi-modal learning for predicting coolant pump failures in heavy duty vehicles. In: PHM Society Asia-Pacific Conference, vol. 4, no. 1 (2023)

12. Kharazian, Z., Lindgren, T., Magnússon, S., Steinert, O., Reyna, O.A.: SCANIA component X dataset: a real-world multivariate time series dataset for predictive maintenance. arXiv preprint arXiv:2401.15199 (2024)

13. SCANIA component dataset X for predictive maintenance. In: 22nd International Symposium on Intelligent Data Analysis (IDA 2024), Industrial Challenge Repository (2024). https://ida2024.org/industrial-challenge/

14. Pedregosa, F., et al.: Scikit-learn: machine learning in Python. J. Mach. Learn. Res. **12**, 2825–2830 (2011). https://scikit-learn.org/stable/

15. PyTorch Development Team: WeightedRandomSampler. PyTorch Documentation (2024). https://pytorch.org/docs/stable/data.html

16. Rengasamy, D., Jafari, M., Rothwell, B., Chen, X., Figueredo, G.P.: Deep learning with dynamically weighted loss function for sensor-based prognostics and health management. Sensors **20**(3), 723 (2020). https://doi.org/10.3390/s20030723

17. Rumelhart, D.E., Hinton, G.E., Williams, R.J.: Learning representations by back-propagating errors. Nature **323**(6088), 533–536 (1986). https://doi.org/10.1038/323533a0

18. Krizhevsky, A., Sutskever, I., Hinton, G.E.: ImageNet classification with deep convolutional neural networks. In: Advances in Neural Information Processing Systems, vol. 25 (2012)

19. LeCun, Y., Bengio, Y., Hinton, G.: Deep learning. Nature **521**(7553), 436–444 (2015). https://doi.org/10.1038/nature14539

20. Hochreiter, S., Schmidhuber, J.: Long short-term memory. Neural Comput. **9**(8), 1735–1780 (1997). https://doi.org/10.1162/neco.1997.9.8.1735

21. Jain, L.C., Medsker, L.R.: Recurrent Neural Networks: Design and Applications. CRC Press, Boca Raton (1999)

22. Wang, Y., Huang, M., Zhu, X., Zhao, L.: Attention-based LSTM for aspect-level sentiment classification. In: Proceedings of the 2016 Conference on Empirical Methods in Natural Language Processing, pp. 606–615 (2016)

Author Index

I. Miliou et al. (Eds.): IDA 2024, LNCS 14642, pp. 277–278, 2024.
https://doi.org/10.1007/978-3-031-58553-1

Printed in the United States
by Baker & Taylor Publisher Services